Synthesis of TiO$_2$ Nanoparticles and Their Catalytic Activity

Synthesis of TiO$_2$ Nanoparticles and Their Catalytic Activity

Editor

Wei Zhou

Basel • Beijing • Wuhan • Barcelona • Belgrade • Novi Sad • Cluj • Manchester

Editor
Wei Zhou
Chemistry and Chemical
Engineering
Qilu University of Technology
Jinan
China

Editorial Office
MDPI
St. Alban-Anlage 66
4052 Basel, Switzerland

This is a reprint of articles from the Special Issue published online in the open access journal *Nanomaterials* (ISSN 2079-4991) (available at: www.mdpi.com/journal/nanomaterials/special_issues/TiO2_nanoparticles).

For citation purposes, cite each article independently as indicated on the article page online and as indicated below:

Lastname, A.A.; Lastname, B.B. Article Title. *Journal Name* **Year**, *Volume Number*, Page Range.

ISBN 978-3-0365-8753-0 (Hbk)
ISBN 978-3-0365-8752-3 (PDF)
doi.org/10.3390/books978-3-0365-8752-3

© 2023 by the authors. Articles in this book are Open Access and distributed under the Creative Commons Attribution (CC BY) license. The book as a whole is distributed by MDPI under the terms and conditions of the Creative Commons Attribution-NonCommercial-NoDerivs (CC BY-NC-ND) license.

Contents

About the Editor vii

Preface ix

Tingting Hu, Panpan Feng, Liping Guo, Hongqi Chu and Fusheng Liu
Construction of Built-In Electric Field in $TiO_2@Ti_2O_3$ Core-Shell Heterojunctions toward Optimized Photocatalytic Performance
Reprinted from: *Nanomaterials* **2023**, *13*, 2125, doi:10.3390/nano13142125 1

Wanting Wang, Yuanting Wu, Long Chen, Chenggang Xu, Changqing Liu and Chengxin Li
Fabrication of Z-Type TiN@(A,R)TiO_2 Plasmonic Photocatalyst with Enhanced Photocatalytic Activity
Reprinted from: *Nanomaterials* **2023**, *13*, 1984, doi:10.3390/nano13131984 17

Christine Joy Querebillo
A Review on Nano Ti-Based Oxides for Dark and Photocatalysis: From Photoinduced Processes to Bioimplant Applications
Reprinted from: *Nanomaterials* **2023**, *13*, 982, doi:10.3390/nano13060982 33

Mohammed Alyami
Ultra-Violet-Assisted Scalable Method to Fabricate Oxygen-Vacancy-Rich Titanium-Dioxide Semiconductor Film for Water Decontamination under Natural Sunlight Irradiation
Reprinted from: *Nanomaterials* **2023**, *13*, 703, doi:10.3390/nano13040703 76

Yuwei Wang, Kelin Xu, Liquan Fan, Yongwang Jiang, Ying Yue and Hongge Jia
B-Doped g-C_3N_4/Black TiO_2 Z-Scheme Nanocomposites for Enhanced Visible-Light-Driven Photocatalytic Performance
Reprinted from: *Nanomaterials* **2023**, *13*, 518, doi:10.3390/nano13030518 86

Lijun Liao, Mingtao Wang, Zhenzi Li, Xuepeng Wang and Wei Zhou
Recent Advances in Black TiO_2 Nanomaterials for Solar Energy Conversion
Reprinted from: *Nanomaterials* **2023**, *13*, 468, doi:10.3390/nano13030468 97

Emilian Chifor, Ion Bordeianu, Crina Anastasescu, Jose Maria Calderon-Moreno, Veronica Bratan and Diana-Ioana Eftemie et al.
Bioactive Coatings Based on Nanostructured TiO_2 Modified with Noble Metal Nanoparticles and Lysozyme for Ti Dental Implants
Reprinted from: *Nanomaterials* **2022**, *12*, 3186, doi:10.3390/nano12183186 124

Hieu Minh Ngo, Amol Uttam Pawar, Jun Tang, Zhongbiao Zhuo, Don Keun Lee and Kang Min Ok et al.
Synthesis of Uniform Size Rutile TiO_2 Microrods by Simple Molten-Salt Method and Its Photoluminescence Activity
Reprinted from: *Nanomaterials* **2022**, *12*, 2626, doi:10.3390/nano12152626 141

Zunfu Hu, Qi Gong, Jiajia Wang, Xiuwen Zheng, Aihua Wang and Shanmin Gao
Construction of Spindle-Shaped Ti^{3+} Self-Doped TiO_2 Photocatalysts Using Triethanolamine-Aqueous as the Medium and Its Photoelectrochemical Properties
Reprinted from: *Nanomaterials* **2022**, *12*, 2298, doi:10.3390/nano12132298 153

Zhenzi Li, Decai Yang, Hongqi Chu, Liping Guo, Tao Chen and Yifan Mu et al.
Efficient Charge Transfer Channels in Reduced Graphene Oxide/Mesoporous TiO$_2$ Nanotube Heterojunction Assemblies toward Optimized Photocatalytic Hydrogen Evolution
Reprinted from: *Nanomaterials* **2022**, *12*, 1474, doi:10.3390/nano12091474 . . . . **165**

About the Editor

Wei Zhou

Wei Zhou received his Ph.D. degree in 2009 from Jilin University, China. He subsequently worked at Heilongjiang University, and he became a full professor in 2015. From 2020, he worked at Qilu University of Technology as an independent Principal Investigator. His research interests include mesoporous materials, semiconductor nanomaterials for solar energy conversion, photocatalysis, photothermal and photoelectrochemical performance. He has published over 150 peer-reviewed SCI papers as first author or corresponding author with over 15000 citations and has an H-index of 70.

Preface

Solar light utilization technologies are largely promoted by the rapid development of material science, which have generated various applications in energy conversion and catalysis. In the past decades, we have experienced the rapid development of solar energy conversion research, resulting in a remarkable increase in energy conversion efficiency and photocatalytic performance. Due to their high stability, low cost, decent activity, and non-toxic properties, TiO_2 nanomaterials have received considerable attention in photocatalytic water splitting, CO_2 conversion, organic synthesis, etc., in terms of efficient utilization of abundant solar energy. Besides the photocatalytic energy conversions, TiO_2 can also be deposited onto substances as bioactive coatings for bioimplant applications to effectively improve the generation of reactive oxygen species for antibacterial. Thus, the advances in catalytic applications by TiO_2 nanomaterials could facilitate the utilization of photocatalysts in real conditions. We are delighted to introduce this Special Issue of "Synthesis of TiO_2 Nanoparticles and Their Catalytic Activity", which includes 10 research and review articles that explore cutting-edge advances in the synthesis and applications of TiO_2 nanomaterials in various fields. We would like to thank all the authors and reviewers for their contributions to this Special Issue. We also hope that the articles presented here are helpful and inspiring for their designs, concepts, and perspectives and can arouse novel understandings in the field of nanomaterials and catalysis.

Wei Zhou
Editor

Article

Construction of Built-In Electric Field in TiO$_2$@Ti$_2$O$_3$ Core-Shell Heterojunctions toward Optimized Photocatalytic Performance

Tingting Hu [1,2], Panpan Feng [3,*], Liping Guo [2], Hongqi Chu [2] and Fusheng Liu [1,*]

[1] State Key Laboratory Base for Eco-Chemical Engineering, College of Chemical Engineering, Qingdao University of Science and Technology, Qingdao 266042, China; hutingting_1981@163.com

[2] Shandong Provincial Key Laboratory of Molecular Engineering, School of Chemistry and Chemical Engineering, Qilu University of Technology (Shandong Academy of Sciences), Jinan 250353, China; guoliping@qlu.edu.cn (L.G.); hqchu@qlu.edu.cn (H.C.)

[3] School of Chemistry and Pharmaceutical Engineering, Shandong First Medical University & Shandong Academy of Medical Sciences, Jinan 250117, China

* Correspondence: fpanpanlhz@163.com (P.F.); liufusheng63@sina.com (F.L.)

Abstract: A series of Ti$_2$O$_3$@TiO$_2$ core-shell heterojunction composite photocatalysts with different internal electric fields were synthesized using simple heat treatment methods. The synthesized Ti$_2$O$_3$@TiO$_2$ core-shell heterojunction composites were characterized by means of SEM, XRD, PL, UV–Vis, BET, SPV, TEM and other related analytical techniques. Tetracycline (TC) was used as the degradation target to evaluate the photocatalytic performance of the synthesized Ti$_2$O$_3$@TiO$_2$ core-shell heterojunction composites. The relevant test results show that the photocatalytic performance of the optimized materials has been significantly enhanced compared to Ti$_2$O$_3$, while the photocatalytic degradation rate has increased from 28% to 70.1%. After verification via several different testing and characterization techniques, the excellent catalytic performance is attributed to the efficient separation efficiency of the photogenerated charge carriers derived from the built-in electric field formed between Ti$_2$O$_3$ and TiO$_2$. When the recombination of electrons and holes is occupied, more charges are generated to reach the surface of the photocatalyst, thereby improving the photocatalytic degradation efficiency. Thus, this work provides a universal strategy to enhance the photocatalytic performance of Ti$_2$O$_3$ by coupling it with TiO$_2$ to build an internal electric field.

Keywords: photocatalysis; Ti$_2$O$_3$@TiO$_2$ core-shell heterojunction; built-in electric field

1. Introduction

Tetracycline (TC) has stable properties due to its four-complex benzene cycloalkyl groups [1,2]. It is highly susceptible to accumulation in the environment, and it is a difficult-to-degrade organic pollutant. Tetracycline (TC) in the environment is persistent, easy to accumulate, and difficult to degrade, which causes serious harm to the food chain and human health [3,4]. In addition, due to its long application, drug resistance is a serious concern. It can be absorbed from the gastrointestinal tract, albeit not completely. About 60%~70% of the dose is residual in the environment. Therefore, the long-term residue of TC in the aquatic environment harms human health and the ecological environment. To alleviate this serious situation, applying photocatalysis technology to the degradation of water pollutants as soon as possible will greatly help with the purification of the water environment [5,6]. Compared with traditional chemical, physical, and biological technologies, semiconductor photocatalytic technology can efficiently degrade organic pollutants in wastewater, such as antibiotic wastewater, into small-molecule substances [7–9].

As the ninth most abundant element on Earth, titanium has been widely studied for its high thermal stability, low cost, and light-responsive ability in metal oxide composites such as titanium dioxide (TiO$_2$) [10–12]. TiO$_2$ can be stimulated by light to produce strong oxidizing h$^+$ and strong reducing e$^-$, which makes TiO$_2$ highly efficient in terms of its photocatalytic performance [13,14]. However, due to its inherent large bandgap (≈3 eV), typical titanium dioxide only responds to ultraviolet radiation with a wavelength of <378 nm [15–17]. In the past few decades, researchers have considerably reduced the bandgap to achieve titanium dioxide absorption of solar energy in the visible spectral range [18]. Research in the literature confirms that titanium dioxide's bandgap width can be reduced to 1.5 eV, while its optical response can be extended to 800 nm, achieving a utilization rate of 40% for total solar energy [19]. However, to design an efficient full-spectrum solar converter with high solar energy utilization, further modification of titanium oxide is needed to reduce the bandgap to less than 0.5 eV.

Magnéli phase titanium oxides were only identified as conductive compounds in the 1950s, with the general formula Ti_nO_{2n-1} (3 < n < 10). This non-stoichiometric semiconductor titanium oxide, which contains structural vacancies in both the titanium and oxygen sublattices, exhibits excellent photocatalytic performance. Due to the limitations of the methods used to manufacture Magnéli phase materials, it is difficult to prepare individual pure phases [20,21]. Therefore, the Magnéli phase materials synthesized by heating TiO_2 or Ti at different temperatures typically contain mixed phases. In addition, the reliability of the electrical and chemical properties of the phases obtained through separation methods is relatively low. Even the most common methods (such as lame synthesis methods, both electrochemical and chemical) for synthesizing titanium suboxides still have many problems, such as the low active surface of the obtained material and the difficulty in adjusting the size and structure. For example, processing must be carried out in a high-temperature and oxygen-free atmosphere. In addition, even at temperatures ranging from 425 to 525 °C, Ti_nO_{2n-1} is unstable and decomposes into various upper structures. However, the enhanced part of the photocatalytic performance with regard to photocatalytic degradation comes from the nanostructures on the surface of bulk titanium suboxides. Ti^{3+} ions play a significant role in reducing the bandgap and effectively boosting the adsorption and activation of reactants, which markedly enhances the light-responsive ability [22]. The relatively small bandgap of Ti_2O_3 enables it to absorb solar energy within the full spectral range [23,24]. In addition, Ti_2O_3 is one of the most representative examples of the Mott insulation system, and its metal–insulator transition characteristics are derived from the electronic correlation effect [25,26]. Due to its narrow bandgap, Ti_2O_3 exhibits high theoretical photocurrent density and high carrier mobility. After modification, the light absorption ability of traditional Ti_2O_3 can be further enhanced. Among many methods, constructing an internal electric field at the interface of the photocatalysts as a driving force for charge separation is considered an efficient method for achieving maximum carrier separation and driving target surface reactions in photocatalytic systems [27]. Specifically, photogenerated electrons will undergo reverse transfer driven by an electric field, greatly accelerating the separation of electron–hole pairs [28,29]. For example, Cui et al. constructed an internal electric field by preparing p–n homojunction perovskite solar cells, which directionally promoted the transport of photogenerated carriers, greatly reducing the carrier recombination losses and improving the power conversion efficiency [30].

As titanium suboxides are usually prepared via heat treatment under an inert or reductive atmosphere, there is a demand for methods to prepare nanostructured titanium suboxides via hydrothermal or other methods. During these synthesis processes, the phase structure of the material is not easily controlled, making it even more difficult to construct controllable heterojunctions in an orderly manner. Research has shown that introducing TiO_2 (anatase)-based symbionts into the low oxidation state structure can significantly improve the conductivity of the composite and the properties related to photoluminescence [31]. Therefore, in this work, we combined the TiO_2 and Ti_2O_3 to construct an orderly core-shell heterojunction, creating an internal electric field. In addition, the band structures of $Ti_2O_3@TiO_2$ heterojunctions were investigated by means of UV–Vis spectroscopy, the Mott–Schottky curve, and surface photovoltage (SPV). Constructing an internal electric field has been proven to promote the transfer efficiency of photogenerated charges and improve photocatalytic activity. The correlation between built-in electric fields and photocatalytic performance was revealed based on multiple test characterizations. The possible enhanced photocatalytic degradation mechanism of tetracycline (TC) was proposed. This work provides a simple and universal method for regulating the built-in electric fields in photocatalysts that can be applied to environmental pollution purification.

2. Materials and Methods

2.1. Chemicals

The Ti_2O_3 chemicals and tetracycline (TC) were purchased from Aladdin. The ethanol and Nafion solution were purchased from Sigma–Aldrich. Deionized water was used in the experiments. The materials used were all of analytical grade (99%) and had not undergone purification treatment.

2.2. Methods

Synthesis of the $Ti_2O_3@TiO_2$ Heterojunction

The $Ti_2O_3@TiO_2$ heterojunction was synthesized via a simple heat treatment method. First, using commercial Ti_2O_3 as the research object, a $Ti_2O_3@TiO_2$ heterojunction system was constructed in an orderly fashion using heat treatment methods. Then, 1 g of Ti_2O_3 was dispersed in a porcelain boat. Program heating was used to raise the temperature at 2 °C min^{-1} from room temperature to different predetermined temperatures. The temperature was kept constant for 2 h and cooled to room temperature. The reaction conditions for the preparation of the $TiO_2@Ti_2O_3$ heterojunction were explored. Reaction temperatures of 100 °C, 200 °C, 300 °C, 400 °C, 500 °C, 550 °C, 600 °C, and 700 °C were selected to explore the photocatalytic activity of $TiO_2@Ti_2O_3$ toward TC decomposition. The prepared catalysts were named Ti_2O_3-100, Ti_2O_3-200, Ti_2O_3-300, Ti_2O_3-400, Ti_2O_3-500, Ti_2O_3-550, Ti_2O_3-600, and Ti_2O_3-700, respectively.

2.3. Characterizations

The crystalline phase structures of the $Ti_2O_3@TiO_2$ samples were determined via X-ray diffraction (XRD) (Rigaku, Tokyo, Japan) with Cu Kα irradiation. Field emission scanning electron microscopy (SEM) (HITACHI, SU8010, Tokyo, Japan) and transmission electron microscopy (TEM) (JEM-ARM200F, Tokyo, Japan) were performed to reveal the surface topography of the $Ti_2O_3@TiO_2$ samples. The adsorption isotherm of the $Ti_2O_3@TiO_2$ samples was used to measure the specific surface area, and a pore size distribution analyzer (BET, Micromeritics, ASAP2460, Norcross, GA, USA) with N_2 as the adsorption medium was used in combination with analysis via the Brunauer–Emmett–Teller (BET) method. The carrier recombination was evaluated based on the photoluminescence (PL) spectra (Edinburgh, Livingston, FLS 980, Scotland, UK) with using a Xenon lamp (excitation wavelength 375 nm). UV–Vis diffuse reflectance spectroscopy was used to analyze the spectral response range of the $Ti_2O_3@TiO_2$ (DRS, Shimadzu, SolidSpec-3700, Tokyo, Japan). The energy gap (Eg) of the $Ti_2O_3@TiO_2$ and Ti_2O_3 was obtained via the Tauc plot method. Thermogravimetric analysis (TGA) was used to monitor the thermal stability performance of the $Ti_2O_3@TiO_2$ and Ti_2O_3 on an SDT Q600 (TA Instruments, New Castle, DE, USA). The SPV transient was recorded using a tunable Nd:YAG laser (EKSPLA, NT 342/1/UVE) excited by pulses with a duration of 5 ns and wavelengths between 420 and 720 nm, and a sampling oscilloscope (GAGE, CS14200, sampling rate of 100 Mm/s) was used by applying logarithmic readings without averaging to avoid potential accuracy loss in a short period. An electrochemical station (CHI660E, Chenhua, Shanghai, China) with a conventional three-electrode system was employed to carry out the electrochemical measurements. The three-electrode system contained working, counter (carbon rod) and reference (Ag/AgCl) electrodes in 1 M KOH alkaline medium. The steps for making the working electrodes were as follows: 4 mg of catalysts was dispersed in 1 mL of mixed solution (water:ethanol = 3:1) before 10 µL was taken and added dropwise to the surface of 1×1 cm^{-2} ITO and then left to dry naturally.

2.4. Measurement of Photocatalytic Activity

The photocatalytic performances of the Ti_2O_3@TiO_2 and Ti_2O_3 were mainly confirmed by examining the photocatalytic degradation under simulated visible light. First, 100 mg of catalyst was added to 100 mL of TC solution (5 mg/L). A mechanical stirrer was used to stir the solution with a speed range of 0–1500 rpm. Before irradiation, a 20 min dark reaction was performed to achieve adsorption–desorption equilibrium. Then, the reaction solution was irradiated using a 300 W Xenon lamp with AM 1.5 (the current was 1.8 A; PLS-SXE300D, Perfect Light, Beijing, China). During the photocatalytic reaction, 4 mL of the solution was withdrawn every 20 min and the concentration of TC was measured. Correspondingly, the absorbance of TC in the supernatant solution was measured at a detection wavelength of 357 nm using a UV–Vis spectrophotometer corresponding to the maximum adsorption for the solution. The photocatalytic activities of the Ti_2O_3@TiO_2 and Ti_2O_3 were studied by analyzing the degradation curves of the tetracycline.

3. Results and Discussion

3.1. The Results of the TGA

To investigate the changes in the quality and heat flux of the materials at different temperatures, TG tests were conducted on the samples. From the results (Figure 1), it can be seen that when the temperature reaches above 500 °C, the material mass gradually increases, which is due to the gradual oxidation of Ti_2O_3 to form TiO_2. When the temperature reaches 750 °C, the mass of the sample tends to stabilize. The sample undergoes phase transformation at different temperatures, so it is highly feasible to construct a Ti_2O_3@TiO_2 heterojunction system in an orderly manner using heat treatment methods.

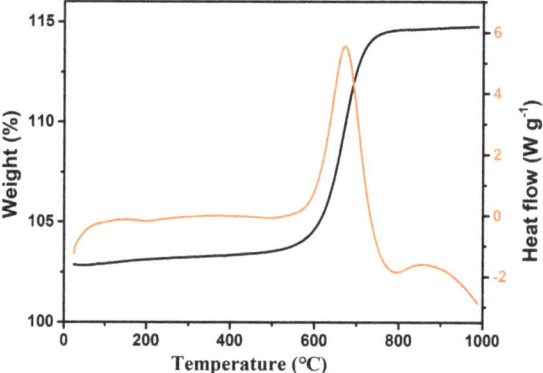

Figure 1. Optical photos of the Ti_2O_3@TiO_2 heterojunction at different heat treatment temperatures.

As shown in Figure 2, after treatment at different temperatures, the color of the Ti_2O_3 changed significantly, with the pure Ti_2O_3 turning black. When the temperature was below 600 °C, the color of the sample became lighter as the temperature increased. When the temperature was above 600 °C, the sample color turned yellow. Therefore, to further investigate the effect of temperature on the microstructures of the materials, scanning electron microscopy (SEM) was used to study the changes in the microstructure of the Ti_2O_3 under different heat treatment temperatures. Based on the results in Figures 1 and 2, the Ti_2O_3, Ti_2O_3-400, Ti_2O_3-500, Ti_2O_3-550, Ti_2O_3-600, and Ti_2O_3-700 were selected for the SEM characterization.

Figure 2. Optical photos of Ti_2O_3 at different heat treatment temperatures.

3.2. The Results of the SEM and XRD Analyses

From the results in Figure 3, it can be seen that the Ti_2O_3 before heat treatment has an irregular 3D particle structure with a smooth surface. When the temperature is below 500 °C, there is no significant change in the microstructure and size of the sample. When the temperature reaches 550 °C, the surface roughness of the sample significantly increases and small particles are generated on the outer surface. As the temperature further increases, the outer particle diameter continuously increases and agglomeration occurs.

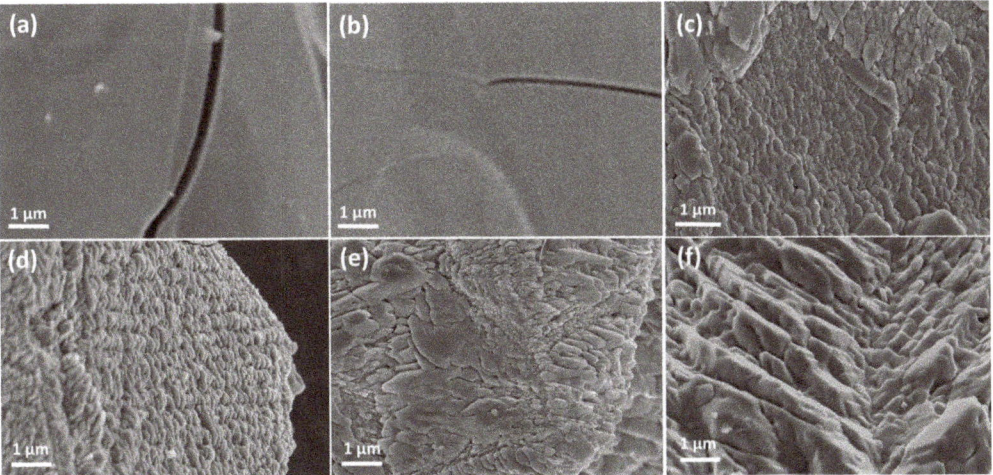

Figure 3. SEM images of Ti_2O_3 under different heat treatment temperatures: (**a**) Ti_2O_3; (**b**) Ti_2O_3-400; (**c**) Ti_2O_3-500; (**d**) Ti_2O_3-550; (**e**) Ti_2O_3-600; and (**f**) Ti_2O_3-700.

In order to investigate the crystal form changes in Ti$_2$O$_3$ after the heat treatment reaction, the samples were characterized by measn of XRD. As shown in Figure 4, the diffraction peaks detected at 23.8°, 33.0°, 34.8°, 40.2°, 48.8°, and 53.7° are related to the Ti$_2$O$_3$ phase (JCPDS No. 43-1033) and can be attributed to the (012), (104), (110), (113), (024), and (116) crystal planes, respectively [32]. When the heat treatment temperature reaches 400 °C, new diffraction peaks begin to appear at 27.4° (110) and 36.0° (101), which belong to the rutile phase TiO$_2$ (JCPDS No. 21-1276) [33]. With the further increase in temperature, the diffraction peaks attributed to the rutile phase TiO$_2$ further appear at 27.4° (110), 36.0° (101), 39.2° (200), 41.2° (111), 44.0° (210), 54.3° (211), 56.6° (220), and 64.1° (310), and the peak intensity also increases with the thermal treatment temperature. When the temperature reaches 600 °C, all the Ti$_2$O$_3$ is converted into the TiO$_2$ rutile phase. In addition, no diffraction peaks of impurities are observed. Therefore, Ti$_2$O$_3$ gradually undergoes phase transition into TiO$_2$ during the heat treatment process. Combining XRD and SEM shows that the Ti$_2$O$_3$-450, Ti$_2$O$_3$-500 and Ti$_2$O$_3$-550 are composed of Ti$_2$O$_3$ and TiO$_2$, forming a core-shell heterojunction. The experimental results show that the progress of TiO$_2$ conversion can be controlled by changing the heat treatment temperature, thereby achieving the orderly construction of a Ti$_2$O$_3$@TiO$_2$ heterojunction system.

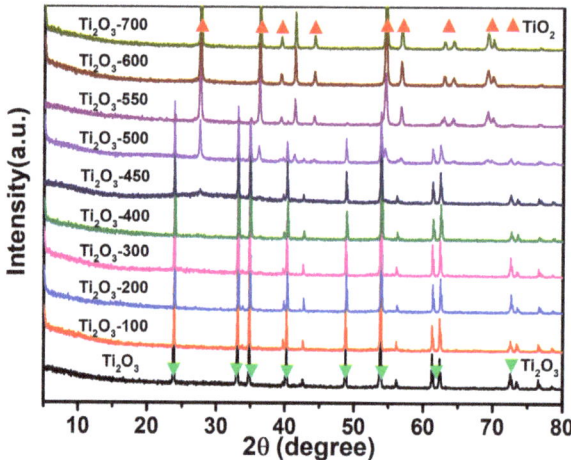

Figure 4. XRD spectra of Ti$_2$O$_3$, Ti$_2$O$_3$-100, Ti$_2$O$_3$-200, Ti$_2$O$_3$-300, Ti$_2$O$_3$-400, Ti$_2$O$_3$-450, Ti$_2$O$_3$-500, Ti$_2$O$_3$-550, Ti$_2$O$_3$-600, and Ti$_2$O$_3$-700.

3.3. The Results of the TEM Analyses

In order to further investigate the microstructure and crystal plane changes of the Ti$_2$O$_3$-based samples during the high-temperature reaction, we conducted TEM testing on the Ti$_2$O$_3$ samples after high-temperature treatment at 550 °C. Figure 5a shows that after the high-temperature treatment, there are significant differences between the microstructure of the outer layer and the internal structure of the Ti$_2$O$_3$@TiO$_2$ heterostructure, such as the particle accumulation and size. As shown in Figure 5b, electron diffraction can further indicate that the particulate matter comprises Ti$_2$O$_3$ and TiO$_2$. The high-resolution transmission test results show that the lattice fringes are d = 0.27 nm and 0.373 nm, belonging to the (104) and (0.12) of Ti$_2$O$_3$, and d = 0.219 nm, belonging to the (111) of TiO$_2$. The TEM results are consistent with the previous XRD and scanning test results, once again proving that the heat treatment method can construct Ti$_2$O$_3$@TiO$_2$ heterojunctions.

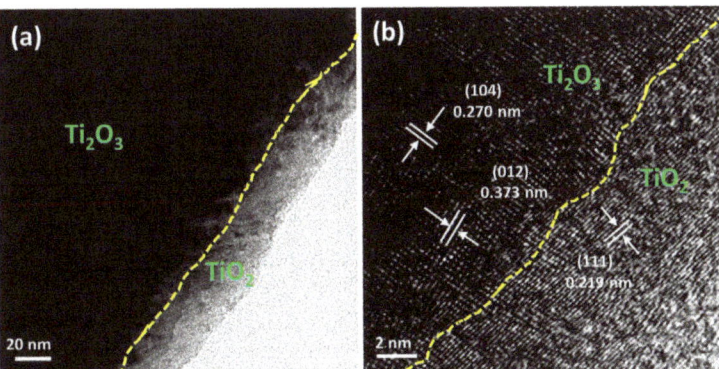

Figure 5. (a,b) High-resolution TEM image of the Ti_2O_3-550 after high-temperature heat treatment.

3.4. The Results of the N_2 Adsorption Analyses

N_2 adsorption–desorption experiments were conducted to investigate the effects of different temperatures on the specific surface area and pore size of the composite materials. Figure 6a shows that the specific surface area of the Ti_2O_3-550 is the largest, approximately 0.796 m^2/g, which is about 12 times that of the original Ti_2O_3 (approximately 0.067 m^2/g). The pore volume of the Ti_2O_3 significantly increased from 0.000467 cm^3/g to 0.0058 cm^3/g (Ti_2O_3-550), demonstrating the reconstruction of the microstructure of the material during the transition from Ti_2O_3 to TiO_2. In addition, in the Ti_2O_3@TiO_2 heterojunction system, the Ti_2O_3-700 bulk phase has the highest content of TiO_2, although its specific surface area is not the largest. This phenomenon proves that the expansion effect of the specific surface area in the system does not originate from the TiO_2. Therefore, the research results indicate that the temperature has significant control over the microstructure of the system. In addition, from the pore size distribution map (Figure 6b), it can be seen that the pore size of the original Ti_2O_3 is mainly distributed around 37.5 nm and 50 nm. With the increase in the heat treatment temperature, the Ti_2O_3@TiO_2 heterojunctions generate new pore structures in the bulk phase after retaining the original pore structure of the material, with the pore sizes mainly being concentrated in the range of 18–20 nm. The results suggest that the microstructure of the Ti_2O_3 did not undergo significant changes during the heat treatment process.

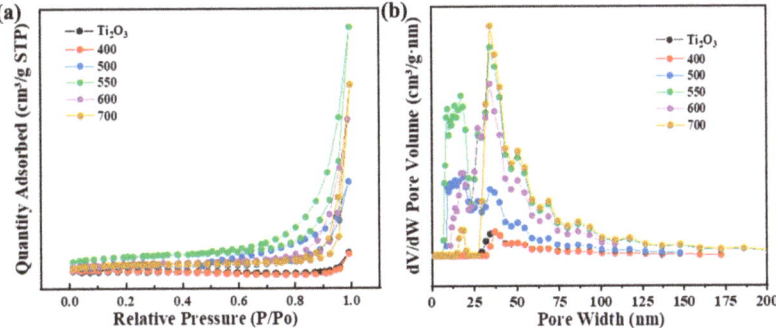

Figure 6. (a) N_2 adsorption–desorption isotherms of Ti_2O_3 at different heat treatment temperatures. (b) Pore size distribution curve of Ti_2O_3 at different heat treatment temperatures.

3.5. The Results of the UV–Vis Analyses

UV visible absorption spectroscopy was used to study the regulatory effect of the built-in electric field on the optical properties of the Ti_2O_3 and Ti_2O_3@TiO_2 heterojunctions. As shown in Figure 7, both the Ti_2O_3 and Ti_2O_3@TiO_2 heterojunctions exhibit strong absorption from 200 to 500 nm. It is worth noting that the Ti_2O_3-550 sample exhibits excellent absorption capacity throughout the entire process and significant tail absorption under the irradiation of visible light (λ > 600 nm). Compared with the as-received Ti_2O_3, the absorption ability of the Ti_2O_3-550 sample in the visible light absorption region remains unchanged, or even slightly enhanced. The existence of an internal electric field can promote the absorption ability of the Ti_2O_3-550 with regard to visible light. In addition, the bandgaps of the Ti_2O_3 and Ti_2O_3@TiO_2 heterojunctions can be estimated using the formula of $\alpha h\nu = A(h\nu - E_g)^{1/2}$ (E_g is the bandgap energy, α is the absorption coefficient, A is a constant, ν is the optical frequency, and h is the Planck constant) [34]. In the graph of photon energy, the bandgap energy value is the intersection point between the extended dashed line and the horizontal axis of the coordinate system (the horizontal axis value is hν = 1240/wavelength, while the vertical axis value is $(\alpha H\nu)^{1/2}$) [35,36]. As shown in Figure 7b, the calculated bandgaps of the pure Ti_2O_3 and Ti_2O_3-550, Ti_2O_3-600, and Ti_2O_3-700 are 0.4 eV, 0.08 eV, 1.82 eV, and 1.89 eV, respectively, indicating that the bandgaps of semiconductor materials can be effectively adjusted and optimized by constructing a reasonable internal electric field. Compared with other reported materials, the synthesized materials have relatively small bandgaps (Table 1).

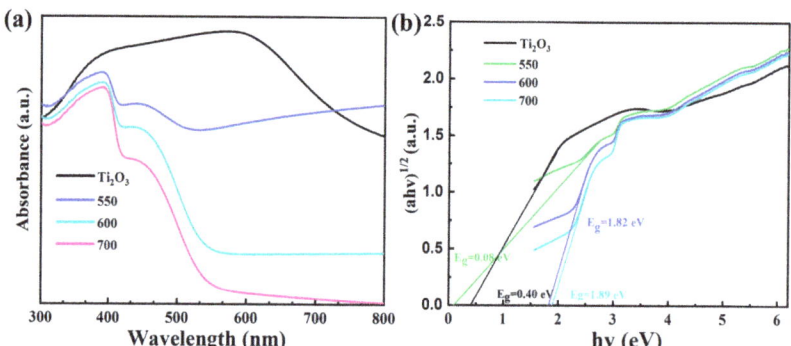

Figure 7. (a) Ultraviolet visible absorption spectra of the Ti_2O_3 and Ti_2O_3/TiO_2 heterojunctions (Ti_2O_3-550, Ti_2O_3-600, and Ti_2O_3-700) and (b) the corresponding Kubelka–Munk conversion reflection spectra.

Table 1. Comparison of the bandgaps of Ti_2O_3@TiO_2 and other reported studies.

Number	Titanium-Based Oxides	Bandgap	References
1	α-Ti_2O_3	1–2 eV	[37]
2	csp-Ti_2O_3	1–2 eV	[37]
3	Anatase TiO_2	3.2 eV	[38]
4	TiO_2	3.3 eV	[39]
5	α-Ti_2O_3	≈0.1 eV	[23]
6	TiO_2	3.23 eV	[40]
7	Ti_2O_3	3.18 eV	[40]
8	T-723 (TiO_x@anatase)	3.04 eV	[41]
9	T-810 (TiO_x@anatase)	3.1 eV	[41]
10	Ti_2O_3@TiO_2	0.08 eV	This work

3.6. The Results of the Optoelectronic Performance Analyses

Photoelectrochemical measurements in a typical three-electrode battery were used to study the electron generation and migration characteristics of the Ti_2O_3 and $Ti_2O_3@TiO_2$ heterojunctions. As shown in Figure 8a, among the pure Ti_2O_3 and $Ti_2O_3@TiO_2$ heterojunction systems, the Ti_2O_3-550 has the strongest photocurrent density, much higher than the original Ti_2O_3, confirming its excellent solar energy utilization ability. The above measurements intuitively prove that the material prepared at 550 °C has the best photoelectric performance. In addition, the relationship between the samples prepared at different heat treatment temperatures and the photocatalytic performance was systematically studied. The catalytic performance of the Ti_2O_3 and $Ti_2O_3@TiO_2$ heterojunctions (Ti_2O_3-400, Ti_2O_3-500, Ti_2O_3-550, Ti_2O_3-600) under visible light was investigated using TC as the substrate. As shown in Figure 8b, after stirring under dark conditions for 20 min, the matrix adsorption–desorption equilibrium of all the samples was observed. The photocatalytic performance of the Ti_2O_3 and $Ti_2O_3@TiO_2$ heterojunction composite materials was evaluated by measuring the degradation of the tetracycline in the sample under simulated visible light irradiation. From the results, it can be seen that Ti_2O_3 significantly enhances its ability to degrade tetracycline by constructing a $Ti_2O_3@TiO_2$ heterojunction, with the degradation rate increasing from 28% to 70.1% (35.05 mg of tetracycline degraded/100 mg of photocatalyst). Compared with other reported materials, the synthesized materials have relatively good photocatalytic degradation efficiency (Table 2). As the temperature further increases, the catalytic efficiency decreases. The experimental results indicate that the excellent photocatalytic performance of the Ti_2O_3-550 can be attributed to the narrower bandgap, reasonable band structure, and larger specific surface area. More importantly, the proper construction of the heterojunction may result in the redistribution of the electron density to build an internal electric field in the $Ti_2O_3@TiO_2$ and inhibit the recombination process of the electron–hole pairs. Therefore, by constructing heterojunctions reasonably and achieving precise regulation of the built-in electric fields, the photocatalytic activity of semiconductor materials can be optimized.

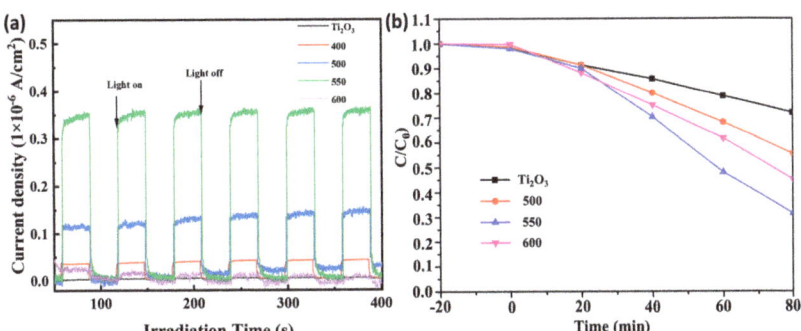

Figure 8. (a) Photogenerated current of the pure Ti_2O_3 and Ti_2O_3-400, Ti_2O_3-500, Ti_2O_3-550, and Ti_2O_3-600, and (b) the photocatalytic degradation performance of tetracycline.

Table 2. Comparison of the degradation efficiency of $Ti_2O_3@TiO_2$ and other reported studies.

Number	Titanium-Based Oxides	Degradation Efficiency (mg of Tetracycline Degraded/mg of Photocatalyst)	References
1	TiO_2	5.67/20	[1]
2	$MnTiO_3/Ag/gC_3N_4$	6.1/10	[42]
3	TBM0.05-5	0.66/10	[8]
4	Ag/ZnO@BC	0.703/10	[43]
5	Bi_3O_4Br	0.622/20	[44]

Table 2. Cont.

Number	Titanium-Based Oxides	Degradation Efficiency (mg of Tetracycline Degraded/mg of Photocatalyst)	References
6	CTF-Bi$_2$WO$_6$	0.771/20	[45]
7	2.0%Au/BiOCOOH	1.13/20	[46]
8	g-C$_3$N$_{4-x}$/g-C$_3$N$_4$	1.56/50	[47]
9	Pristine TiO$_2$	0.63/10	[8]
10	Ti$_2$O$_3$@TiO$_2$	35.05/100	This work

3.7. Discussion on the Photocatalytic Mechanism

In order to investigate the built-in electric field effect in the Ti$_2$O$_3$ and Ti$_2$O$_3$@TiO$_2$ heterojunctions in depth, the surface photovoltage (SPV) was used to investigate the separation and migration process of the photogenerated charge carriers in the materials [48]. Figure 9a,b show that the SPV response signals of the original Ti$_2$O$_3$ and Ti$_2$O$_3$-400 were not detected. This may be due to the materials' low separation efficiency or the high recombination efficiency of the photogenerated carriers. When the heat treatment temperature was further increased to 500 and 550 °C (Figure 9c,d), significant photovoltage response signals were observed at 275–400 nm for the Ti$_2$O$_3$-500 and Ti$_2$O$_3$-550, which can be attributed to the electronic transitions between the valence band and conduction band [49,50]. Moreover, the SPV response is positive, indicating that the upward energy band bends toward the surface/interface [51]. It can be clearly seen that the SPV response values have increased to 4.1 µV and 2.33 mV, respectively. This increase is caused by the built-in electric field constructed by the Ti$_2$O$_3$ and TiO$_2$, which continuously adjusts with the composition. With the increase in the heat treatment temperature, the amount of TiO$_2$ in the outer layer from the conversion of the Ti$_2$O$_3$ gradually increases and the built-in electric field effect is continuously enhanced. Thus, the separation of the photoexcited charge carriers is promoted, resulting in changes in the surface photovoltage. When the direction of the applied electric field is changed, there is no significant change in the photogenerated voltage of the Ti$_2$O$_3$-500 (Figure 9c), indicating that the direction of the built-in electric field of the Ti$_2$O$_3$-500 is opposite to the direction of the positive applied electric field. As shown in Figure 9d, the photogenerated voltage of the Ti$_2$O$_3$-550 significantly increases with the applied electric field. When the direction of the applied electric field is changed, the photogenerated voltage of the Ti$_2$O$_3$-500 is suppressed and reduced, indicating that the direction of the built-in electric field of the Ti$_2$O$_3$-550 is the same as the direction of the positive applied electric field. However, when the heat treatment temperature is increased to 600 °C (Figure 9e), the outer TiO$_2$ content further increases. The SPV response of the Ti$_2$O$_3$-600 decreases to 0.75 mV, indicating that in the presence of excessive TiO$_2$, the built-in electric field effect decreases and the separation of the photoexcited charge carriers and the recombination of the photogenerated electron–hole pairs are suppressed. In addition, the photogenerated voltage of the Ti$_2$O$_3$-600 increases significantly with the increase in the applied electric field. When the direction of the applied electric field changes, the photogenerated voltage of the Ti$_2$O$_3$-600 is suppressed and reduced, indicating that the direction of the built-in electric field of the Ti$_2$O$_3$-600 is the same as that of the positive applied electric field. Photogenerated charge carriers can recombine through radiation or non-radiation decay. Therefore, by comparing the changes in the SPV response with the heat treatment temperature in the Ti$_2$O$_3$ and Ti$_2$O$_3$@TiO$_2$ heterojunctions, it can be inferred that the Ti$_2$O$_3$-550 has the best built-in electric field effect, which can effectively suppress non-radiative decay, allowing for the more effective separation of the photogenerated charge carriers and transport of the photogenerated electrons to the catalyst surface to facilitate the degradation of the tetracycline [52,53].

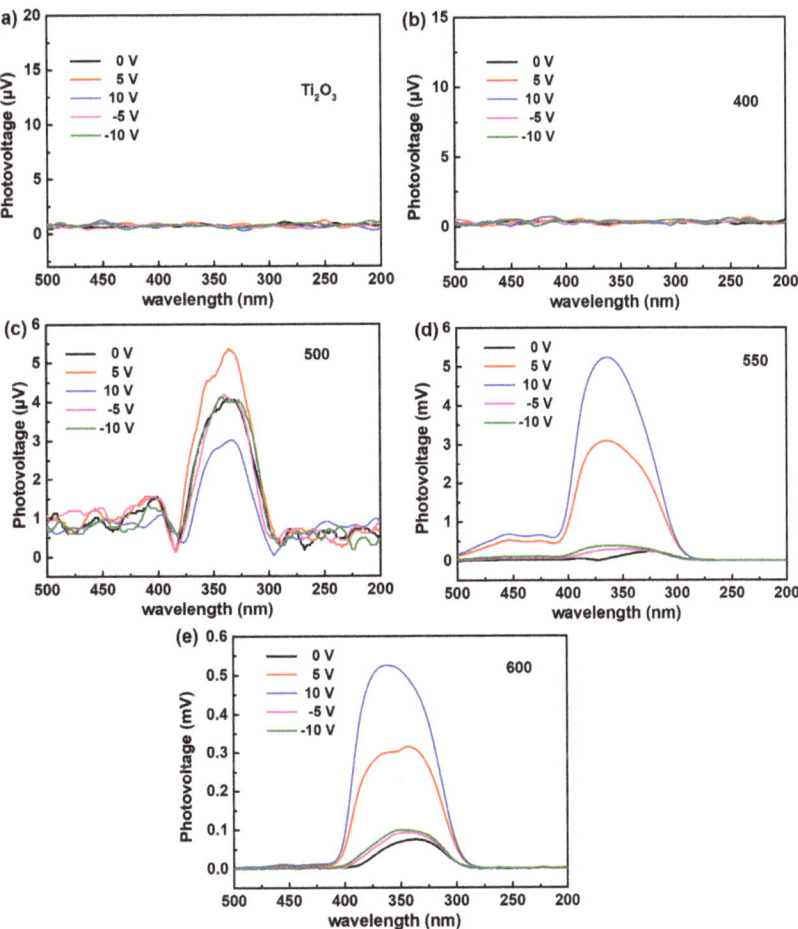

Figure 9. Surface photovoltage (SPV) diagrams of the Ti_2O_3 and Ti_2O_3/TiO_2 heterojunctions: (a) Ti_2O_3, (b) Ti_2O_3-400, (c) Ti_2O_3-500, (d) Ti_2O_3-550, and (e) Ti_2O_3-600.

As shown in Figure 10, the slope of the straight-line portion of the Mott–Schottky curve of the Ti_2O_3 and $Ti_2O_3@TiO_2$ heterojunctions (Ti_2O_3-500, Ti_2O_3-550, and Ti_2O_3-600) within the passivation zone is positive, indicating that the semiconductor film formed by the composite materials in this solution is an n-type semiconductor [54,55]. When the $Ti_2O_3@TiO_2$ heterojunction is not in contact with the solution, the material's Fermi energy level is higher than the solution's chemical potential. When the $Ti_2O_3@TiO_2$ heterojunction enters the solution, the carriers, i.e., electrons, in the bulk phase will spontaneously transfer from the high energy level to the low energy level, that is, from the side of the $Ti_2O_3@TiO_2$ to the solution, to balance the Fermi energy levels [56]. A space charge layer is formed on one side of the $Ti_2O_3@TiO_2$ film, causing the bulk phase to carry opposite charges to the solution. Excess charges are distributed within the space charge layer. Due to the lower concentration of electrons in the $Ti_2O_3@TiO_2$ bulk phase compared to the solution, electrons continue to enter the solution. Therefore, in the surface area on the semiconductor side, electrons are consumed, leaving only positive charges. This process causes the energy ratio near the surface to be corrected internally in the semiconductor, resulting in the energy bands of the Ti_2O_3 and $Ti_2O_3@TiO_2$ heterojunctions bending upwards in this region. As electrons continue to propagate, the energy band continues to bend upwards. If an external voltage is continuously applied to the Ti_2O_3 and $Ti_2O_3@TiO_2$ heterojunctions, the charge distribution at the interface changes with the electron input and the band bending changes. When the electrons that flood into the space charge layer neutralize the excess positive charge there, the energy band of the semiconductor returns to the same level as when it was not in contact with the solution. At this point, the required voltage is the flat band voltage. Suppose there is a built-in electric field in the semiconductor. In that case, it will decrease the necessary external voltage required to determine the potential size of the built-in electric field.

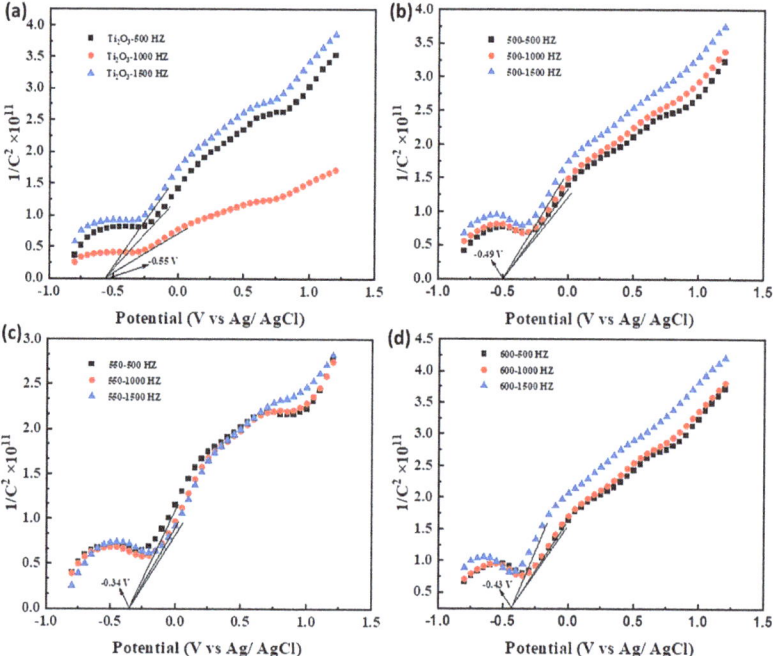

Figure 10. Mott–Schottky diagrams of the Ti_2O_3 and Ti_2O_3/TiO_2 heterojunctions: (**a**) Ti_2O_3, (**b**) Ti_2O_3-500, (**c**) Ti_2O_3-550, and (**d**) Ti_2O_3-600.

As shown by the results, the flat band potentials of the Ti_2O_3, Ti_2O_3-500, Ti_2O_3-550, and Ti_2O_3-600 are −0.55 V, −0.49 V, −0.34 V, and −0.43 V, respectively. The results indirectly prove that in the $Ti_2O_3@TiO_2$ heterojunction system, the built-in electric field of the Ti_2O_3-550 has the highest potential. In addition, the results of the Mott–Schottky curve indicate that the Ti_2O_3-550 has the highest carrier concentration, which is consistent with the conclusion obtained via the SPV. The Ti_2O_3-550 has a high Fermi energy level, which can endow the $Ti_2O_3@TiO_2$ heterostructure with a more excellent electronic reservoir capability and accelerate the carrier separation and conversion.

These experimental results indicate that under visible light irradiation, the photocatalytic reaction of the $Ti_2O_3@TiO_2$ heterojunction is direct photocatalysis, where electrons are excited by visible light from the valence band of the TiO_2 to the transition band of the Ti_2O_3, which is lower than the conduction band of the TiO_2. Then, electrons can easily transfer from the Ti (III) site to the adjacent Ti (IV) sites through the valence excitation process [22]. Under visible light irradiation, photogenerated holes are injected into the conduction band of the photocatalyst by the built-in electric field. Subsequently, electrons are captured to produce reactive substances and induce degradation reactions (Figure 11).

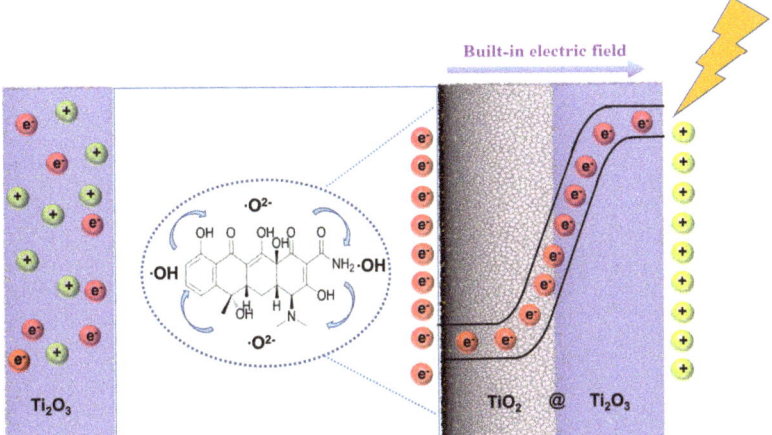

Figure 11. Schematic diagram of photocatalytic TC degradation over the Ti$_2$O$_3$@TiO$_2$ heterojunctions.

4. Conclusions

In summary, Ti$_2$O$_3$@TiO$_2$ heterojunctions have been constructed via a sample heat treatment method as an efficient TC degradation system. The photocatalytic activity of the Ti$_2$O$_3$@TiO$_2$ heterojunctions for TC degradation under visible light was investigated. Compared with the pure Ti$_2$O$_3$, the photocatalytic performance of the Ti$_2$O$_3$@TiO$_2$ heterojunctions was significantly improved. After the heterojunction structure was constructed, the photodegradation rate increased from 28% to 70.1% (35.05 mg of tetracycline degraded/100 mg of photocatalyst). The relevant experimental results indicated that the excellent photocatalytic performance of the Ti$_2$O$_3$-550 can be attributed to the narrow bandgap and reasonable band structure derived from the built-in electric field. By using the SPV and Mott–Schottky methods, it was revealed that the construction of the heterojunctions in the Ti$_2$O$_3$@TiO$_2$ resulted in the construction of an internal electric field, which suppressed the recombination process of the photogenerated electron–hole pairs. The most significant aspect here is that the built-in electric field can be controlled by changing the content of Ti$_2$O$_3$ and TiO$_2$. Overall, this work provides a simple method to precisely regulate the built-in electric field and optimize the photocatalytic activity of semiconductor materials by reasonably constructing heterojunctions.

Author Contributions: Conceptualization, F.L., P.F. and H.C.; methodology, T.H. and L.G.; formal analysis, T.H., F.L., P.F. and H.C.; investigation, T.H. and L.G.; writing—original draft preparation, T.H.; writing—review and editing, F.L., P.F. and H.C.; visualization, P.F.; supervision, F.L., P.F. and H.C. All authors have read and agreed to the published version of the manuscript.

Funding: We gratefully acknowledge the support of the National Natural Science Foundation of China (22205124), the Shandong Province Natural Science Foundation (ZR2021QB070), Basic Research Projects for the Pilot Project of Integrating Science and Education and Industry of Qilu University of Technology (Shandong Academy of Sciences) (2023PY024), the Development Plan of the Youth Innovation Team in Colleges and Universities of Shandong Province, and Basic Research Projects for the Pilot Project of Integrating Science and Education and Industry of Qilu University of Technology (Shandong Academy of Sciences) (2023PX108).

Data Availability Statement: Not applicable.

Conflicts of Interest: The authors declare no conflict of interest.

References

1. Wu, S.; Hu, H.; Lin, Y.; Zhang, J.; Hu, Y.H. Visible light photocatalytic degradation of tetracycline over TiO_2. *Chem. Eng. J.* **2020**, *382*, 122842. [CrossRef]
2. Xiao, Z.-J.; Feng, X.-C.; Shi, H.-T.; Zhou, B.-Q.; Wang, W.-Q.; Ren, N.-Q. Why the cooperation of radical and non-radical pathways in PMS system leads to a higher efficiency than a single pathway in tetracycline degradation. *J. Hazard. Mater.* **2022**, *424*, 127247. [CrossRef] [PubMed]
3. Zhang, Q.; Jiang, L.; Wang, J.; Zhu, Y.; Pu, Y.; Dai, W. Photocatalytic degradation of tetracycline antibiotics using three-dimensional network structure perylene diimide supramolecular organic photocatalyst under visible-light irradiation. *Appl. Catal. B Environ.* **2020**, *277*, 119122. [CrossRef]
4. Li, R.; Zhang, Y.; Deng, H.; Zhang, Z.; Wang, J.J.; Shaheen, S.M.; Xiao, R.; Rinklebe, J.; Xi, B.; He, X.; et al. Removing tetracycline and Hg(II) with ball-milled magnetic nanobiochar and its potential on polluted irrigation water reclamation. *J. Hazard. Mater.* **2020**, *384*, 121095. [CrossRef]
5. Guo, J.; Wang, L.; Wei, X.; Alothman, Z.A.; Albaqami, M.D.; Malgras, V.; Yamauchi, Y.; Kang, Y.; Wang, M.; Guan, W.; et al. Direct Z-scheme $CuInS_2/Bi_2MoO_6$ heterostructure for enhanced photocatalytic degradation of tetracycline under visible light. *J. Hazard. Mater.* **2021**, *415*, 125591. [CrossRef]
6. Yang, D.; Xu, Y.; Pan, K.; Yu, C.; Wu, J.; Li, M.; Yang, F.; Qu, Y.; Zhou, W. Engineering surface oxygen vacancy of mesoporous CeO_2 nanosheets assembled microspheres for boosting solar-driven photocatalytic performance. *Chin. Chem. Lett.* **2022**, *33*, 378–384. [CrossRef]
7. Shen, Q.; Wei, L.; Bibi, R.; Wang, K.; Hao, D.; Zhou, J.; Li, N. Boosting photocatalytic degradation of tetracycline under visible light over hierarchical carbon nitride microrods with carbon vacancies. *J. Hazard. Mater.* **2021**, *413*, 125376. [CrossRef]
8. Wang, Y.; Rao, L.; Wang, P.; Guo, Y.; Guo, X.; Zhang, L. Porous oxygen-doped carbon nitride: Supramolecular preassembly technology and photocatalytic degradation of organic pollutants under low-intensity light irradiation. *Environ. Sci. Pollut. Res.* **2019**, *26*, 15710–15723. [CrossRef]
9. Fang, B.; Xing, Z.; Sun, D.; Li, Z.; Zhou, W. Hollow semiconductor photocatalysts for solar energy conversion. *Adv. Powder Mater.* **2022**, *1*, 100021. [CrossRef]
10. Cui, Y.; Zheng, J.; Wang, Z.; Li, B.; Yan, Y.; Meng, M. Magnetic induced fabrication of core-shell structure $Fe_3O_4@TiO_2$ photocatalytic membrane: Enhancing photocatalytic degradation of tetracycline and antifouling performance. *J. Environ. Chem. Eng.* **2021**, *9*, 106666. [CrossRef]
11. Low, J.; Dai, B.; Tong, T.; Jiang, C.; Yu, J. In situ irradiated X-ray photoelectron spectroscopy investigation on a direct Z-scheme TiO_2/CdS composite film photocatalyst. *Adv. Mater.* **2019**, *31*, 1802981. [CrossRef]
12. Zhou, W.; Li, W.; Wang, J.-Q.; Qu, Y.; Yang, Y.; Xie, Y.; Zhang, K.; Wang, L.; Fu, H.; Zhao, D. Ordered mesoporous black TiO_2 as highly efficient hydrogen evolution photocatalyst. *J. Am. Chem. Soc.* **2014**, *136*, 9280–9283. [CrossRef]
13. Wang, W.; Han, Q.; Zhu, Z.; Zhang, L.; Zhong, S.; Liu, B. Enhanced photocatalytic degradation performance of organic contaminants by heterojunction photocatalyst $BiVO_4/TiO_2/RGO$ and its compatibility on four different tetracycline antibiotics. *Adv. Powder Technol.* **2019**, *30*, 1882–1896. [CrossRef]
14. Zhou, W.; Sun, F.; Pan, K.; Tian, G.; Jiang, B.; Ren, Z.; Tian, C.; Fu, H. Well-ordered large-pore mesoporous anatase TiO_2 with remarkably high thermal stability and improved crystallinity: Preparation, characterization, and photocatalytic performance. *Adv. Funct. Mater.* **2011**, *21*, 1922–1930. [CrossRef]
15. Guo, Q.; Zhou, C.; Ma, Z.; Yang, X. Fundamentals of TiO_2 photocatalysis: Concepts, mechanisms, and challenges. *Adv. Mater.* **2019**, *31*, 1901997. [CrossRef]
16. Li, Z.; Li, Z.; Zuo, C.; Fang, X. Application of nanostructured TiO_2 in UV photodetectors: A review. *Adv. Mater.* **2022**, *34*, 2109083. [CrossRef]
17. Schneider, J.; Matsuoka, M.; Takeuchi, M.; Zhang, J.; Horiuchi, Y.; Anpo, M.; Bahnemann, D.W. Understanding TiO_2 photocatalysis: Mechanisms and materials. *Chem. Rev.* **2014**, *114*, 9919–9986. [CrossRef]
18. Li, Z.; Li, H.; Wang, S.; Yang, F.; Zhou, W. Mesoporous black $TiO_2/MoS_2/Cu_2S$ hierarchical tandem heterojunctions toward optimized photothermal-photocatalytic fuel production. *Chem. Eng. J.* **2022**, *427*, 131830. [CrossRef]
19. Boscaro, P.; Cacciaguerra, T.; Cot, D.; Fajula, F.; Hulea, V.; Galarneau, A. C, N-doped TiO_2 monoliths with hierarchical macro-/mesoporosity for water treatment under visible light. *Microporous Mesoporous Mater.* **2019**, *280*, 37–45. [CrossRef]
20. Jagminas, A.; Ramanavičius, S.; Jasulaitiene, V.; Šimėnas, M. Hydrothermal synthesis and characterization of nanostructured titanium monoxide films. *RSC Adv.* **2019**, *9*, 40727–40735. [CrossRef]
21. Walsh, F.; Wills, R. The continuing development of Magnéli phase titanium sub-oxides and Ebonex® electrodes. *Electrochim. Acta* **2010**, *55*, 6342–6351. [CrossRef]
22. Liu, H.; Yang, W.; Ma, Y.; Yao, J. Extended visible light response of binary TiO_2-Ti_2O_3 photocatalyst prepared by a photo-assisted sol–gel method. *Appl. Catal. A Gen.* **2006**, *299*, 218–223. [CrossRef]
23. Li, Y.; Zhu, Y.; Wang, M.; Zhao, M.; Xue, J.; Chen, J.; Wu, T. Recent Progress on Titanium Sesquioxide: Fabrication, Properties, and Applications. *Adv. Funct. Mater.* **2022**, *32*, 2203491. [CrossRef]
24. Cai, Y.; Zhu, H.; Shi, Q.; Cheng, Y.; Chang, L.; Huang, W. Photothermal conversion of Ti_2O_3 film for tuning terahertz waves. *iScience* **2022**, *25*, 103661. [CrossRef] [PubMed]

25. Li, Y.; Weng, Y.; Yin, X.; Yu, X.; Kumar, S.S.; Wehbe, N.; Wu, H.; Alshareef, H.N.; Pennycook, S.J.; Breese, M.B.; et al. Orthorhombic Ti$_2$O$_3$: A Polymorph-Dependent Narrow-Bandgap Ferromagnetic Oxide. *Adv. Funct. Mater.* **2018**, *28*, 1705657. [CrossRef]
26. Chen, H.; Liang, J.; Li, L.; Zheng, B.; Feng, Z.; Xu, Z.; Luo, Y.; Liu, Q.; Shi, X.; Liu, Y.; et al. Ti$_2$O$_3$ nanoparticles with Ti^{3+} sites toward efficient NH$_3$ electrosynthesis under ambient conditions. *ACS Appl. Mater. Interfaces* **2021**, *13*, 41715–41722. [CrossRef]
27. Luo, Z.; Ye, X.; Zhang, S.; Xue, S.; Yang, C.; Hou, Y.; Xing, W.; Yu, R.; Sun, J.; Yu, Z.; et al. Unveiling the charge transfer dynamics steered by built-in electric fields in BiOBr photocatalysts. *Nat. Commun.* **2022**, *13*, 2230. [CrossRef]
28. Zhang, H.; Chen, X.; Zhang, Z.; Yu, K.; Zhu, W.; Zhu, Y. Highly-crystalline triazine-PDI polymer with an enhanced built-in electric field for full-spectrum photocatalytic phenol mineralization. *Appl. Catal. B Environ.* **2021**, *287*, 119957. [CrossRef]
29. Li, Z.; Wang, S.; Wu, J.; Zhou, W. Recent progress in defective TiO$_2$ photocatalysts for energy and environmental applications. *Renew. Sustain. Energy Rev.* **2022**, *156*, 111980. [CrossRef]
30. Cui, P.; Wei, D.; Ji, J.; Huang, H.; Jia, E.; Dou, S.; Wang, T.; Wang, W.; Li, M. Planar p–n homojunction perovskite solar cells with efficiency exceeding 21.3%. *Nat. Energy* **2019**, *4*, 150–159. [CrossRef]
31. Ramanavicius, S.; Jagminas, A.; Ramanavicius, A. Gas sensors based on titanium oxides. *Coatings* **2022**, *12*, 699. [CrossRef]
32. Chen, Y.; Mao, J. Sol–gel preparation and characterization of black titanium oxides Ti$_2$O$_3$ and Ti$_3$O$_5$. *J. Mater. Sci. Mater. Electron.* **2014**, *25*, 1284–1288. [CrossRef]
33. Liu, X.; Li, X.; Zhu, L.; Wang, X. Preparation of molecularly imprinted Ag-TiO$_2$ for photocatalytic removal of ethyl paraben. *Environ. Sci. Pollut. Res.* **2021**, *29*, 10308–10318. [CrossRef]
34. Lai, M.T.L.; Lee, K.M.; Yang, T.C.K.; Lai, C.W.; Chen, C.Y.; Johan, M.R.; Juan, J.C. Highly effective interlayer expanded MoS$_2$ coupled with Bi$_2$WO$_6$ as pn heterojunction photocatalyst for photodegradation of organic dye under LED white light. *J. Alloys Compd.* **2023**, *953*, 169834. [CrossRef]
35. Thill, A.S.; Lobato, F.O.; Vaz, M.O.; Fernandes, W.P.; Carvalho, V.E.; Soares, E.A.; Poletto, F.; Teixeira, S.R.; Bernardi, F. Shifting the band gap from UV to visible region in cerium oxide nanoparticles. *Appl. Surf. Sci.* **2020**, *528*, 146860. [CrossRef]
36. Jian, S.; Tian, Z.; Hu, J.; Zhang, K.; Zhang, L.; Duan, G.; Yang, W.; Jiang, S. Enhanced visible light photocatalytic efficiency of La-doped ZnO nanofibers via electrospinning-calcination technology. *Adv. Powder Mater.* **2022**, *1*, 100004. [CrossRef]
37. Zhao, X.; Selcuk, S.; Selloni, A. Formation and stability of reduced TiO$_x$ layers on anatase TiO$_2$ (101): Identification of a novel Ti$_2$O$_3$ phase. *Phys. Rev. Mater.* **2018**, *2*, 015801. [CrossRef]
38. Velempini, T.; Prabakaran, E.; Pillay, K. Recent developments in the use of metal oxides for photocatalytic degradation of pharmaceutical pollutants in water—A review. *Mater. Today Chem.* **2021**, *19*, 100380. [CrossRef]
39. Yu, X.; Li, Y.; Hu, X.; Zhang, D.; Tao, Y.; Liu, Z.; He, Y.; Haque, A.; Liu, Z.; Wu, T.; et al. Narrow bandgap oxide nanoparticles coupled with graphene for high performance mid-infrared photodetection. *Nat. Commun.* **2018**, *9*, 4299. [CrossRef]
40. Abdel-Aziz, M.; Yahia, I.; Wahab, L.; Fadel, M.; Afifi, M. Determination and analysis of dispersive optical constant of TiO$_2$ and Ti$_2$O$_3$ thin films. *Appl. Surf. Sci.* **2006**, *252*, 8163–8170. [CrossRef]
41. Zhang, H.; Zhao, Y.; Chen, S.; Yu, B.; Xu, J.; Xu, H.; Haoa, L.; Liu, Z. Ti^{3+} self-doped TiO$_x$@anatase core–shell structure with enhanced visible light photocatalytic activity. *J. Mater. Chem. A* **2013**, *1*, 6138–6144. [CrossRef]
42. Thinley, T.; Yadav, S.; Prabagar, J.S.; Hosakote, A.; Kumar, K.A.; Shivaraju, H.P. Facile synthesis of MnTiO$_3$/Ag/gC$_3$N$_4$ nanocomposite for photocatalytic degradation of tetracycline antibiotic and synthesis of ammonia. *Mater. Today Proc.* **2023**, *75*, 24–30. [CrossRef]
43. Li, J.; Wang, B.; Pang, Y.; Sun, M.; Liu, S.; Fang, W.; Chen, L. Fabrication of 0D/1D Bi$_2$MoO$_6$/Bi/TiO$_2$ heterojunction with effective interfaces for boosted visible-light photo-catalytic degradation of tetracycline. *Colloids Surf. A Physicochem. Eng. Asp.* **2022**, *638*, 128297. [CrossRef]
44. Zhang, W.; Peng, Y.; Yang, Y.; Zhang, L.; Bian, Z.; Wang, H. Bismuth-rich strategy intensifies the molecular oxygen activation and internal electrical field for the photocatalytic degradation of tetracycline hydrochloride. *Chem. Eng. J.* **2021**, *430*, 132963. [CrossRef]
45. Hosny, M.; Fawzy, M.; Eltaweil, A.S. Green synthesis of bimetallic Ag/ZnO@Biohar nanocomposite for photocatalytic degradation of tetracycline, antibacterial and antioxidant activities. *Sci. Rep.* **2022**, *12*, 7316. [CrossRef]
46. Ma, C.; Wei, J.; Jiang, K.; Yang, Z.; Yang, X.; Yang, K.; Zhang, Y.; Zhang, C. Self-assembled micro-flowers of ultrathin Au/BiOCOOH nanosheets photocatalytic degradation of tetracycline hydrochloride and reduction of CO$_2$. *Chemosphere* **2021**, *283*, 131228. [CrossRef]
47. Feng, C.; Ouyang, X.; Deng, Y.; Wang, J.; Tang, L. A novel g-C$_3$N$_4$/g-C$_3$N$_{4-x}$ homojunction with efficient interfacial charge transfer for photocatalytic degradation of atrazine and tetracycline. *J. Hazard. Mater.* **2023**, *441*, 129845. [CrossRef]
48. Zhang, Y.; Xie, T.; Jiang, T.; Wei, X.; Pang, S.; Wang, X.; Wang, D. Surface photovoltage characterization of a ZnO nanowire array/CdS quantum dot heterogeneous film and its application for photovoltaic devices. *Nanotechnology* **2009**, *20*, 155707. [CrossRef]
49. Hao, L.; Huang, H.; Guo, Y.; Zhang, Y. Multifunctional Bi$_2$O$_2$(OH)(NO$_3$) nanosheets with {001} active exposing facets: Efficient photocatalysis, dye-sensitization, and piezoelectric-catalysis. *ACS Sustain. Chem. Eng.* **2018**, *6*, 1848–1862. [CrossRef]
50. Kronik, L.; Shapira, Y. Surface photovoltage phenomena: Theory, experiment, and applications. *Surf. Sci. Rep.* **1999**, *37*, 1–206. [CrossRef]
51. Chen, R.; Fan, F.; Dittrich, T.; Li, C. Imaging photogenerated charge carriers on surfaces and interfaces of photocatalysts with surface photovoltage microscopy. *Chem. Soc. Rev.* **2018**, *47*, 8238–8262. [CrossRef] [PubMed]

52. Chen, R.; Fan, F.; Li, C. Unraveling charge-separation mechanisms in photocatalyst particles by spatially resolved surface photovoltage techniques. *Angew. Chem. Int. Ed.* **2022**, *134*, e202117567. [CrossRef]
53. Zhang, Z.; Chen, X.; Zhang, H.; Liu, W.; Zhu, W.; Zhu, Y. A highly crystalline perylene imide polymer with the robust built-in electric field for efficient photocatalytic water oxidation. *Adv. Mater.* **2020**, *32*, 1907746. [CrossRef] [PubMed]
54. Forghani, M.; McCarthy, J.; Cameron, A.P.; Davey, S.B.; Donne, S.W. Semiconductor properties of electrodeposited manganese dioxide for electrochemical capacitors: Mott-schottky analysis. *J. Electrochem. Soc.* **2021**, *168*, 020508. [CrossRef]
55. Sun, B.; Zhou, W.; Li, H.; Ren, L.; Qiao, P.; Li, W.; Fu, H. Synthesis of particulate hierarchical tandem heterojunctions toward optimized photocatalytic hydrogen production. *Adv. Mater.* **2018**, *30*, 1804282. [CrossRef]
56. Xu, D.; Zhang, S.-N.; Chen, J.-S.; Li, X.-H. Design of the Synergistic Rectifying Interfaces in Mott–Schottky Catalysts. *Chem. Rev.* **2022**, *123*, 1–30. [CrossRef]

Disclaimer/Publisher's Note: The statements, opinions and data contained in all publications are solely those of the individual author(s) and contributor(s) and not of MDPI and/or the editor(s). MDPI and/or the editor(s) disclaim responsibility for any injury to people or property resulting from any ideas, methods, instructions or products referred to in the content.

Article

Fabrication of Z-Type TiN@(A,R)TiO₂ Plasmonic Photocatalyst with Enhanced Photocatalytic Activity

Wanting Wang [1], Yuanting Wu [1], Long Chen [1], Chenggang Xu [1], Changqing Liu [1,2,*] and Chengxin Li [2]

[1] Shaanxi Key Laboratory of Green Preparation and Functionalization for Inorganic Materials, School of Material Science and Engineering, Shaanxi University of Science & Technology, Xi'an 710021, China
[2] State Key Laboratory for Mechanical Behavior of Materials, School of Materials Science and Engineering, Xi'an Jiaotong University, Xi'an 710049, China
* Correspondence: liuchangqing@sust.edu.cn

Abstract: Plasmonic effect-enhanced Z-type heterojunction photocatalysts comprise a promising solution to the two fundamental problems of current TiO_2-based photocatalysis concerning low-charge carrier separation efficiency and low utilization of solar illumination. A plasmonic effect-enhanced TiN@anatase-TiO_2/rutile-TiO_2 Z-type heterojunction photocatalyst with the strong interface of the N–O chemical bond was synthesized by hydrothermal oxidation of TiN. The prepared photocatalyst shows desirable visible light absorption and good visible-light-photocatalytic activity. The enhancement in photocatalytic activities contribute to the plasma resonance effect of TiN, the N–O bond-connected charge transfer channel at the TiO_2/TiN heterointerface, and the synergistically Z-type charge transfer pathway between the anatase TiO_2 (A-TiO_2) and rutile TiO_2 (R-TiO_2). The optimization study shows that the catalyst with a weight ratio of A-TiO_2/R-TiO_2/TiN of approximately 15:1:1 achieved the best visible light photodegradation activity. This work demonstrates the effectiveness of fabricating plasmonic effect-enhanced Z-type heterostructure semiconductor photocatalysts with enhanced visible-light-photocatalytic activities.

Keywords: Z-type system; LSPR; photocatalyst; TiO_2

Citation: Wang, W.; Wu, Y.; Chen, L.; Xu, C.; Liu, C.; Li, C. Fabrication of Z-Type TiN@(A,R)TiO₂ Plasmonic Photocatalyst with Enhanced Photocatalytic Activity. *Nanomaterials* 2023, 13, 1984. https://doi.org/10.3390/nano13131984

Academic Editor: Chiara Maccato

Received: 15 May 2023
Revised: 26 June 2023
Accepted: 27 June 2023
Published: 30 June 2023

Copyright: © 2023 by the authors. Licensee MDPI, Basel, Switzerland. This article is an open access article distributed under the terms and conditions of the Creative Commons Attribution (CC BY) license (https://creativecommons.org/licenses/by/4.0/).

1. Introduction

Green nanotechnology driven by solar energy has attracted great interest in alleviating the environmental hazards of pesticides, organic dyes, toxic gases, and industrial wastewater [1–6]. Since Carey et al. [7,8] used semiconductors to degrade pollutants in 1976, TiO_2 has proven to be a material that can be used for environmental purification. However, pure TiO_2 possesses a wide band gap (about 3.2 eV) [9,10]. Due to this limitation, it only responds to UV light and has low solar energy utilization (about 4%). Furthermore, the recombination rate of photogenerated charge carriers generated after TiO_2 excitation is much higher than that of interfacial charge transfer, resulting in low activity even under UV light [11]. Therefore, promoting solar utilization and charge carrier separation is the key to improving the photocatalytic performance of the catalysts.

Combining the localized surface plasmon resonance (LSPR) effect with semiconductor photocatalysts is a promising method to promote both the charge carrier separation efficiency and the responsive solar illumination range [12]. Till now, most plasmonic photocatalysts relied on noble metal nanostructures (such as Au or Ag) [13]. However, their potential for practical applications is limited due to their rarity, high cost, low thermal stability, and easy dissolution upon the exposure to air or humidity. Thus, novel plasmonic photocatalysts without noble metal components should be developed to overcome these problems.

Recently, TiN has emerged as an attractive competitor in photocatalytic applications due to its plasmonic resonance absorption properties [14]. In addition, the work function of TiN is about 4 eV versus vacuum [15], which is greater than or equal to the electron affinity of most semiconductor metal oxide photocatalysts, including TiO_2. Therefore, TiN tends to form a favorable energetic alignment to promote hot carrier-enhanced solar energy conversion [16]. Naldoni et al. [17] explored the plasmonic-enhanced TiO_2 photocatalysts by coupling with TiN, demonstrating that the LSPR effect of the TiN introduced an enhanced photocurrent generation and photocatalytic activity. Fakhouri et al. [18] demonstrated a significant photoactive improvement to bilayered RF magnetron-sputtering TiN/TiO_2 thin films due to enhanced charge separation at the heterojunction. Clatworthy et al. [19] demonstrated enhanced photocatalytic activity of $TiN-TiO_2$ nanoparticle composites and proposed that hot electrons migration can be promoted due to TiO_2 photovoltage by combining visible light with UV light. However, there is usually a certain lattice mismatch between different semiconductors, therefore constructed heterostructures usually result in large lattice defects and interface resistance [20]. These lattice defects often form the capturing center of photogenerated carriers [21], and the interface resistance would restrict charge transfer and affect their stability [22], thus greatly affecting the efficiency of charge carrier separation. Therefore, a novel plasmonic photocatalyst without noble metal components could be developed if a nanostructured TiN/TiO_2 composite with good contact could be created. Zhu et al. [23] found that the epitaxial growth of different semiconductors on conductive precursors and the regulation of growth conditions can significantly reduce the interface contact resistance, which can solve the challenge of building heterostructures to obtain high photogenerated charge separation characteristics. Li et al. [24] fabricated a TiN/TiO_2 plasmonic photocatalyst by in situ growing TiO_2 on TiN nanoparticles, demonstrating good visible light photocatalytic performance. Furthermore, Zhang et al. [25] significantly reduced the interface resistance and greatly improved their ability to photoelectrochemical decompose water by forming a strong interface contact of the S–O covalent bond at the interface of the Cu_2S/Fe_2O_3 heterostructure. In our previous work [26], the Ti–O–Zr bonded $TiO_2/ZrTiO_4$ heterointerface was constructed by growing $ZrTiO_4$ in situ on TiO_2 to enhance the transport of photogenerated carriers. However, the possibility of forming a chemically bonded TiO_2/TiN heterostructure and its synergistic enhancement of photocatalytic activities with the LSPR effect of TiN have yet to be explored.

In addition, fabrication of the direct Z-type heterojunction is an alternative strategy to obtain a semiconductor photocatalyst with high performance due to its advantage in charge carrier separation and utilizing the high-redox properties of each component [27]. For direct Z-type heterojunctions with staggered band structure, the photogenerated electrons on lower CB and the holes on the higher VB recombine. Meanwhile, the electrons and holes with stronger redox abilities are retained [28,29]. Thus, the charge carrier separation can be enhanced, and the highest redox potential of the heterojunction can be retained, thus contributing to the promoted photocatalytic activities. In our previous work, we successfully constructed a direct Z-type A-TiO_2/R-TiO_2 heterojunction by synergistically mediating oxygen vacancy contents and the band structure of the catalysts through B-doping [30]. However, to our best knowledge, there is no report concerning the possibility of combining TiN plasmonic enhancement with direct Z-type TiO_2-based heterojunction.

In this study, to solve the two fundamental problems of current TiO_2-based photocatalysis on low charge carrier separation efficiency and low utilization of solar illumination, we optimized a unique plasmonic effect-enhanced Z-type TiN@A-TiO_2/R-TiO_2 photocatalyst with a strong interface of the N–O chemical bond through hydrothermal in situ oxidation of TiN to (A,R)-TiO_2. In this photocatalyst system, the desirable visible light absorption could be attributed to the LSPR effect of the TiN component. The charge carrier separation efficiency could be enhanced by the Z-type charge transfer mode at the interface of the A-TiO_2/R-TiO_2 heterojunction. The obtained TiN@A-TiO_2/R-TiO_2 photocatalyst showed a distinct enhancement in visible light absorption, photocurrent generation, and photodegradation activities, demonstrating a simple way to promote the photoactive properties of semiconductor photocatalysts by fabricating plasmonic effect-enhanced Z-type heterostructures.

2. Experimental Section

2.1. Chemicals

Commercial titanium nitride (TiN, AR) were procured from Aladdin Reagent Co., Ltd., Shanghai, China. Ethanol (C_2H_5OH, CP); hydrogen peroxide (H_2O_2, 35 wt%, AR), rhodamine B (RhB, AR), and concentrated sulfuric acid (H_2SO_4, AR) were procured from the National Reagent company, Beijing, China. All reagents were used as received.

2.2. Preparation of the Catalyst

All samples were prepared by a simple hydrothermal process. Firstly, TiN powder was dispersed in 40 mL of deionized water and sonicated for 15 min to obtain TiN suspension. A total of 1 mL of H_2SO_4 (1 M) and a certain amount of H_2O_2 was added dropwise and stirred for 2 h. The suspension was then hydrothermally treated at 180 °C for 5 h, then washed and dried at 60 °C for 24 h to obtain the target samples. The hydrolysis degree of TiN was determined by the amount of H_2O_2 added. In this work, the mass fraction of added H_2O_2 is 0%, 0.5%, 1.0%, 2.5%, and 5.0%. The obtained catalysts were labeled as TiN, sample 1 (S1), sample 2 (S2), sample 3 (S3), and sample 4 (S4), respectively.

2.3. Characterization Methods

Compositions were recorded on a D/max-2200PC powder X-ray diffraction (XRD), with Cu Kα radiation over a 2θ ranging from 10° to 70°. Morphologies and microstructures were recorded by SEM (FEI Verios 460, Hillsboro, OR, USA), TEM, and HRTEM (FEI Tecnai G2 F20 S-TWIN, Hillsboro, OR, USA). XPS was studied on an X-ray photoelectron spectroscope (XPS, AXIS SUPRA, Manchester, UK) with a monochromatic Al Kα source. Comparing with the standard binding energy of adsorbed carbon (284.6 eV), charge correction was applied after the peak fitting using the CasaXPS analysis software. UV–Vis diffuse reflectance spectra (DRS) were tested on a UV–Vis-NIR spectrophotometer (Cary 5000, Santa Clara, CA, USA). Photoluminescence (PL) spectra were obtained by a fluorescence spectrophotometer (F-4600, Rigaku, Japan) with the excitation source at 345 nm. EPR spectra were conducted by a Bruker A300 spectrometer, during which DMPO was applied using a 300 W Xe lamp as the light source. The FT-IR spectrum (4000–500 cm^{-1}) was obtained on Vertex70 Bruker FT-IR Spectroscopy.

Photoelectrochemical analysis was performed on a CHI760D electrochemical workstation equipped with a 300 W Xe lamp and a cut off filter (>420 nm), in which 20 µL of catalyst slurry on an FTO substrate of 2 cm × 2 cm was used as the working electrode. The slurry was prepared by dispersing 5 mg catalyst powder into polyvinylidene difluoride N-methyl pyrrolidone solution (0.5 g, 2 wt%) through ultrasonic vibration. A total of 0.5 M Na_2SO_4 solution, platinum, and Ag/AgCl were used as the electrolyte solution, counter, and reference electrode, respectively.

2.4. Photocatalytic Performance

The photocatalytic activity of various catalysts was evaluated by RhB photodegradation. Firstly, 30 mL of RhB solution (10 mg/L) was prepared, then 30 mg of catalyst was added under stirring. Then, the above suspension was illuminated under visible light for 120 min, collected, centrifuged, and measured at regular intervals of 30 min. The peak absorbency of the centrifuged RhB solution at 554 nm was applied to analyze its concentration using a UV–Vis spectrophotometer.

3. Results and Discussion

3.1. Structural Characterization of the Photocatalysts

Figure 1 shows the XRD results of samples obtained with various H_2O_2 content. Except for the single-phase TiN sample, all other samples show diffraction peaks ascribed to three phases—that is, A-TiO_2, R-TiO_2, and TiN [31]. Moreover, as the H_2O_2 content increases, the peak intensity of TiN decreases and that of TiO_2 increases. The content of each phase was determined by the Rietveld method and shown in Table 1. With an increase in the H$_2$O$_2$ content, the phase proportion of TiN gradually decreases accompanied by an increase in the TiO$_2$ content, indicating the hydrolysis of TiN and its conversion into TiO$_2$. Furthermore, the ratio of A-TiO$_2$ to R-TiO$_2$ increases accordingly. Specifically, for sample S2, the weight ratio of the three phases (A-TiO$_2$:R-TiO$_2$:TiN) is about 15:1:1.

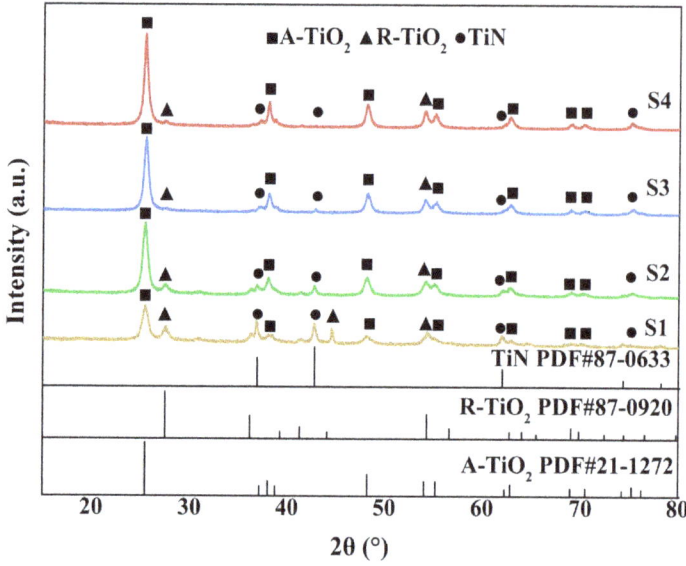

Figure 1. XRD results of the prepared samples obtained with various H$_2$O$_2$ content.

Table 1. Compositions and ratios of the A/R-TiO$_2$ and TiN in the prepared samples.

Samples	Content of H$_2$O$_2$ (wt%)	TiO$_2$/wt%			TiN/wt%
		A-TiO$_2$	R-TiO$_2$	A-TiO$_2$:R-TiO$_2$	
TiN	0	-	-	-	100%
S1	0.5	80.9%	19.1%	4.25	21.1%
S2	1.0	93.8%	6.2%	15.13	7.2%
S3	2.5	97.3%	2.7%	36.40	1.7%
S4	5.0	98.3%	1.7%	57.82	0.2%

Figure 2 shows the FT-IR spectra of sample TiN and S2. In the spectra, the wide absorption band at 3440 cm^{-1} and the peaks around 1633 cm^{-1} are ascribed to the adsorbed water and hydroxyl groups [32], respectively. The NO$_x$-determined peaks appeared at 1382 cm^{-1} and 1346 cm^{-1} [33]. Furthermore, peaks between 500–800 cm^{-1} are believed to be caused by the stretching vibration of Ti–O–Ti bonds [34]. Compared to the TiN sample, the maximum strength of the Ti–O–Ti bond increased significantly, and a new peak of NO$_x$ appeared, indicating the formation of TiO$_2$ and the possible existence of a newly formed N–O bond in sample S2.

Figures S1–S3 show the SEM images, particle size distributions, and BET surface areas of all the samples. As can be seen, all the samples show uniform and fine particle distribution with an average particle size of about 50 nm, except for sample TiN and S2, which show a slightly smaller size (around 35 nm). Moreover, except for the comparison samples (P25 and TiN), all the other samples have close BET surface areas, suggesting that surface area is not the reason for the photocatalytic performance difference between various samples. Microstructure was studied through TEM analysis. In the TEM results (Figure 3a), uniformly distributed irregular nanoparticles including polygonal, spherical, and rod-shaped particles can be observed. The length of rod particles is 50–200 nm, while the particle size of polygon and spherical particles is about 25 nm. In the HRTEM results (Figure 3b), spacings of 0.210, 0.324, and 0.352 nm of the lattice fringes, correspond to the (200) plane of TiN, the (110) plane of R-TiO$_2$, and the (101) plane of A-TO$_2$, respectively [35,36]. Moreover, the (A,R)-TiO$_2$ are identified on the surface of TiN, and all three phases are in close contact. Furthermore, the energy spectra (Figure 3c–f) show the evenly distributed Ti, N, and O elements, indicating the potential formation of the TiN/(A,R)-TiO$_2$ heterointerface. Therefore, it can be deduced that the in situ oxidation of TiN and growth of (A,R)-TiO$_2$ could create heterojunctions with an intimate contact interface, improving charge transfer efficiency.

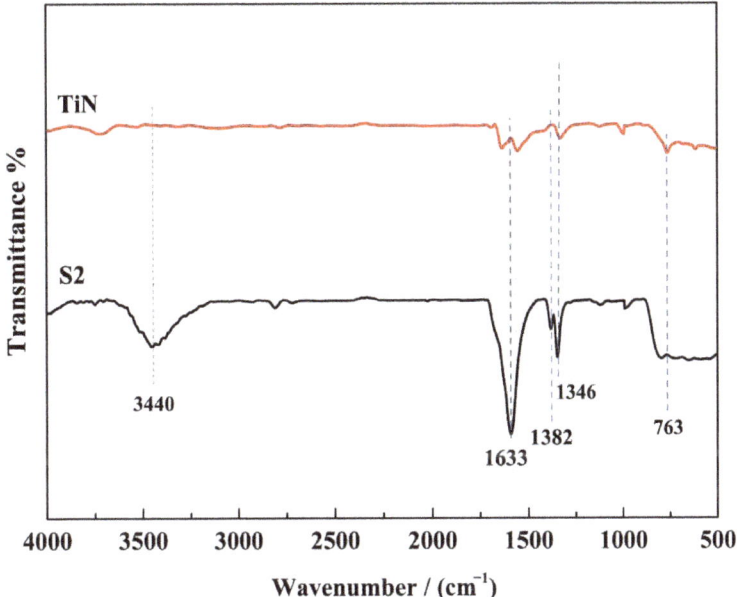

Figure 2. FT-IR spectra of TiN and sample S2.

Figure 4a is the XPS survey spectrum of S2, demonstrating the presence of C, O, and Ti elements. The N element was not detected, suggesting that it may appear in the interior of the particles. The Ti 2p XPS spectrum was fitted into four peaks. The Ti 2p3/2 (458.1 eV) and the Ti 2p1/2 (464.1 eV) were for TiO$_2$ [37–40]. The Ti 2p3/2 (457.3 eV) and Ti 2p1/2 (462.5 eV) were for partially oxidized TiN [41]. This observation confirms the creation of TiO$_2$ from the oxidation of TiN and suggests the possibility of forming a chemical contact interface between TiN and TiO$_2$. By further analysis of the O 1s spectrum of S2 (Figure 4c), four peaks can be fitted at 533.09, 531.54, 529.71, and 529.15 eV, which could be attributed to adsorbed H$_2$O (A$_O$), Ti–O in Ti$_2$O$_3$ suggests the existence of oxygen vacancies (V$_O$), and oxygen in Ti–O–N and Ti–O lattices [42–44], respectively. Considering the relatively high area ratio of the Vo XPS peak, it can be deduced that there is a high content of V$_O$ in the sample. Based on the analysis of the O1s spectrum, it is clear that N–O bonds exist between TiN and its oxidation products TiO$_2$, which contribute to the good contact interface of the formed TiN/TiO$_2$ heterostructure.

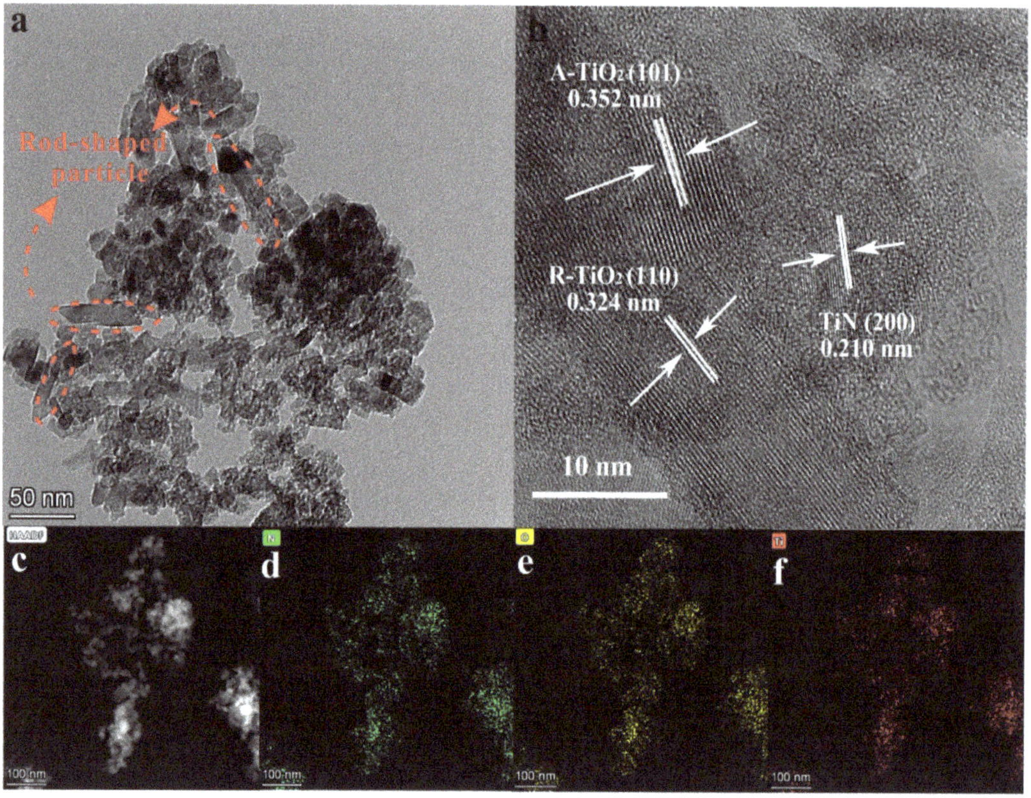

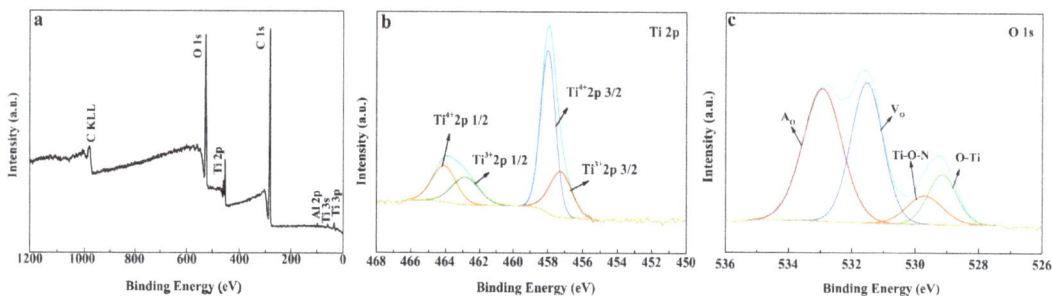

Figure 3. (a) TEM; (b) HRTEM; (c) HADDF images and EDS mappings of the elements N (d); O (e); Ti (f) for S2.

Figure 4. XPS spectra of S2: (a) full spectrum; (b) Ti 2p; and (c) O1s.

3.2. Photodegradation Performance

Figure 5 shows the photodegradation performance of the catalysts. As can be seen, all samples showed no obvious adsorption in the dark reaction. The sample TiN did not have a degrading effect on RhB, indicating that it is not the main catalytic carrier in the photocatalytic reaction process but a cocatalyst. Compared to TiN samples, the hydrolyzed samples showed obvious degrading behavior on RhB. With the deeper degree of hydrolysis, i.e., the decrease in TiN content and the increase in weight ratio of A-TiO_2 to R-TiO_2, the photodegradation rate increases first and then decreases. Among them, the S2 sample has the best degradation efficiency, reaching more than 97% in 90 min. As for the kinetics of RhB degradation, the degradation curves are well-fitted by a mono-exponential curve, indicating that the photodegradation experiments follow the first-order kinetics [35]. Figure 5b shows the relationship between ln (C_0/C) and t for all experiments using different samples, where C_0 is the initial RhB content and C is the RhB concentration at reaction time t. By regression analysis of the linear curve in the graph, the value of the apparent first-order rate constant can be directly obtained, in which the value of sample S2 is the highest 0.02272 min^{-1}. In addition, the cycling experiment (Figure 5c) shows that the sample can maintain a degradation efficiency of more than 90% after five cycles, showing good stability. From the XRD analysis in Figure 5d, no detectable differences can be seen between the as-prepared and cycled S2, indicating a well-preserved crystalline structure of the catalyst after multiple photocatalytic cycles. Moreover, a few studies on the RhB photodegradation performance of TiO_2-based photocatalysts are summarized in Table 2. The table shows that the visible-light degradation performance of RhB over the catalyst prepared in this work was enhanced, indicating that the prepared TiN@(A,R)TiO_2 is a promising visible-light photocatalyst.

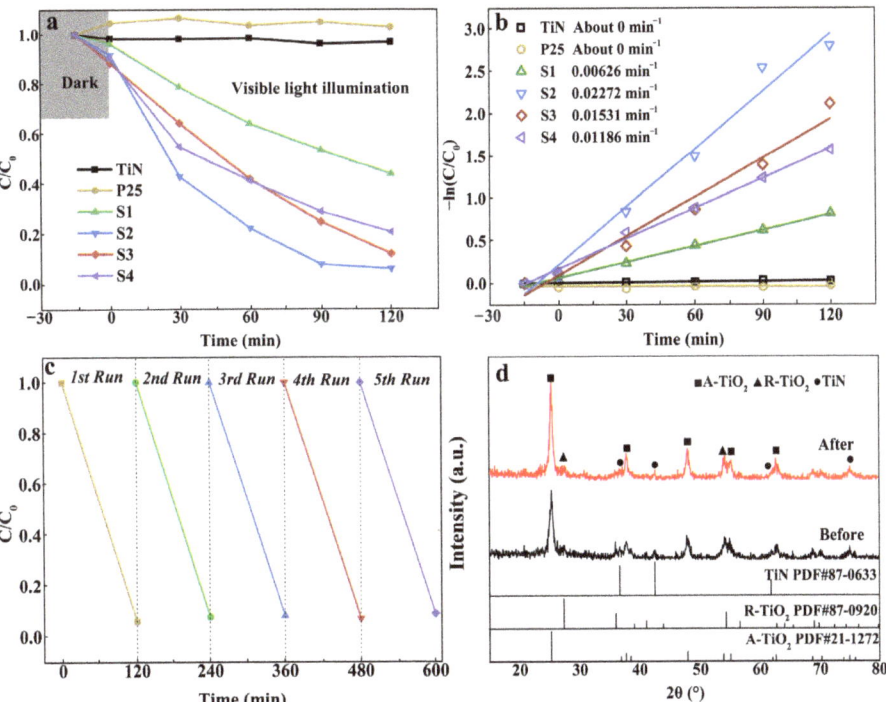

Figure 5. (a) Photodegradation performance; (b) kinetics of all prepared catalysts; (c) cycling experiments of sample S2; and (d) XRD patterns of the as-prepared and cycled sample S2.

Figure S4 shows the result of the free radical capture experiment. By adding IPA, TEOA, BQ, and $AgNO_3$ as the capture agents of ·OH, h^+, ·O_2^-, and e^-, respectively, the effect of free radicals on photocatalysis was investigated [45]. The addition of TEOA and BQ has the greatest impact on the photodegradation rate, suggesting that the corresponding h^+ and ·O_2^- may play the main role in the photodegradation process. EPR test was carried out on sample S2, and the result is shown in Figure 6. In the O_2^- free radical detection, the DMPO-·O_2^- signal peak of 1:1:1:1 was detected, and its intensity increased with prolonged irradiation time, confirming that the O_2^- radical is the main active species, whereas for

·OH radicals, no DMPO-·OH signal peak of 1:2:2:1 can be detected, suggesting that no ·OH radical can be produced during light irradiation. This result indirectly verified that h$^+$ might participate in the following photodegradation reaction without conversion into ·OH.

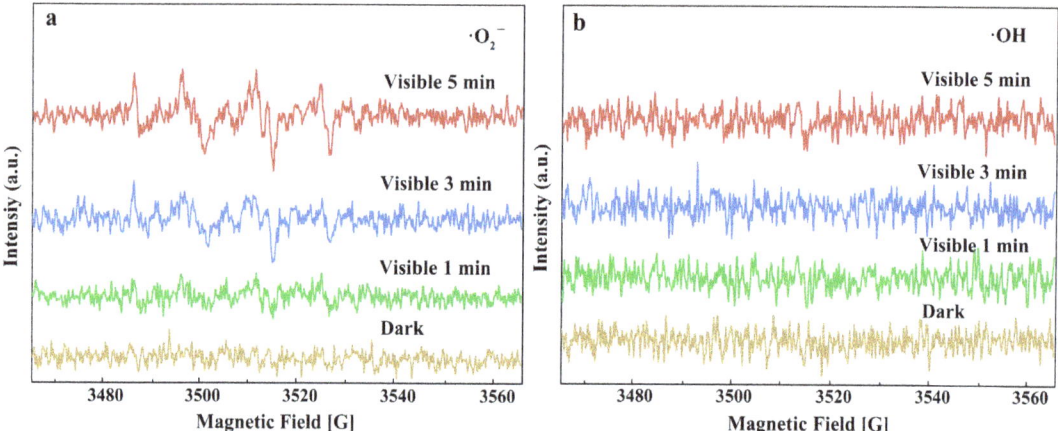

Figure 6. EPR results of (**a**) DMPO-•O_2^-; (**b**) DMPO-•OH with sample S2.

Table 2. Summary of recent relative works on the RhB photodegradation performance of TiO_2-based heterojunction photocatalysts.

Photocatalyst	C_0 (mg/L)	Dosage (mg)	Light Source	Degradation Rate	Time (min)	Kinetic Rate (min^{-1})	Ref.
Ag@TiO_2	10	100	150 W Xe lamp	98.2%	120	0.0188	[46]
TiO_2 hollow boxes	100	50	Visible light	96.5%	240	0.0025	[47]
Ag_2O/TiO_2	4.79	40	UV light	87.7%	80	0.0277	[48]
$Ag/ZnO/AgO/TiO_2$	10	30	350 W Xe lamp	99.3%	100	0.0230	[49]
$Pt/A/R-TiO_2$	-	-	UV light	92.4%	90	0.0280	[50]
$Bi_2WO_6/TiO_2/Pt$	20	100	UV light	60.0%	40	0.0210	[51]
$g-C_3N_4/TiO_2$	50	5	Visible light	87.0%	300	0.0115	[52]
$A/R-TiO_2$	10	25	UV light	About 100%	50	-	[53]
$Au/A/R-TiO_2$	-	-	UV light	97%	60	0.0470	[54]
TiN@(A,R)TiO_2	10	30	Visible light	97.0%	90	0.0227	This work

3.3. Photocatalytic Mechanism

Figure 7a shows the UV–Vis DRS spectra of all prepared catalysts. TiN shows full spectrum absorption characteristics similar to those of metals. Compared to P25 and sample S4 with little TiN content showing no obvious visible light absorption, the other samples show obvious light absorption in the entire visible light region (390–780 nm). Moreover, with increasing hydrolysis degree of TiN, the light absorption intensity gradually decreases, confirming that component TiN plays a decisive role in the light absorption ability of the prepared photocatalyst. The result is consistent with the report that the presence of TiN contributed to improving the material's entire solar light absorption capability [31].

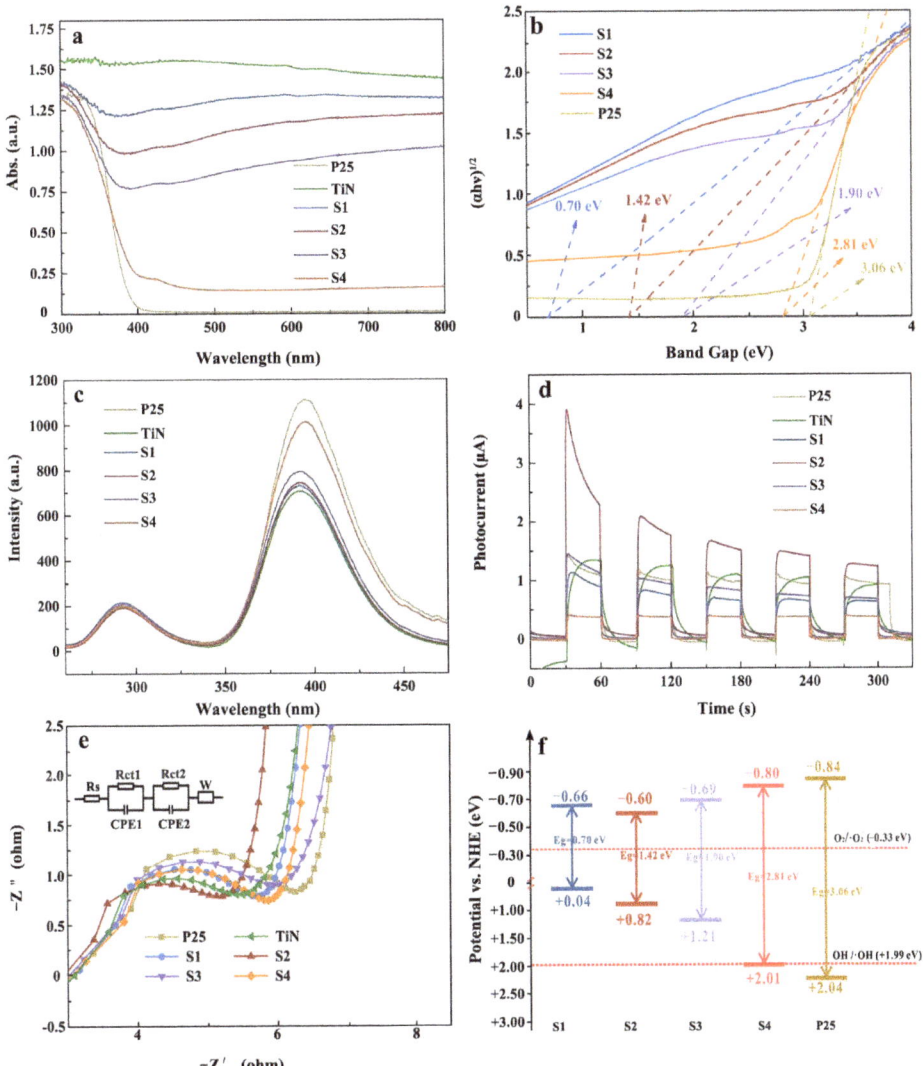

Figure 7. (**a**) UV–Vis DRS; (**b**) Plot of $(\alpha h\nu)^{1/2}$ versus $h\nu$; (**c**) PL spectra; (**d**) TP curves; (**e**) EIS plots; (**f**) band structures of the samples S1 (blue), S2 (red), S3 (purple), S4 (orange), P25 (yellowish-brown) and TiN (green).

The band gap (Eg) is further obtained through the conversion of Formula (1) [55]:

$$\alpha h\nu = A(h\nu - E_g)^{n/2}, \tag{1}$$

where A is a constant, n = 1 for indirect semiconductors [56], and α and h are the absorption coefficient and photon energy, respectively.

With the decrease in the TiN content, the Eg of the sample gradually increases from 0.70 eV (S1) to 3.06 eV (P25). Therefore, the presence of TiN can effectively reduce the band gap of the sample, thus significantly improving the capability of light absorbance and utilization.

Figure 7c shows the PL spectral analysis of the samples, in which a higher fluorescence intensity represents a higher carrier recombination rate [57]. TiN and P25 showed the lowest and strongest fluorescence intensity, respectively. The fluorescence intensity of the others gradually increased with increasing TiO$_2$ phase content. In particular, sample S2 also maintains low fluorescence intensity, indicating its outstanding charge separation ability. From the instantaneous photocurrent results in Figure 7d, the highest transient photocurrent signal can be observed for sample S2. The photocurrent signal decreases remarkably when the TiO$_2$ phase content is further increased. The above results demonstrate the photocurrent enhancement effect of TiN on TiO$_2$.

Figure 7e shows the electrochemical impedance spectroscopy (EIS) analysis of the samples. In addition, the resultant Nyquist plots (insert in Figure 7e) were fitted with an equivalent circuit using Zman software. As is shown, the equivalent circuit consists of internal resistance (Rs), charge transfer resistance (Rct1, Rct2), Warburg impedance (W), and double-layer capacitance (CPE1, CPE2) [58]. Compared to P25, the charge transfer resistance of the other samples (Figure 7e) is reduced to a certain extent, suggesting that the presence of TiN could improve the conductivity of the samples [59]. In particular, the charge transfer resistance of sample S2 is the lowest, demonstrating the greatest charge transfer rate. The above results demonstrate that the best charge carries separation and transfer can be obtained in sample S2. The reason can be explained by its proper phase proportion, and the formation of the N–O bond at the interface of (A,R)-TiO$_2$ and TiN, which can effectively reduce the interface contact resistance of the heterostructure.

The calculated flat band potentials (E_{fb}) are shown in Figure S5. The Mott–Schottky curves and calculation process of E_{fb} can be found in Figures S5 and S6. Considering that the E_{CB} of the n-type semiconductor is about 0.1 eV higher than its E_{fb}, the E_{CB} of the samples can be further deduced [60]. Combined with the band gap values (Eg), their energy band structures can be obtained (Figure 7f) through the following formula (2) [61]:

$$E_{VB} = E_{CB} + Eg. \qquad (2)$$

It can be seen in Figure 7f that the presence of TiN can significantly decrease the Eg of the samples by improving the valence band (VB) potential. For sample S2, the Eg was reduced from 3.06 eV to 1.42 eV with the VB position changing from +2.04 to +0.82 eV, and CB position was slightly changed compared to P25. With a narrow band gap, sample S2 is more conducive to generating e$^-$ and h$^+$ charge carriers, while it shows no ability to produce ·OH active species due to its high valence band position (VB), which is in good agreement with the result of the EPR test.

Considering the staggered band structures of the A-TiO$_2$/R-TiO$_2$, Type-II or Z-type charge transfer modes may occur in the heterojunction, as shown in Figure 8. The values of the CB and VB for (A,R)-TiO$_2$ are obtained from the literature [62]. If Type-II mode is formed, electrons transfer to the CB of A-TiO$_2$, the reduction potential of which is weak and cannot further reduce surface-adsorbed oxygen to generate ·O$_2^-$ for the following photodegradation process, and this situation is inconsistent with our experimental results. Therefore, the Z-type charge transfer pathway is preferred for the heterojunction constructed in our work. Specifically, according to the literature, for partially reduced samples, the VB and CB positions of R-TiO$_2$ are higher than that of A-TiO$_2$ and the work function of R-TiO$_2$ ($\varphi \approx 4.3$ eV) is smaller than that of A-TiO$_2$ ($\varphi \approx 4.7$ eV) [63,64]. When they are in contact, free electrons spontaneously flow from R-TiO$_2$ to A-TiO$_2$ to obtain their Fermi energy levels to reach equilibrium. At this time, there are a large number of negatively charged electrons near the A-TiO$_2$ interface. In contrast, positive charges are gathered at the R-TiO$_2$ interface, generating a built-in electric field. Due to the shift in the Fermi energy level, R-TiO$_2$ will generate an upward band bending, while A-TiO$_2$ will generate a downward band bending [65]. The Z-type electron transfer path is generated due to the formed electric field and the energy band bending. To further prove the formed heterojunction is a Z-type photocatalyst, the Ag nanoparticles were photo-deposited on the catalyst to track where the electrons flow to. Figure 9 shows the EDS, TEM, and HRTEM images of the photodeposition of Ag nanoparticles on sample S2. It shows the uniform distribution of Ag, N, O, and Ti elements, and the Ag nanoparticles were isolated on R-TiO$_2$ and apart from A-TiO$_2$. The results suggest that the electrons were left on R-TiO$_2$, confirming the Z-type charge transfer pathway in the formed heterojunction.

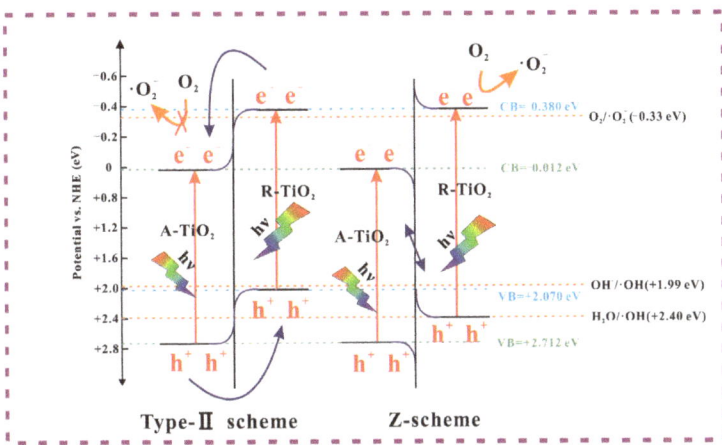

Figure 8. Schematic illustration of the possible charge carrier transfer mode in Type-II and direct Z-type photocatalysts.

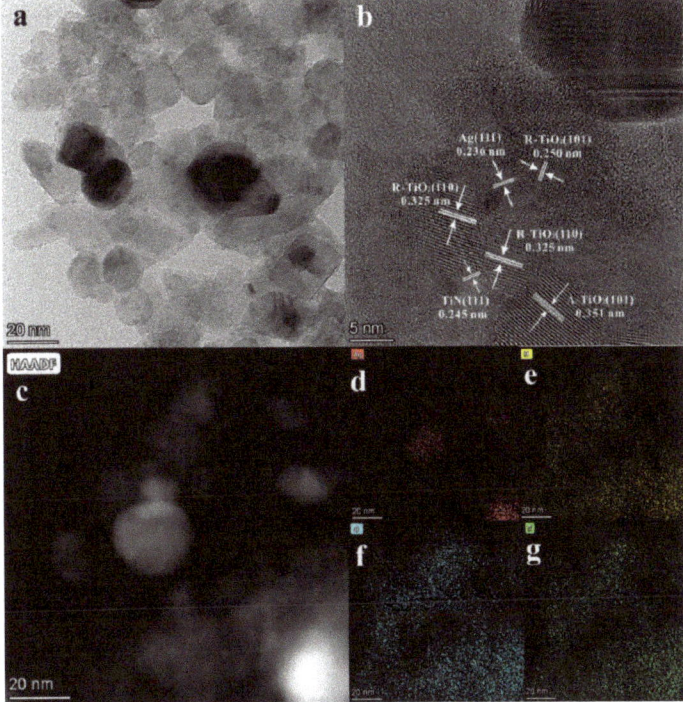

Figure 9. (**a**) TEM; (**b**) HRTEM; (**c**) HADDF images and EDS mappings of the elements Ag (**d**); N (**e**); O (**f**); Ti (**g**) for S2 with photo-deposited Ag nanoparticles.

In this work, the plasmonic component TiN can broaden the absorbed light range and generate hot electrons due to the LSPR effect. Since nanostructured TiO$_2$ was obtained in situ from TiN and charge transfer channel N–O bonds were formed between TiN and TiO$_2$, the resulting intimate contacted interface benefits the electron transfer between them. Furthermore, the work function of TiN is ~3.7 eV (φ_m), and the electron affinity of TiO$_2$ is ~4.2 eV (φ_s). Considering the barrier energy (the lowest energy required for an electron in the metal to be injected into the semiconductor) can be calculated as $\varphi = \varphi_m - \varphi_s$ [66], a negative value (−0.5 eV) can be obtained, suggesting the quick injecting of the hot electrons into TiO$_2$. Therefore, the improved photocatalytic performance of TiN@A-TiO$_2$/R-TiO$_2$ heterojunction can be concluded and shown in Figure 10. First, the plasmonic properties of TiN greatly broaden the light absorption range, generating and injecting hot electrons into TiO$_2$. Furthermore, the N–O bond contacted TiO$_2$/TiN heterointerface can significantly reduce the contact resistance of the interface and improve the charge transfer efficiency. Moreover, the optimized three-phase ratio and the formed Z-type A-TiO$_2$/R-TiO$_2$ heterojunction with an intimate interface contribute to the charge carrier separation and retain its high redox capacity. Thus, more active species will participate in the following photodegradation activities.

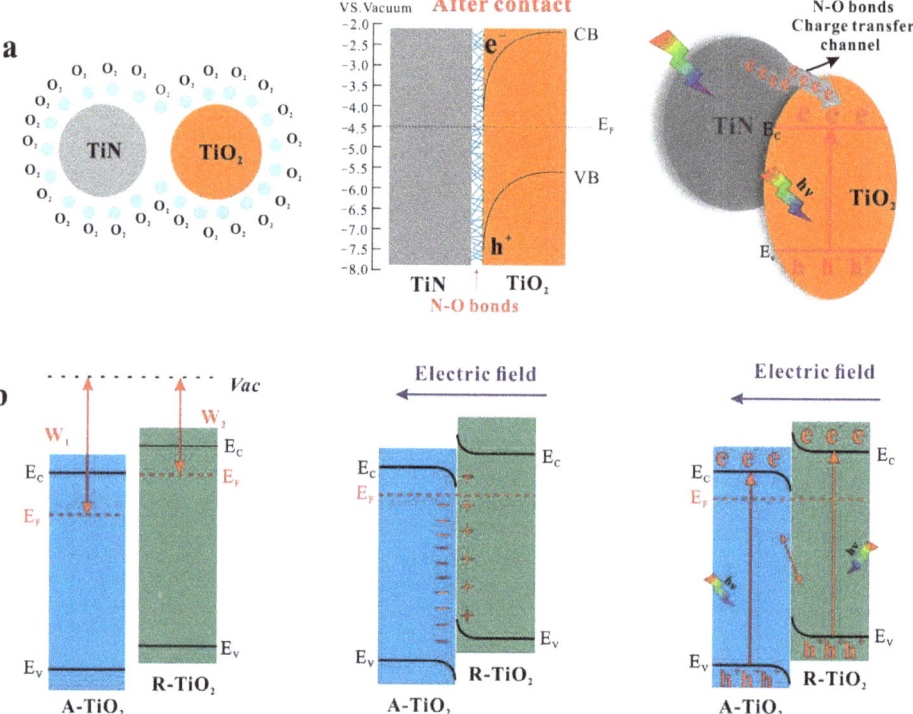

Figure 10. (**a**,**b**) Schematic illustration of the formation and a possible photoinduced catalytic mechanism of the TiN@(A,R)TiO$_2$ heterojunction.

4. Conclusions

In this work, a plasmonic effect-enhanced TiN@A-TiO$_2$/R-TiO$_2$ direct Z-type heterojunction was fabricated through the simple hydrothermal reaction process. By regulating the amount of H$_2$O$_2$ oxidant, the proportion of TiN, anatase TiO$_2$, and rutile TiO$_2$ contents can be successfully adjusted and the interface charge transfer channel (N–O bond) has been constructed. Due to the Z-type charge transfer path between A-TiO$_2$ and R-

TiO$_2$, the N–O bond connected charge transfer channel at the TiN/TiO$_2$ interface, and the synergistic plasma resonance effect of TiN, the optimized photocatalyst shows a distinct increment in visible light absorption, photocurrent generation, and photocatalytic performance, demonstrating an effective approach to promote the photoactive properties of semiconductor photocatalysts.

Supplementary Materials: The following supporting information can be downloaded at: https://www.mdpi.com/article/10.3390/nano13131984/s1, Figure S1: SEM images of (a) TiN, (b) S1, (c) S2, (d) S3, (e) S4. Figure S2: Particle size distribution of all samples (a) TiN, (b) S1, (c) S2, (d) S3, (e) S4. Figure S3: N2 adsorption/desorption isotherm plots and pore size distribution of all samples. Figure S4: Free radical capture experiment of sample S2. Figure S5: Mott-Schottky curves of samples (a) P25, (b) TiN, (c) S1, (d) S2, (e) S3, (f) S4. Figure S6: Schematic diagram of flat band potential of the prepared samples.

Author Contributions: Conceptualization, C.L. (Changqing Liu) and C.L. (Chengxin Li); Data curation, W.W. and L.C.; Formal analysis, C.X.; Funding acquisition, Y.W. and C.L. (Changqing Liu); Investigation, W.W.; Methodology, L.C.; Project administration, Y.W. and C.L. (Chengxin Li); Resources, Y.W. and C.L. (Chengxin Li); Supervision, Y.W. and C.L. (Changqing Liu); Validation, C.L. (Changqing Liu); Writing—original draft, W.W.; Writing—review & editing, C.L. (Changqing Liu). All authors have read and agreed to the published version of the manuscript.

Funding: This work has been supported by the National Natural Science Foundation of China (Grant No. 51702194 and 52173214), the Natural Science Foundation of Shaanxi Province (Grant No. 2023-JC-YB-384), and the Youth Innovation Team of Shaanxi Universities (2022-70).

Data Availability Statement: The data presented in this study are available on request from the corresponding author.

Conflicts of Interest: The authors declare no conflict of interest.

References

1. Masudy-Panah, S.; Siavash Moakhar, R.; Chua, C.S.; Kushwaha, A.; Dalapati, G.K. Stable and Efficient CuO Based Photocathode through Oxygen-Rich Composition and Au–Pd Nanostructure Incorporation for Solar-Hydrogen Production. *ACS Appl. Mater. Interfaces* **2017**, *9*, 27596–27606. [CrossRef] [PubMed]
2. Chong, M.N.; Jin, B.; Chow, C.W.K.; Saint, C. Recent Developments in Photocatalytic Water Treatment Technology: A Review. *Water Res.* **2010**, *44*, 2997–3027. [CrossRef] [PubMed]
3. Takata, T.; Domen, K. Particulate Photocatalysts for Water Splitting: Recent Advances and Future Prospects. *ACS Energy Lett.* **2019**, *4*, 542–549. [CrossRef]
4. Tong, H.; Ouyang, S.; Bi, Y.; Umezawa, N.; Oshikiri, M.; Ye, J. Nano-Photocatalytic Materials: Possibilities and Challenges. *Adv. Mater.* **2012**, *24*, 229–251. [CrossRef]
5. Hunge, Y.M.; Yadav, A.A.; Kang, S.-W.; Mohite, B.M. Role of Nanotechnology in Photocatalysis Application. *Recent Pat. Nanotechnol.* **2023**, *17*, 5–7. [CrossRef]
6. Yadav, A.A.; Hunge, Y.M.; Kang, S.-W.; Fujishima, A.; Terashima, C. Enhanced Photocatalytic Degradation Activity Using the V$_2$O$_5$/RGO Composite. *Nanomaterials* **2023**, *13*, 338. [CrossRef]
7. Carey, J.H.; Lawrence, J.; Tosine, H.M. Photodechlorination of PCB's in the Presence of Titanium Dioxide in Aqueous Suspensions. *Bull. Environ. Contam. Toxicol.* **1976**, *16*, 697–701. [CrossRef]
8. Poudel, M.B.; Kim, A.A. Silver Nanoparticles Decorated TiO$_2$ Nanoflakes for Antibacterial Properties. *Inorg. Chem. Commun.* **2023**, *152*, 110675. [CrossRef]
9. Hou, J.; Yang, C.; Wang, Z.; Jiao, S.; Zhu, H. Bi$_2$O$_3$ Quantum Dots Decorated Anatase TiO$_2$ Nanocrystals with Exposed {001} Facets on Graphene Sheets for Enhanced Visible-Light Photocatalytic Performance. *Appl. Catal. B Environ.* **2013**, *129*, 333–341. [CrossRef]
10. Kment, S.; Riboni, F.; Pausova, S.; Wang, L.; Wang, L.; Han, H.; Hubicka, Z.; Krysa, J.; Schmuki, P.; Zboril, R. Photoanodes Based on TiO$_2$ and α-Fe$_2$O$_3$ for Solar Water Splitting—Superior Role of 1D Nanoarchitectures and of Combined Heterostructures. *Chem. Soc. Rev.* **2017**, *46*, 3716–3769. [CrossRef]
11. Reddy, N.L.; Kumar, S.; Krishnan, V.; Sathish, M.; Shankar, M.V. Multifunctional Cu/Ag Quantum Dots on TiO$_2$ Nanotubes as Highly Efficient Photocatalysts for Enhanced Solar Hydrogen Evolution. *J. Catal.* **2017**, *350*, 226–239. [CrossRef]
12. Zhang, L.; Ding, N.; Lou, L.; Iwasaki, K.; Wu, H.; Luo, Y.; Li, D.; Nakata, K.; Fujishima, A.; Meng, Q. Localized Surface Plasmon Resonance Enhanced Photocatalytic Hydrogen Evolution via Pt@Au NRs/C$_3$N$_4$ Nanotubes under Visible-Light Irradiation. *Adv. Funct. Mater.* **2019**, *29*, 1806774. [CrossRef]

13. Boerigter, C.; Campana, R.; Morabito, M.; Linic, S. Evidence and Implications of Direct Charge Excitation as the Dominant Mechanism in Plasmon-Mediated Photocatalysis. *Nat. Commun.* **2016**, *7*, 10545. [CrossRef]
14. Guler, U.; Kildishev, A.V.; Boltasseva, A.; Shalaev, V.M. Plasmonics on the Slope of Enlightenment: The Role of Transition Metal Nitrides. *Faraday Discuss.* **2015**, *178*, 71–86. [CrossRef]
15. Lima, L.P.B.; Diniz, J.A.; Doi, I.; Godoy Fo, J. Titanium Nitride as Electrode for MOS Technology and Schottky Diode: Alternative Extraction Method of Titanium Nitride Work Function. *Microelectron. Eng.* **2012**, *92*, 86–90. [CrossRef]
16. Al-Hamdi, A.M.; Rinner, U.; Sillanpää, M. Tin Dioxide as a Photocatalyst for Water Treatment: A Review. *Process Saf. Environ. Prot.* **2017**, *107*, 190–205. [CrossRef]
17. Mascaretti, L.; Barman, T.; Bricchi, B.R.; Münz, F.; Li Bassi, A.; Kment, Š.; Naldoni, A. Controlling the Plasmonic Properties of Titanium Nitride Thin Films by Radiofrequency Substrate Biasing in Magnetron Sputtering. *Appl. Surf. Sci.* **2021**, *554*, 149543. [CrossRef]
18. Fakhouri, H.; Arefi-Khonsari, F.; Jaiswal, A.K.; Pulpytel, J. Enhanced Visible Light Photoactivity and Charge Separation in TiO_2/TiN Bilayer Thin Films. *Appl. Catal. Gen.* **2015**, *492*, 83–92. [CrossRef]
19. Clatworthy, E.B.; Yick, S.; Murdock, A.T.; Allison, M.C.; Bendavid, A.; Masters, A.F.; Maschmeyer, T. Enhanced Photocatalytic Hydrogen Evolution with TiO_2 –TiN Nanoparticle Composites. *J. Phys. Chem. C* **2019**, *123*, 3740–3749. [CrossRef]
20. Cherevan, A.S.; Gebhardt, P.; Shearer, C.J.; Matsukawa, M.; Domen, K.; Eder, D. Interface Engineering in Nanocarbon–Ta_2O_5 Hybrid Photocatalysts. *Energy Env. Sci* **2014**, *7*, 791–796. [CrossRef]
21. Bai, S.; Ge, J.; Wang, L.; Gong, M.; Deng, M.; Kong, Q.; Song, L.; Jiang, J.; Zhang, Q.; Luo, Y.; et al. A Unique Semiconductor-Metal-Graphene Stack Design to Harness Charge Flow for Photocatalysis. *Adv. Mater.* **2014**, *26*, 5689–5695. [CrossRef]
22. Maarisetty, D.; Baral, S.S. Defect Engineering in Photocatalysis: Formation, Chemistry, Optoelectronics, and Interface Studies. *J. Mater. Chem. A* **2020**, *8*, 18560–18604. [CrossRef]
23. Zhu, H.; Yang, Y.; Kang, Y.; Niu, P.; Kang, X.; Yang, Z.; Ye, H.; Liu, G. Strong Interface Contact between $NaYF_4$:Yb,Er and CdS Promoting Photocatalytic Hydrogen Evolution of $NaYF_4$:Yb,Er/CdS Composites. *J. Mater. Sci. Technol.* **2022**, *102*, 1–7. [CrossRef]
24. Li, J.; Gao, L.; Sun, J.; Zhang, Q.; Guo, J.; Yan, D. Synthesis of Nanocrystalline Titanium Nitride Powders by Direct Nitridation of Titanium Oxide. *J. Am. Ceram. Soc.* **2001**, *84*, 3045–3047. [CrossRef]
25. Zhang, Y.; Huang, Y.; Zhu, S.; Liu, Y.; Zhang, X.; Wang, J.; Braun, A. Covalent S-O Bonding Enables Enhanced Photoelectrochemical Performance of Cu_2S/Fe_2O_3 Heterojunction for Water Splitting. *Small* **2021**, *17*, 2100320. [CrossRef] [PubMed]
26. Ning, Q.; Zhang, L.; Liu, C.; Li, X.; Xu, C.; Hou, X. Boosting Photogenerated Carriers for Organic Pollutant Degradation via In-Situ Constructing Atom-to-Atom $TiO_2/ZrTiO_4$ Heterointerface. *Ceram. Int.* **2021**, *47*, 33298–33308. [CrossRef]
27. Guan, Y.; Hu, S.; Li, P.; Zhao, Y.; Wang, F.; Kang, X. In-Situ Synthesis of Highly Efficient Direct Z-Scheme Cu_3P/g-C_3N_4 Heterojunction Photocatalyst for N_2 Photofixation. *Nano* **2019**, *14*, 1950083. [CrossRef]
28. Liu, X.; Zhang, Q.; Ma, D. Advances in 2D/2D Z-Scheme Heterojunctions for Photocatalytic Applications. *Sol. RRL* **2021**, *5*, 2000397. [CrossRef]
29. Xu, Q.; Zhang, L.; Yu, J.; Wageh, S.; Al-Ghamdi, A.A.; Jaroniec, M. Direct Z-Scheme Photocatalysts: Principles, Synthesis, and Applications. *Mater. Today* **2018**, *21*, 1042–1063. [CrossRef]
30. Liu, C.; Li, X.; Xu, C.; Wu, Y.; Hu, X.; Hou, X. Boron-Doped Rutile TiO_2/Anatase $TiO_2/ZrTiO_4$ Ternary Heterojunction Photocatalyst with Optimized Phase Interface and Band Structure. *Ceram. Int.* **2020**, *46*, 20943–20953. [CrossRef]
31. Kaur, M.; Shinde, S.L.; Ishii, S.; Jevasuwan, W.; Fukata, N.; Yu, M.-W.; Li, Y.; Ye, J.; Nagao, T. Marimo-Bead-Supported Core–Shell Nanocomposites of Titanium Nitride and Chromium-Doped Titanium Dioxide as a Highly Efficient Water-Floatable Green Photocatalyst. *ACS Appl. Mater. Interfaces* **2020**, *12*, 31327–31339. [CrossRef] [PubMed]
32. Zhou, X.; Peng, F.; Wang, H.; Yu, H.; Yang, J. Preparation of Nitrogen Doped TiO_2 Photocatalyst by Oxidation of Titanium Nitride with H_2O_2. *Mater. Res. Bull.* **2011**, *46*, 840–844. [CrossRef]
33. Huang, D.G.; Liao, S.J.; Zhou, W.B.; Quan, S.Q.; Liu, L.; He, Z.J.; Wan, J.B. Synthesis of Samarium- and Nitrogen-Co-Doped TiO_2 by Modified Hydrothermal Method and Its Photocatalytic Performance for the Degradation of 4-Chlorophenol. *J. Phys. Chem. Solids* **2009**, *70*, 853–859. [CrossRef]
34. Divyasri, Y.V.; Lakshmana Reddy, N.; Lee, K.; Sakar, M.; Navakoteswara Rao, V.; Venkatramu, V.; Shankar, M.V.; Gangi Reddy, N.C. Optimization of N Doping in TiO_2 Nanotubes for the Enhanced Solar Light Mediated Photocatalytic H_2 Production and Dye Degradation. *Environ. Pollut.* **2021**, *269*, 116170. [CrossRef] [PubMed]
35. Liu, C.; Xu, C.; Wang, W.; Chen, L.; Li, X.; Wu, Y. Oxygen Vacancy Mediated Band-Gap Engineering via B-Doping for Enhancing Z-Scheme A-TiO_2/R-TiO_2 Heterojunction Photocatalytic Performance. *Nanomaterials* **2023**, *13*, 794. [CrossRef] [PubMed]
36. Al-Dhaifallah, M.; Abdelkareem, M.A.; Rezk, H.; Alhumade, H.; Nassef, A.M.; Olabi, A.G. Co-decorated Reduced Graphene/Titanium Nitride Composite as an Active Oxygen Reduction Reaction Catalyst with Superior Stability. *Int. J. Energy Res.* **2021**, *45*, 1587–1598. [CrossRef]
37. Wang, W.-K.; Chen, J.-J.; Gao, M.; Huang, Y.-X.; Zhang, X.; Yu, H.-Q. Photocatalytic Degradation of Atrazine by Boron-Doped TiO_2 with a Tunable Rutile/Anatase Ratio. *Appl. Catal. B Environ.* **2016**, *195*, 69–76. [CrossRef]
38. Zhang, X.; Zhang, Y.; Yu, Z.; Wei, X.; Wu, W.D.; Wang, X.; Wu, Z. Facile Synthesis of Mesoporous Anatase/Rutile/Hematite Triple Heterojunctions for Superior Heterogeneous Photo-Fenton Catalysis. *Appl. Catal. B Environ.* **2020**, *263*, 118335. [CrossRef]

39. Hashemizadeh, I.; Golovko, V.B.; Choi, J.; Tsang, D.C.W.; Yip, A.C.K. Photocatalytic Reduction of CO_2 to Hydrocarbons Using Bio-Templated Porous TiO_2 Architectures under UV and Visible Light. *Chem. Eng. J.* **2018**, *347*, 64–73. [CrossRef]
40. Hao, H.; Shi, J.-L.; Xu, H.; Li, X.; Lang, X. N-Hydroxyphthalimide-TiO_2 Complex Visible Light Photocatalysis. *Appl. Catal. B Environ.* **2019**, *246*, 149–155. [CrossRef]
41. Cheng, X.; Yu, X.; Xing, Z.; Yang, L. Synthesis and Characterization of N-Doped TiO_2 and Its Enhanced Visible-Light Photocatalytic Activity. *Arab. J. Chem.* **2016**, *9*, S1706–S1711. [CrossRef]
42. Sarkar, A.; Khan, G.G. The Formation and Detection Techniques of Oxygen Vacancies in Titanium Oxide-Based Nanostructures. *Nanoscale* **2019**, *11*, 3414–3444. [CrossRef] [PubMed]
43. Zhang, Y.; Chen, J.; Hua, L.; Li, S.; Zhang, X.; Sheng, W.; Cao, S. High Photocatalytic Activity of Hierarchical SiO_2@C-Doped TiO_2 Hollow Spheres in UV and Visible Light towards Degradation of Rhodamine B. *J. Hazard. Mater.* **2017**, *340*, 309–318. [CrossRef]
44. Poudel, M.B.; Awasthi, G.P.; Kim, H.J. Novel Insight into the Adsorption of Cr(VI) and Pb(II) Ions by MOF Derived Co-Al Layered Double Hydroxide @hematite Nanorods on 3D Porous Carbon Nanofiber Network. *Chem. Eng. J.* **2021**, *417*, 129312. [CrossRef]
45. Nosaka, Y.; Nosaka, A.Y. Generation and Detection of Reactive Oxygen Species in Photocatalysis. *Chem. Rev.* **2017**, *117*, 11302–11336. [CrossRef] [PubMed]
46. Zhang, F.; Cheng, Z.; Cui, L.; Duan, T.; Anan, A.; Zhang, C.; Kang, L. Controllable Synthesis of Ag@TiO_2 Heterostructures with Enhanced Photocatalytic Activities under UV and Visible Excitation. *RSC Adv.* **2016**, *6*, 1844–1850. [CrossRef]
47. Zhao, X.; Du, Y.; Zhang, C.; Tian, L.; Li, X.; Deng, K.; Chen, L.; Duan, Y.; Lv, K. Enhanced Visible Photocatalytic Activity of TiO_2 Hollow Boxes Modified by Methionine for RhB Degradation and NO Oxidation. *Chin. J. Catal.* **2018**, *39*, 736–746. [CrossRef]
48. Liu, G.; Wang, G.; Hu, Z.; Su, Y.; Zhao, L. Ag_2O Nanoparticles Decorated TiO_2 Nanofibers as a p-n Heterojunction for Enhanced Photocatalytic Decomposition of RhB under Visible Light Irradiation. *Appl. Surf. Sci.* **2019**, *465*, 902–910. [CrossRef]
49. Bian, H.; Zhang, Z.; Xu, X.; Gao, Y.; Wang, T. Photocatalytic Activity of Ag/ZnO/AgO/TiO_2 Composite. *Phys. E Low-Dimens. Syst. Nanostructures* **2020**, *124*, 114236. [CrossRef]
50. Wang, W.-K.; Chen, J.-J.; Zhang, X.; Huang, Y.-X.; Li, W.-W.; Yu, H.-Q. Self-Induced Synthesis of Phase-Junction TiO_2 with a Tailored Rutile to Anatase Ratio below Phase Transition Temperature. *Sci. Rep.* **2016**, *6*, 20491. [CrossRef]
51. Lu, Y.; Zhao, K.; Zhao, Y.; Zhu, S.; Yuan, X.; Huo, M.; Zhang, Y.; Qiu, Y. Bi_2WO_6/TiO_2/Pt Nanojunction System: A UV–Vis Light Responsive Photocatalyst with High Photocatalytic Performance. *Colloids Surf. Physicochem. Eng. Asp.* **2015**, *481*, 252–260. [CrossRef]
52. Li, Y.; Lv, K.; Ho, W.; Dong, F.; Wu, X.; Xia, Y. Hybridization of Rutile TiO_2 (R-TiO_2) with g-C_3N_4 Quantum Dots (CN QDs): An Efficient Visible-Light-Driven Z-Scheme Hybridized Photocatalyst. *Appl. Catal. B Environ.* **2017**, *202*, 611–619. [CrossRef]
53. Zhang, X.; Lin, Y.; He, D.; Zhang, J.; Fan, Z.; Xie, T. Interface Junction at Anatase/Rutile in Mixed-Phase TiO_2: Formation and Photo-Generated Charge Carriers Properties. *Chem. Phys. Lett.* **2011**, *504*, 71–75. [CrossRef]
54. Yu, Y.; Wen, W.; Qian, X.-Y.; Liu, J.-B.; Wu, J.-M. UV and Visible Light Photocatalytic Activity of Au/TiO_2 Nanoforests with Anatase/Rutile Phase Junctions and Controlled Au Locations. *Sci. Rep.* **2017**, *7*, 41253. [CrossRef]
55. Xu, X.; Ding, X.; Yang, X.; Wang, P.; Li, S.; Lu, Z.; Chen, H. Oxygen Vacancy Boosted Photocatalytic Decomposition of Ciprofloxacin over Bi_2MoO_6: Oxygen Vacancy Engineering, Biotoxicity Evaluation and Mechanism Study. *J. Hazard. Mater.* **2019**, *364*, 691–699. [CrossRef]
56. Baran, T.; Wojtyła, S.; Minguzzi, A.; Rondinini, S.; Vertova, A. Achieving Efficient H_2O_2 Production by a Visible-Light Absorbing, Highly Stable Photosensitized TiO_2. *Appl. Catal. B Environ.* **2019**, *244*, 303–312. [CrossRef]
57. Lakshmanareddy, N.; Navakoteswara Rao, V.; Cheralathan, K.K.; Subramaniam, E.P.; Shankar, M.V. Pt/TiO_2 Nanotube Photocatalyst—Effect of Synthesis Methods on Valance State of Pt and Its Influence on Hydrogen Production and Dye Degradation. *J. Colloid Interface Sci.* **2019**, *538*, 83–98. [CrossRef]
58. Poudel, M.B.; Kim, A.A.; Lohani, P.C.; Yoo, D.J.; Kim, H.J. Assembling Zinc Cobalt Hydroxide/Ternary Sulfides Heterostructure and Iron Oxide Nanorods on Three-Dimensional Hollow Porous Carbon Nanofiber as High Energy Density Hybrid Supercapacitor. *J. Energy Storage* **2023**, *60*, 106713. [CrossRef]
59. Liu, Y.; Shen, S.; Zhang, J.; Zhong, W.; Huang, X. $Cu_{2-x}Se$/CdS Composite Photocatalyst with Enhanced Visible Light Photocatalysis Activity. *Appl. Surf. Sci.* **2019**, *478*, 762–769. [CrossRef]
60. Yin, W.; Bai, L.; Zhu, Y.; Zhong, S.; Zhao, L.; Li, Z.; Bai, S. Embedding Metal in the Interface of a p-n Heterojunction with a Stack Design for Superior Z-Scheme Photocatalytic Hydrogen Evolution. *ACS Appl. Mater. Interfaces* **2016**, *8*, 23133–23142. [CrossRef]
61. Zhang, S.; Liu, Y.; Ma, R.; Jia, D.; Wen, T.; Ai, Y.; Zhao, G.; Fang, F.; Hu, B.; Wang, X. Molybdenum (VI)-oxo Clusters Incorporation Activates g-C_3N_4 with Simultaneously Regulating Charge Transfer and Reaction Centers for Boosting Photocatalytic Performance. *Adv. Funct. Mater.* **2022**, *32*, 2204175. [CrossRef]
62. Guan, Y.; Liu, W.; Zuo, S.; Yuan, K.; Wu, F.; Ji, J.; Yao, C. Double Z-Scheme TiO_2 (R)/C-TiO_2 (A) Heterojunction Greatly Enhanced Efficiency of Photocatalytic Desulfurization under Sunlight. *J. Mater. Sci. Mater. Electron.* **2020**, *31*, 22297–22311. [CrossRef]
63. Lin, Y.; Jiang, Z.; Zhu, C.; Hu, X.; Zhang, X.; Zhu, H.; Fan, J.; Lin, S.H. C/B Codoping Effect on Band Gap Narrowing and Optical Performance of TiO_2 Photocatalyst: A Spin-Polarized DFT Study. *J. Mater. Chem. A* **2013**, *1*, 4516. [CrossRef]
64. Kashiwaya, S.; Morasch, J.; Streibel, V.; Toupance, T.; Jaegermann, W.; Klein, A. The Work Function of TiO_2. *Surfaces* **2018**, *1*, 73–89. [CrossRef]

65. Xu, Q.; Zhang, L.; Cheng, B.; Fan, J.; Yu, J. S-Scheme Heterojunction Photocatalyst. *Chem* **2020**, *6*, 1543–1559. [CrossRef]
66. Li, C.; Yang, W.; Liu, L.; Sun, W.; Li, Q. In Situ Growth of TiO_2 on TiN Nanoparticles for Non-Noble-Metal Plasmonic Photocatalysis. *RSC Adv.* **2016**, *6*, 72659–72669. [CrossRef]

Disclaimer/Publisher's Note: The statements, opinions and data contained in all publications are solely those of the individual author(s) and contributor(s) and not of MDPI and/or the editor(s). MDPI and/or the editor(s) disclaim responsibility for any injury to people or property resulting from any ideas, methods, instructions or products referred to in the content.

 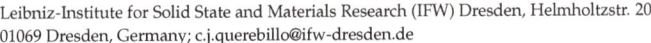

Review

A Review on Nano Ti-Based Oxides for Dark and Photocatalysis: From Photoinduced Processes to Bioimplant Applications

Christine Joy Querebillo

Leibniz-Institute for Solid State and Materials Research (IFW) Dresden, Helmholtzstr. 20, 01069 Dresden, Germany; c.j.querebillo@ifw-dresden.de

Abstract: Catalysis on TiO_2 nanomaterials in the presence of H_2O and oxygen plays a crucial role in the advancement of many different fields, such as clean energy technologies, catalysis, disinfection, and bioimplants. Photocatalysis on TiO_2 nanomaterials is well-established and has advanced in the last decades in terms of the understanding of its underlying principles and improvement of its efficiency. Meanwhile, the increasing complexity of modern scientific challenges in disinfection and bioimplants requires a profound mechanistic understanding of both residual and dark catalysis. Here, an overview of the progress made in TiO_2 catalysis is given both in the presence and absence of light. It begins with the mechanisms involving reactive oxygen species (ROS) in TiO_2 photocatalysis. This is followed by improvements in their photocatalytic efficiency due to their nanomorphology and states by enhancing charge separation and increasing light harvesting. A subsection on black TiO_2 nanomaterials and their interesting properties and physics is also included. Progress in residual catalysis and dark catalysis on TiO_2 are then presented. Safety, microbicidal effect, and studies on Ti-oxides for bioimplants are also presented. Finally, conclusions and future perspectives in light of disinfection and bioimplant application are given.

Keywords: reactive oxygen species; photocatalytic efficiency; charge separation; light harvesting; black TiO_2; titanium alloys; residual disinfection; dark catalysis; antibacterial; inflammation

Citation: Querebillo, C.J. A Review on Nano Ti-Based Oxides for Dark and Photocatalysis: From Photoinduced Processes to Bioimplant Applications. *Nanomaterials* **2023**, *13*, 982. https://doi.org/10.3390/nano13060982

Academic Editors: Wei Zhou and Antonino Gulino

Received: 30 January 2023
Revised: 13 February 2023
Accepted: 24 February 2023
Published: 8 March 2023

Copyright: © 2023 by the author. Licensee MDPI, Basel, Switzerland. This article is an open access article distributed under the terms and conditions of the Creative Commons Attribution (CC BY) license (https://creativecommons.org/licenses/by/4.0/).

1. Introduction

Titanium dioxide (TiO_2), or titania, occurs naturally and forms spontaneously when bare Ti is exposed to air [1,2]. At room temperature, TiO_2 is commonly considered an n-type semiconductor [3,4] with a bandgap (E_g) of around 3 eV (3.2 eV for anatase, 3 eV for rutile) [2,4]. Since the discovery of its photoelectrochemical (PEC) water-splitting ability by Fujishima and Honda [3,5], TiO_2 remains today as the standard photocatalyst. The excitation of its electrons from the valence band (VB) to the conduction band (CB) upon exposure to light with energy $E \geq E_g$, i.e., UV (to violet, 413 nm) light for the bulk form, results in photogenerated charge carriers (electrons and holes) which can be utilized in various processes, resulting in its photocatalytic activity. In addition, TiO_2 exhibits desirable material properties, such as its wide availability, biocompatibility, chemical stability in an aqueous environment, and affordability [6–14], resulting in the application of TiO_2 in different fields, such as catalysis, alternative/clean energy technologies, environmental cleaning, pollutant degradation, medicine, pharmaceuticals, disinfection, and biomedical implants [1–4,6–21].

The mechanism of photocatalysis on TiO_2 depends on the type of catalytic reaction that is being investigated, though it primarily involves interfacial (and bulk) processes of the photoinduced electrons and holes, which occur at different time scales (Figure 1) [22–26]. The faster charge carrier generation due to photon absorption (1) compared to electron–hole recombination (2) enables the possibility of having CB electrons and VB holes that can be tapped for reductive (3) and oxidative (4) processes, respectively. However, the recombination process lies at the same time scale as these interfacial charge transfer (CT) processes, and therefore efforts to further delay the recombination process can improve the photocatalytic efficiency of TiO$_2$.

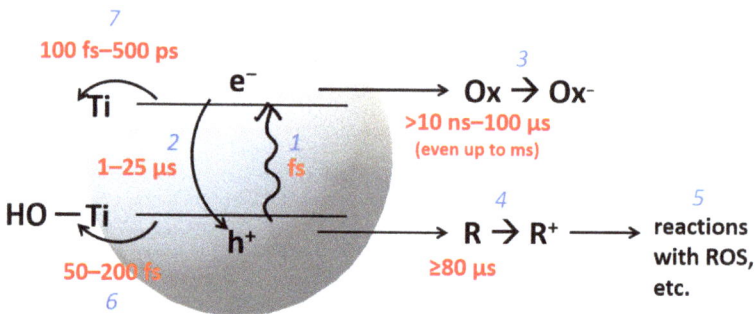

Figure 1. The core steps in photocatalytic mechanism include (1) absorption of light which forms charge carriers, (2) recombination of charge carriers, (3) reductive process with electrons in the conduction band (CB), (4) oxidative pathway undertaken by a valence-band (VB) hole, (5) further reactions (hydrolysis, reactions with reactive oxygen species (ROS), etc.), (6) trapping a VB hole at a Ti-OH group on the surface, and (7) trapping of CB electron at a Ti(IV) site to produce Ti(III). Redrawn from ref. [26] to include timescales obtained from [22–26]. The timescale in 3 and 4 are mainly from nontrapped electrons and holes (i.e., directly from the CB and VB).

Furthermore, the VB holes can transport quickly to hole trap sites at the surface, such as at surface Ti-OH groups (6), which is often the case for the photogenerated holes because of the fast femtosecond process. Consequently, free VB holes are scarcely present in TiO$_2$ [24], and surface-trapped holes are usually responsible for the oxidation reaction which can occur faster in the ps–ns range [22–26]. The electrons, on the other hand, can get trapped at Ti(IV) sites (7) and form Ti(III), which can also participate in other redox processes [22–26].

Direct recombination of the photogenerated carriers usually does not occur and instead happens upon meeting trapped complementary charge carriers. For example, mobile electrons can recombine with trapped holes, decreasing the former's lifetime and reducing the photocatalytic performance of TiO$_2$. As such, anatase usually performs better as a photocatalyst than rutile due to its longer electron lifetime (>few ms for anatase vs. ~24 ns for rutile) and stronger band bending [27,28]. Further photocatalytic reactions on TiO$_2$ occur on the surface involving reactive oxygen species (ROS) (Figure 1, (5)) [22].

In the past decades, most studies on TiO$_2$ are focused on its photocatalytic activity (both in the bulk and nanomaterial form), and the literature is overflowing with strategies to improve its performance by addressing inherent deterrents, extending the spectral range for which TiO$_2$-based photocatalysis can be used, or enhancing the light utilization through various geometries (nanotextures, structured arrays, etc.). These improvements are based mainly on the key steps in photocatalysis, focusing on light absorption, generation of charge carriers, their separation and transport, catalyst replenishment, and prevention of back and side processes. With the goal and steps clarified, TiO$_2$ photocatalysis still faces a number of challenges considering that modern materials require multiple functionalities and therefore a balance of its properties. Hence, the influence of different TiO$_2$ properties on its photocatalytic performance is also a major focus of research.

Almost overshadowed by the numerous works on photocatalysis but persisting mainly due to the excellent biocompatibility and oxidative bleaching ability of TiO$_2$, research on TiO$_2$ in the "dark" or in the absence of light has also been growing steadily over the years. Titanium, from which TiO$_2$ can be grown, exhibits excellent mechanical properties which helped in establishing its place in the field of biomedical implants. For example, titanium and its alloys are standard materials used for bone implants, and strong research in medicine and engineering is focused on continuously improving their physicochemical properties, mechanical properties, and designability/processability [29–33], on top of added functionalities desired in modern biomaterials, such as its antimicrobial and regenerative properties [34–39]. The last two are also attributed to its catalytic ROS-forming ability, which should then be considered in bioimplant applications. After implantation surgery, i.e., in the postoperative phase, implant material surfaces are devoid of light. Understanding the mechanism of "dark" catalysis on TiO_2 in the physiological condition is therefore important in addressing the current challenges and limitations not only for Ti and TiO_2 but also for more advanced Ti alloys and Ti-based materials and their oxides for implant application.

Considering the vast use of TiO_2, it is not surprising that the published works on TiO_2 catalysis come from many different disciplines of different perspectives. Modern scientific and engineering challenges then often entail a multidisciplinary approach to answer the increasingly becoming more complex questions. Yet, if we look at the literature, it seems that there is still a need to consolidate this immense knowledge of TiO_2 catalysis from a multidisciplinary perspective.

The different knowledge and experiences we gain by working on different topics involving titania can help us develop the skill of understanding catalysis on TiO_2 nanomaterials from a more inclusive viewpoint. Different fields working on titania, for instance, electromagnetic field enhancement on semiconductors [40–50], dye photodegradation (see Section 2.2), and bioimplant applications [31,36,51–56] to name a few, may all involve catalytic properties of TiO_2 nanomaterials, yet they also need a nuanced understanding of TiO_2 in light of specific, targeted applications. Such background and experience can certainly help us easily understand the literature though regularly immersing oneself in various literature on TiO_2 catalysis easily available to us nowadays, and constantly discussing with colleagues and peers can also help us be familiarized and updated with the progress on TiO_2 catalysis. As such, despite the experience and background of the author, which certainly helped in the reading and analysis of the literature for this review, the method of regularly keeping up to date with the literature, engaging in scientific discussions, and a period of intensive gathering and reading of the literature on TiO_2 catalysis was therefore adapted in preparing this review. Many excellent works and reviews helped in the preparation of a general survey, with emphasis on certain points of interest—namely, the advancements in photocatalytic enhancement, the photocatalytic and dark bactericidal activities, and the use of Ti (and Ti-based) oxide nanomaterials for bioimplants. In some topics, the readers are referred to excellent reviews available in the literature, and the scope is limited to oxides of Ti (mainly TiO_2). On the other hand, the preparation of this review was conducted with the awareness that the answers to present-day complex scientific questions may still not be available in the literature, and due to the multidisciplinary nature of these questions and the explosion of the available literature on the internet, not all available review materials on the internet and in print can be included in this review. Nevertheless, inspired by the present challenges that the bioimplant community wishes to address, a review on photo- and dark catalysis of TiO_2 is presented here.

In this review, the catalysis of TiO_2 nanomaterials, both in the presence of light (photocatalysis) and in the dark, is presented to give a general overview of the full spectrum of its catalytic activity. A section discussing the role of ROS in TiO_2 photocatalysis, which is crucial in photo- and dark catalysis on TiO_2, is included. As the understanding of the role of ROS in photocatalysis is more extensive, it is beneficial to look at it from a mechanistic perspective without going into details, as excellent reviews and articles also exist in the literature [22,57,58]. The influence of some properties of TiO_2 on its performance in photocatalysis will then be presented to understand the surface engineering that has been carried out to advance the photocatalysis field. TiO_2 nanoparticles (NPs), having been extensively developed for photocatalysis, also pose some risks, and the safety of using them will also be discussed. Together with this, the other side of the coin, the photocatalytic antibacterial property afforded by TiO_2 NPs, is presented. Then, highlights and advancements in the efforts to boost the photocatalytic efficiency of TiO_2 mainly in terms of charge separation enhancement and improvements in light harvesting will be presented. Black TiO_2 nanomaterials, a current hot topic in the field of TiO_2 photocatalysis, will also be presented in light of their physics and photocatalytic activity.

Many studies on dark catalysis are investigations on the influence of Ti-based implants on the inflammatory response and vice versa. In addition, early studies on dark catalysis are observations carried out mainly as a reference to photocatalytic works and the residual effect after the removal of irradiation. These will be presented to serve as a bridge between photo- and "dark catalysis" and will be followed by studies addressing and contributing to the so-far understanding of the dark catalysis mechanism. Dark catalysis is important when looking at the inflammatory response which yields ROS, resulting in some microbicidal effect of TiO_2 and improving the performance of biomedical implants. As there is a huge scientific community working on biomedical implants who are looking at improving the performance, regenerative ability, and other properties of Ti-based materials, discussions on Ti and Ti-based oxides for biomedical applications, the safety of these materials, and the inflammatory condition are also included. Finally, a conclusion/future perspective in terms of photo- and dark catalysis on Ti-based oxides for disinfection and bioimplant application is given.

2. TiO_2 Photocatalysis

Since photocatalysis is mainly due to redox reactions of photogenerated charge carriers producing reactive surface species and that they mostly occur in the presence of water and/or oxygen, it is important to look at the role of reactive oxygen species (ROS).

2.1. Reactive Oxygen Species in TiO_2 Photocatalysis

ROS can be considered primary intermediates of photocatalytic reactions with these four recognized as the main ones: hydroxyl radical ($\cdot$OH), superoxide anion radical ($\cdot O_2^-$), hydrogen peroxide (H_2O_2), and singlet oxygen ($^1O_2$) [57,59]. ROS seem to form mainly from the interaction of the VB hole with molecules (such as H_2O) or species, oxidizing the latter and typically resulting in $\cdot$OH centers [58,60,61]. This could also happen in hole-trapping processes in TiO_2, such as at bridging O_2^-, resulting in the formation mainly of $\cdot O^-$ ("deprotonated $\cdot$OH") [22,62,63]. Because of the high potential barrier of free $\cdot$OH for desorption, adsorbed $\cdot$OH is considered more favorable and is usually equated to trapped holes due to the adsorption–desorption equilibrium [22,57].

A detailed summary of generating the four major ROS on the TiO_2 surface can be viewed in terms of bridging and terminal OH sites (Figure 2) [58]. The reactions occurring at the anatase and rutile are differentiated by the arrow lines (double lines are restricted to anatase), whereas broken lines refer to adsorption/desorption. At the bridged OH site (Figure 2a), a photogenerated hole attacks the O^{2-} bridge (step a), forming Ti-O$\cdot$ and Ti-OH (step b), which can be reversed by recombining with an electron from the CB. Some surface-trapped holes at the anatase can be released as $\cdot$OH into the solution. At the rutile surface with its suitable distance between adjacent Ti surface atoms, a different scenario occurs. Once another hole is formed in the same trapped hole-containing particle, the hole could migrate and interact with the existing hole resulting in a peroxo-bridged structure at the surface (step c). Further reactions of these structures could then generate other ROS [57,58].

At the terminal OH site (Figure 2b), a photogenerated electron can interact and is trapped at the Ti^{4+} site, transforming it to Ti^{3+} (step b) [64]. The trapped electron in the Ti^{3+} could then reduce oxygen to form an $\cdot O_2^-$ (adsorbed) (step d) (which, with further reduction, could become an adsorbed H_2O_2 or as (Ti)-OOH (step c)). The H_2O_2 that is adsorbed could also be reoxidized to produce an adsorbed $\cdot O_2^-$, which can be desorbed to return to the initial state (step a). As the peroxo bond needs to be dissociated, the production of $\cdot$OH is highly unlikely when the adsorbed H_2O_2 is being reduced [57,58]. These schemes (Figure 2) also show a sensible explanation for the influence of the adsorption of H_2O_2 in forming ROS, which has been well-considered for increasing the photocatalytic performance of TiO_2.

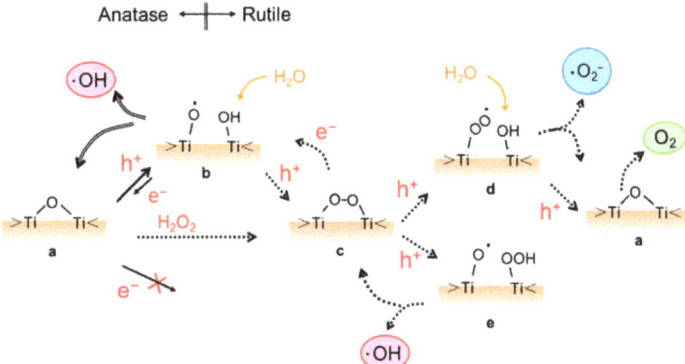

(a) at an anionic bridged OH site of TiO_2

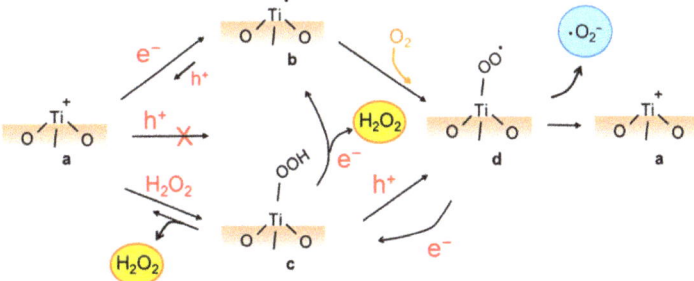

(b) at a cationic terminal OH site of TiO_2

Figure 2. Photocatalytic reaction pathways on TiO_2 proposed (a) at a bridge OH site and (b) at a terminal OH site. Reprinted (adapted) with permission from Nosaka, Y., Nosaka, A. Understanding Hydroxyl Radical (•OH) Generation Processes in Photocatalysis. *ACS Energy Lett.* **2016**, *1*, 356–359, doi:10.1021/acsenergylett.6b00174. Copyright 2016 American Chemical Society [58].

Adsorbed ROS can also have a more direct impact on photocatalytic performance [57]. Anatase and rutile show different reactivity towards forming ·OH and ·O_2^- [28,65], likely due to their H_2O_2 adsorption [63] in addition to their band edge alignment. One-step oxidation of H_2O_2 produces ·O_2^-, which is more remarkable for anatase, whereas one-step reduction produces ·OH, which is only observed for rutile or rutile-containing forms and is believed to be due to the structure of the adsorbed H_2O_2 on rutile vs. anatase [66].

2.2. Nanomorphologies and Structural States of TiO_2

Due to its wide applicability, TiO_2 has been produced via different means, with the resulting TiO_2 structural polymorphs—i.e., anatase, rutile, or brookite, among others [67]—and morphology being highly influenced by the preparation method. The different TiO_2 morphologies add to the variety of properties and performance exhibited by TiO_2. In addition to bulk TiO_2 [68–71], in recent decades, TiO_2 nanomaterials of various morphologies have also been developed, resulting in the current plethora of TiO_2 nanomorphologies (Figure 3). These have been synthesized using different means for various targeted applications, achieving a range of photocatalytic efficiencies (Table 1). Note that some morphologies are preferably prepared using certain procedures (e.g., sol-gel method for nanopowders and anodization for nanotubes), whereas some procedures (e.g.

hydrothermal synthesis) can be used and modified to produce various morphologies (such as nanospindles, nanorhombus, nanorods, or nanosheets).

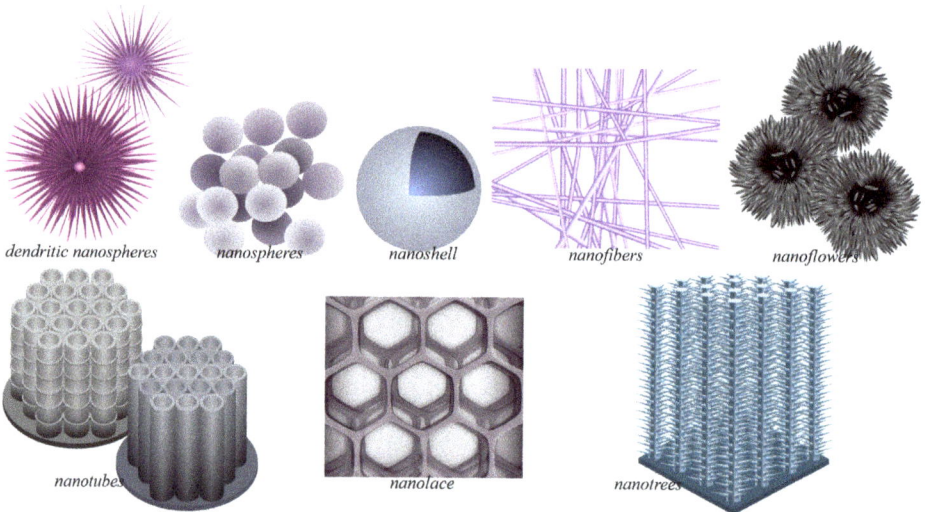

Figure 3. Plethora of TiO$_2$ nanostructures. This includes free-standing nanomaterials, such as TiO$_2$ dendritic nanospheres, nanospheres, nanoshells, nanofibers, and nanoflowers, and nanostructures grown on bulk substrates, such as nanotubes, nanolace, and nanotrees. Nanomaterials present increased catalytic activity due to their interesting properties (increased surface area, enhanced charge separation, light absorption/harvesting), which are desirable in catalytic TiO$_2$ applications.

Table 1. Reported photocatalytic performance of some TiO$_2$ nanomorphologies.

Morphology	Material Dimension	Crystal State	Synthetic Procedure	Photocatalytic Performance §	Targeted Applications	Ref.
Commercially available Degussa P25 nanopowder	30–40 nm	anatase + rutile	either deposited as film or used as dispersion/colloid	Pseudo-first-order rate constant, $k = 0.0085$–0.012 min^{-1} AO7 [72,73], RhB [73,74] degradation; UV light, 100–200 mW HeCd laser with I = 60 or 100 mW cm^{-2}, 254~325 nm	Dye-sensitized solar cells, etc.	[72–74]
Quantum dots	~4.7 nm	anatase	autoclave method (+heating)	~18% MO photodegradation* (Xe lamp with a glass filter, 400 nm cut off, 100 mW cm^{-2}) [75]; *estimated $k \sim 0.0033$ min^{-1}	Energy and environmental applications	[75,76]
Nanocrystals/nanopowder/ nanoparticles (NPs)/ nanospheres	particle size: 8–10.2 nm [77,78], 14–18 nm [79,80], 19–23 nm [81]	anatase	sol-gel method + heat treatment	$k = 0.002$–0.036 min^{-1} (MB degradation) [78,79,81], $k = 0.090$–0.105 min^{-1} (MR degradation) [77] (UV light 250–625 W; max. ~250–368 nm); 20% (NOx degradation after 60 min; UV light-20 W, 287.5 nm) [80], *estimated $k \sim 0.0037$ min^{-1} UV light 250–625 W; max. ~250–368 nm)	Pollutant degradation and self-cleaning [77–79,81], exhaust gas decomposition [80], energy storage [78]	[77–81]
Nanoporous shell (polyimide support)	2.7–2.8 nm pore size, 12–20 nm shell thickness,	anatase	in situ complexation hydrolysis	~80–95% degradation rate* (365 nm, 30-W UV lamp) *estimated $k = 0.04$–0.07 min^{-1}	Air purification and water disinfection	[82]
Nanospindles	~190 nm in length; growth direction along [001]	anatase	hydrothermal synthesis	$k = 0.0306$ min^{-1} (RhB degradation; UV illum.; sim. sunlight AM 1.5 G filter, (100 mW/cm^2))	Dye-sensitized solar cells, etc.	[74]
Nanowires	100 nm diameter × 800 μm	anatase	hydrothermal reaction, proton exchange calcination	$k = 0.0154$ h^{-1} (0 0 000256 min^{-1}) (phenol degradation in water; LED UV lamp (18 W); ca. 12 mW cm^{-2})	Photocatalysis, gas sensors, etc.	[83]
Nanowires/ nanobelts	15 nm diameter	anatase	hydrothermal growth + heat treatment	51.96%* (MO degradation; 350-W Xe lamp); *estimated $k \sim 0.0066$ min^{-1}	Organic pollutants degradation	[84]

Table 1. Cont.

Morphology	Material Dimension	Crystal State	Synthetic Procedure	Photocatalytic Performance §	Targeted Applications	Ref.
Nanobelts	800 μm × 400 nm × 20 nm	anatase	hydrothermal reaction, proton exchange, calcination	$k = 0.0256$ h^{-1} (0.000426 min) (phenol degradation in water; LED UV lamp (18 W); ca. 12 mW cm^{-2})	Photocatalysis, gas sensors, etc.	[83]
Nanorods (NRs)	~1.5 nm diameter and ~8.3 nm in length	rutile	solvothermal reaction	k~0.068 min^{-1} (RhB degradation, 300 W Xe lamp; full spectrum)	Organic pollutant degradation	[85]
Nanofibers/core-shell nanofibers	diameter < 100 nm	anatase, rutile, anatase + rutile	electrospinning/ hydrolysis or alkoxide method	Estimated $k = 0.027$–0.1118 min^{-1} (lower values for rutile, higher values for anatase, mixed/core-shell have values in between, with values closer to the shell structure) (RhB; UV irradiation	Energy and environmental applications	[86]
Nanotubes (bamboo-type)	2.5 μm in length, 100 nm diameter	amorphous	anodization	$k = 0.0045$ min^{-1} (AO7 degradation), 0.0187 min^{-1} (MB degradation) (UV light, 200 mW HeCd laser with I = 60 mWcm^{-2}, 325 nm)	Dye-sensitized solar cells, etc.	[72]
	20–60 nm diameter, 0.5–0.9 μm in length, and ~60 nm interpore distances	anatase + rutile	dynamic anodization + heat treatment	$k = 0.0007$–0.0013 min^{-1} (MB degradation, 9W UV black-light lamp (λ = 365 nm), 1 mW/cm^2 radiation at surface)	Devices; energy and environmental applications	[87]
Nanotubes (smooth)	4.5 μm length, 45 nm diameter	anatase	anodization + heat treatment	$k = 0.0158$ min^{-1} (AO7 degradation), 0.0213 min^{-1} (MB degradation) (UV light, 200 mW HeCd laser with I = 60 mWcm^{-2}, 325 nm)	Dye-sensitized solar cells, etc.	[72]
	50–60 nm diameter, 1.7–2.2 μm in length, and ~60 nm interpore distances	anatase + rutile	dynamic anodization + heat treatment	$k = 0.0024$–0.0049 min^{-1} (MB degradation, 9 W UV black-light lamp (λ = 365 nm), 1 mW/cm^2 radiation at surface)	Devices; energy and environmental applications	[87]

Table 1. Cont.

Morphology	Material Dimension	Crystal State	Synthetic Procedure	Photocatalytic Performance §	Targeted Applications	Ref.
Nanotubes (grown from Ti-6Al-4V)	diameter increases with anodizing potential; thickness ~280 nm	anatase (+rutile)	anodization + heat treatment	~8–42% photodegradation efficiency (depending on the anodization voltage or tube diameter); estimated $k = 0.0005$–0.003 min^{-1} (MB degradation; UV-A lamp, 0.39 W/cm^2)	Organic pollutant degradation	[88]
Nanoribbons	200–300 nm in width; several microns in length	anatase (+rutile)	alkaline hydrothermal treatment	k ~0.05342–0.08164 min^{-1} (RhB degradation; simulated sunlight, 300 W 230 V E27)	Organic pollutants degradation	[89]
Nanoribbons (nanopitted)	width 20–200 nm, length of 1 μm–few μm with pits of dia. 5–15 nm	TiO$_2$-B	alkaline hydrothermal treatment	k ~ 0.0024–0.011 min^{-1} (MB degradation) (natural sunlight)	Dye degradation	[90]
2D nanogrid/nanolaces/inverse-opal-like structure	230–610 nm diameter of holes	anatase	opal-templated sol-gel-based synthesis + heat treatment	k~0.022–0.058 min^{-1} (RhB degradation; 500 W Xe arc lamp, 400 nm cutoff, 25 mW cm^{-2})	Degradation of various environmental contaminants	[91]
	~150 nm diameter holes, lace thickness 10–20 nm	anatase (+rutile)	alternating-voltage anodization + heat treatment/ 2-step anodization	$k = 0.003$ min^{-1} (Cr (VI) photocatalytic reduction; simulated sunlight, 300 W xenon lamp with 100 W cm^{-2} irradiation)	Heavy metal (Cr(VI)) removal from wastewater	[92,93]
	~150 nm diameter holes, lace thickness 10–20 nm	black TiO$_2$	2-step anodization process + heat-treat. in reducing atmosphere	$k = 0.0657$ min^{-1} (Cr (VI) photocatalytic reduction; simulated sunlight, 300 W xenon lamp with 100 W cm^{-2} irradiation)	Heavy metal (Cr(VI)) removal from wastewater	[93]
Nanosheets	thickness < 7 nm	anatase	hydrothermal process	$k = 0.013$ min^{-1} (RhB degradation, mercury lamp (300 W, as UV light source), 2.62 mW/cm^2)	Renewable energy; environment	[75]
Nanoflowers	2–6 nm diameter with ~10 nm thin petals	anatase + rutile	hydrothermal + calcination	$k = 0.03$–0.12 min^{-1} (MB degradation at diff. pH; highest k at pH 4; UV lamp (100 W, 365 nm, 6.5 mW cm^{-2}))	Effluent treatment	[94]

Table 1. Cont.

Morphology	Material Dimension	Crystal State	Synthetic Procedure	Photocatalytic Performance §	Targeted Applications	Ref.
Dendritic nanospheres	2–3 µm diameter of entire structure; nanowire/nanoribbon spikes: 500 nm–1.5 µm long and dia. in nm	rutile	low-temperature hydrothermal method	k–0.018–0.024 min^{-1} (A07/RhB degradation; UVP Mineralight lamp (254 nm, 40 mW cm^{-2}))	Photocatalytic membrane water purification	[73]
Nanotrees	130–180 nm nanowire diameter; 10–20 nm nanoparticle size	anatase	hydrothermal method, (1) proton exchange + calcination, (2) TiO$_2$ NP decoration; sequence variation has an effect	k = 0.021, 0.071 min^{-1} (depending on the lattice parameter) (toluene gas decomposition; UV lamp (PL-L 18W/10/4P, Philips; 365 nm; 8.7 mW cm^{-2})	Photocatalysis, gas sensors, etc.	[95]
	Sheet-on-belt (SOB): 800 µm × 400 nm × 20 nm (trunk), up to 100 nm long (branches). Sheet-on-wire (SOW): 100 nm diameter × 800 µm (trunk) with branches up to 200 nm. Branch thickness: a few nm	mostly anatase (+a bit of rutile)	hydrothermal reaction, proton exchange, calcination, solution combustion synthesis + calcination	Up to k = 0.346 h^{-1} (or 0.00576 min^{-1}) (SOB photocatalytic); up to k = 0.40 h^{-1} (or 0.0067 min^{-1}) (SOW photo-electrocatalytic) (phenol degradation in water) (LED UV lamp (18 W); ca. 12 mW cm^{-2})	Environmental remediation, energy storage, green energy production	[83]
	branched nanowire; 1 µm thick, branch: 10 nm thick, 45 nm long	anatase + rutile	H$_2$O$_2$ oxidation, intermediate calcination, and H$_2$SO$_4$ treatment	k = 0.0007–0.0086 min^{-1} (UV + various organic compounds); 0.0057 min^{-1} (visible + RhB) (18 W UV lamp, 5 mW/cm^2; 500 W Xe-lamp, 420 nm cut off, 200 mW/cm^2)	Photocatalytic water-splitting, environmental remediation	[96]
Nanosheet + quantum dots (QDs)	nanosheet: thickness <7 nm quantum dot: 3–5.6 nm	anatase	nanosheet, hydrothermal; QD, autoclave (+heating) homojunction: grinding	k = 0.027–0.064 min^{-1} (RhB degradation, mercury lamp (300 W, as UV light source); 2.62 mW/cm^2)	Renewable energy technologies, environmental protection	[75]

Table 1. Cont.

Morphology	Material Dimension	Crystal State	Synthetic Procedure	Photocatalytic Performance §	Targeted Applications	Ref.
Nanoparticles, nanobucks, and nanorods	15–25 nm NPs; 100–150 nm nanobuck; 15 nm × 150 nm nanorod	rutile + anatase	(one-step) hydrothermal synthesis	$k = 0.033$ min^{-1} (RhB degradation) (250 W Xe lamp–simulated solar light source)	Wastewater treatment	[97]
Nanorhombus nanocuboids	nanorhombus: 55–80 nm × 35–40 nm; nanocuboids: 20–30 nm × 30–50 nm	anatase	hydrothermal synthesis	$k = 0.0134$–0.0318 min^{-1} (RhB degradation; UV; simulated sunlight AM 1.5 G filter, (100 mW/cm^2))	Dye-sensitized solar cells, etc.	[74]

§ acid orange 7 (AO7), rhodamine B (RhB), methyl orange (MO), methylene blue (MB), methyl red (MR).

The morphologies of nanosized titanium dioxide can be grouped according to their dimension classifications: zero-dimensional (0D) includes nanopowders, nanocrystals, and quantum dots (QD); one-dimensional (1D) includes nanowires, nanofibers, nanotubes, and nanorods; two-dimensional (2D) includes nanosheets; and three-dimensional (3D) includes nanotube arrays. Mixed morphologies, such as nanosheets with QDs, also exist.

Many studies on photocatalysis have been conducted on nanoparticle suspensions [98–100], which are inconvenient and result in practical difficulties [8] because complete photocatalyst recovery is challenging. Therefore, studies have resorted to photocatalyst immobilization [101–104], which, however, requires catalysts of high activity. TiO_2 of different nanomorphologies were used, evaluated, and/or compared in terms of performance [105–107], and some of these are in Table 1 to show the influence of morphology on photocatalysis. The photocatalytic performance reported, usually measured by dye degradation rate, varies and depends on the experimental setup/condition, such as the illumination and probe used, though the typical pseudo-first-order rate constant k is within the 10^{-3}–10^{-1} min^{-1} range. Therefore, studies also sometimes include a reference TiO_2, such as commercially available P25 for benchmarking. Nevertheless, from this summary (Table 1), one also sees that generally, some improvement in photocatalytic performance is brought about by nanomorphology based on the obtained k values being mainly in the 10^{-1}–10^{-2} min^{-1} range for systems with varied morphology, which are at least one order of magnitude better than the usual for those with nanopowders ($k \sim 10^{-2}$–10^{-3} min^{-1}).

Post-synthesis heat treatment (calcination/annealing) mainly dictates the structural state. Heat treatment can be performed to transform the amorphous state to rutile ($\geq 600\ °C$) or anatase (300 °C to 500 °C) or to transform anatase to the more stable rutile [108]. The crystalline state also influences photocatalytic activity. It is commonly agreed that anatase is better than the other states (such as rutile) due to its higher surface affinity (i.e., better adsorption and probably due to the ROS formation as discussed in Section 2.1) and slower recombination rate. However, mixed states (such as the case of P25) also exhibit good photocatalytic activity, though the surface crystalline state seems to play a more crucial role in such cases of mixed states due to the fact that photocatalytic reactions take place at the surface [86].

2.2.1. Safety of TiO_2 Nanoparticles

Though TiO_2 is considered a safe, biologically inert material [109,110], the development of TiO_2 NPs with novel properties and applications resulted in its increased use and production. Hence, it has to be evaluated in terms of its toxicology. TiO_2 NPs are posed as possible carcinogens to humans, though TiO_2 is allowed for use as an additive (E171) in the food and pharmacy industry [111]. A sound basis of why TiO_2 NPs have been scrutinized is the observed appearance of their unique size-dependent properties when inorganic NPs reach the limit of ≤ 30 nm in diameter. In this size range, drastic changes in the behavior of the NPs can appear, enhancing their reactivity at the surface [112]. While the increased reactivity at this size renders their enhanced catalytic effect, undesirable reactivity could also occur. The main adverse effects caused by TiO_2 NPs seem to be due to their ability to induce oxidative stress, resulting in cellular dysfunction and inflammation, among others [113]. At high levels of oxidative stress, cell-damage responses are observed, whereas, at moderate levels, inflammatory responses may kick in due to the activation of ROS-sensitive signaling pathways [114,115].

TiO_2 NP-induced oxidative stress is therefore related to increased formation of ROS and the resulting oxidized products and to the decrease in the cellular antioxidants. The damage and extent caused depend on the physicochemical properties of the titania particles. For instance, ·OH production depends on the TiO_2 NP crystal structure and size and was found to correlate with cytotoxicity, e.g., against hamster ovary cells [116], pointing to ·OH as the main damaging species for UV-irradiated TiO_2 NPs [117]. There are conflicting studies on whether it is the size or the irradiation of TiO_2 that contributes to its ability to induce oxidative stress, and the readers are referred to extensive reviews, such as Skocaj et al.'s [118], on this topic and other TiO$_2$ toxicity related discussions.

2.3. Photocatalytic Disinfection Using TiO$_2$ Nanostructures

The photocatalytic ROS production on TiO$_2$ NPs, while it may pose some health risks, can also provide benefits. As early as 1985, the photocatalytic microbicidal effect of TiO$_2$ has been reported by Matsunaga et al. [119]. More studies have then been carried out on the bacteria-killing action of TiO$_2$ [120–123]. Maness et al. [120] attribute this effect to the lipid peroxidation in the microbe (in their case, in *E. coli*) due to the photocatalytic oxidative property of TiO$_2$ NPs. Upon initiation of lipid peroxidation by ROS, propagation can happen via the generation of peroxy radical intermediates, which can also react with other lipid molecules. Superoxide radical could also be involved, as it can also be photogenerated on TiO$_2$. This can react with an intermediate hydroperoxide to form new reaction chains that can go through the damaged cell membrane. Once the cell wall is broken down, TiO$_2$ NPs themselves could also possibly directly attack the cell membrane [120]. It is important to remember though that the microbe's response to photocatalytic disinfection action can also be influenced by its level of protective enzymes against oxidative stress [122].

It is generally accepted that the photocatalytic antibacterial properties of TiO$_2$ are due to its ROS formation, whereby ·OH is thought to play a crucial role [122]. Yet, novel TiO$_2$-based materials also point to the role and use of other ROS. The development of different TiO$_2$ nanostructures paves the way for advancements in photocatalytic disinfection on TiO$_2$, and some examples, also in relation to the formation of ROS, are presented here.

Nanocomposites made from TiO$_2$ NPs on Si nanostructured surfaces can be used as antibacterial surfaces for dental and orthopedic implants, and the TiO$_2$ NPs themselves can be spray-coated to surfaces for disinfection of microbes upon irradiation [65] (Figure 4). The nanostructures are said to rupture the bacterial cell wall, whereas the ROS from TiO$_2$ NPs can oxidize organic matter (such as bacteria) to prevent bacterial growth [124], which could eventually form biofilms. On the other hand, the free radicals photogenerated on TiO$_2$ can also disrupt and destroy biofilms. This is important since killing bacteria within a biofilm is quite challenging; the biofilm shields the bacteria from antibiotics, antibodies, and immune cells [125,126]. Using TiO$_2$ exposed to UVA, the biofilm formation of *P. aerigunosa* was inhibited via ROS attack to disrupt the bacterial cell membrane (disabling bacteria to form biofilm), and the ε-poly(L-lysine) of the cells already in the biofilm was weakened [65]. Through a plasma electrolytic oxidation process, Nagay and coworkers produced N- and Bi-doped TiO$_2$ coatings that kill bacteria because of the generation of ROS upon visible light exposure [51,127].

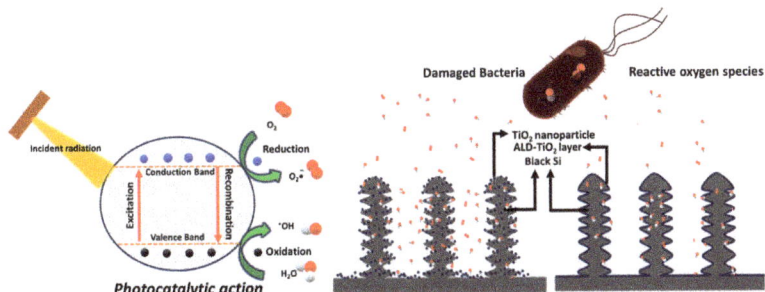

Figure 4. Antibacterial TiO$_2$ nanoparticle hybrid nanostructures. The photocatalytic ROS generation on bacterial-disruptive nanostructures presents a low-cost antibacterial technology. Reprinted with permission from [124]. Copyright 2022, The Authors. Published by American Chemical Society. CC BY-NC-ND 4.0 license (https://creativecommons.org/licenses/by-nc-nd/4.0/).

Efforts to extend photosensitization with TiO_2 were performed by forming composites with materials such as MoS_2 and *l*-arginine and/or doping with Yb and Er [51,128]. However, the broader spectral range or near-IR sensitization is usually brought about by the additional component (and not by TiO_2). On the other hand, other works utilize the photogenerated charge carriers from TiO_2 and enhance the catalytic effect by the addition of other components for antibacterial purposes. For example, TiO_2 combined with graphdiyne (GDY) was synthesized into nanofibers by electrostatic force to produce ROS and prolong the antibacterial effect. When exposed to light, electrons and holes are generated on the TiO_2 and GDY surface, with the photogenerated electrons of TiO_2 being easily transferrable to the GDY surface. There, $\cdot OH$ and $\cdot O^{2-}$ are formed because they can react easily with water and O_2. The extended lifetimes of the charge carriers enhance ROS generation and the resulting bactericidal effect. Overall, these processes inhibit the methicillin-resistant *S. aureus* (MRSA) biofilm formation and promote the regeneration of bone tissues [51,65,129].

In general, the photocatalytic disinfection by TiO_2 nanocomposite antimicrobial coatings entails the incorporation of inorganic metals/nonmetals (such as Ag, Cu, Mn, P, Ca, and F) and/or 2D materials (graphydiyne, MXenes, and metal–organic frameworks, etc.) into TiO_2 to control the porosity of the surface, crystallinity, charge transfer, and disinfecting property against critical pathogens, such as *S. aureus* and *E. coli* but also H1N1, vesicular stomatitis virus (the safe surrogate virus for SARS-CoV-2), and the human coronavirus HCoV-NL63 [65]. Such light-catalyzed coatings could prevent microbes from reactivating to completely destroy them, and with the high mobility of ROS in the air, airborne microbes could also be targeted [130]. The high interest in the inactivation and disinfection of coronaviruses emerged recently due to the recent pandemic. Some of these studies were on the photocatalytic disinfection of coronavirus using TiO_2 NP coatings, with the mechanism attributed to their generation of ROS [65,131,132]. For interest in antimicrobial coatings, readers are referred to Kumaravel et al. [65].

2.4. Efforts to Improve the Photocatalytic Efficiency of TiO_2 Nanomaterials

As seen in Section 2.2, the photocatalytic activity of TiO_2 nanomaterials improves by resorting to different morphologies. Their nanosize alone increases the surface area, providing more active sites. Additionally, due to the interesting properties afforded by its size, the use of nanomaterials also improves photocatalytic efficiency by enhancing charge separation and light harvesting and increasing the surface-to-bulk ratio. These improvements are presented in this section.

2.4.1. Enhanced Charge Separation

The increased photocatalytic performance of TiO_2 nanoparticles cannot be attributed to the increased specific surface area alone but also to the *increase in the surface-to-bulk ratio* with decreasing particle size. The latter results in shorter diffusion pathways that the charge carriers have to traverse to reach the surface, which is the photocatalytic reaction site [133]. Adding adsorbed species further provides electron/hole scavengers that could also improve the charge separation [133–135]. The decrease in the size though also blueshifts the TiO_2 absorption edge [136] and could result in unstable NPs [137]. Therefore, an optimized size is needed to balance the properties in terms of charge separation, light absorption, and stability. During thermal treatment, which is typically needed for good crystallinity and increased photocatalytic performance, aggregation could occur, and this can be prevented by preparing highly-dispersed TiO_2 clusters (such as those synthesized with zeolites [138]), which also shows high photocatalytic activity. These spatially separated TiO_2 species were also prepared as single-site catalysts that also show high photocatalytic electron–hole pair reactivity and selectivity [22]. The high photocatalytic reactivity observed for highly dispersed TiO_2 species is attributed to the highly selective formation of a longer-lived (up to μs), localized charge-transfer excited state compared to that of bulk TiO_2 (ns) [133].

The aggregate formation is not always disadvantageous. The mechanisms portrayed for a single semiconductor NP (e.g., Figures 1, 4 and 5) are simplified, and a more accurate representation would consider that TiO$_2$ NPs have the tendency to self-aggregate in aqueous solution to form a 3D framework. This happens by aligning their atomic planes with each other, allowing for efficient charge carrier transport without interfacial trap interferences in a so-called "antenna effect". Through this, the photogenerated excitons in a nanoparticle will be transported throughout the network until they get trapped individually in a suitable site (e.g., via a redox reaction with an adsorbed electron acceptor/donor on one particle in the network). Charge carriers that are not trapped continue to traverse through the network until they themselves react. Therefore, through forming 3D aggregates, better electron mobility is achieved for TiO$_2$ particles [22,139].

When the aggregates align, they act as if they are an array of nanowires that facilitate efficient CT throughout the network. In fact, 1D morphologies, such as TiO$_2$ nanowires, nanotubes, and nanorods, are hailed for their efficient electron transport since the photoexcited charge carriers could move along the length, increasing delocalization and resulting in long diffusion lengths (>200 nm), which delays the charge recombination and prevents electrons from residing in traps [140–145]. For example, using TiO$_2$ nanotube arrays (TNAs) for PEC water splitting can improve the photocatalytic efficiency and result in a photocurrent of up to 10 times since loss of photogenerated electrons is prevented with the electrons being able to diffuse along the tube towards the collecting substrate. The improved performance using TNA is said to not only come from enhanced charge separation and better electron transport due to the orderly arrangement [140,146–148] of this 3D structure but also due to better light harvesting [133,140], which is discussed in the next section.

Meanwhile, from Table 1, the photocatalytic activity of pure TiO$_2$ nanomaterials is considerably small and requires mostly UV light due to its bandgap. Hence, various attempts were conducted to improve its efficiency and increase its absorption range. *Doping* is one widely used approach, and this has been performed with transition metal ions, such as Fe^{3+}, which increased the efficiency for pollutant photodecomposition likely due to the fact that it decreases the E_g and broadens the absorption range to the visible region [22,149]. Oxygen doping in TiO$_2$ interstices can also improve photocatalytic performance by enhancing charge separation efficiency. For the photodegradation of methyl orange, O-doped or oxidative TiO$_2$ showed a 2~3.7 times higher rate than pristine TiO$_2$, with the former fabricated using KMnO$_4$ to create trap sites to separate charges via bandgap impurity states [150].

Depositing other catalysts has also been another approach. For example, a noble metal, such as Pt, Au, and Ru, can be deposited to increase the photocatalytic activity of TiO$_2$ towards the decomposition of organics and photocatalytic water splitting [22,135,151–153]. This enhanced activity is likely due to improved charge separation as bulk electrons transfer to the metal and therefore to the surface of TiO$_2$ [22,151–154]. Though not all photogenerated electrons transfer from the titania to the metal, the enhanced separation of photogenerated charge carriers increases the electron lifetime in TiO$_2$. The separation is likely enhanced by the surface plasmon resonance (SPR) of the metal and the resulting increased localized EM field caused by the exposure of the metal to light. This induces charge carrier formation near the TiO$_2$ surface, with which carriers can easily reach surface sites and improve charge separation. Loading TiO$_2$ with gold instead of platinum also extends the absorption of TiO$_2$ to the visible range up to near-infrared [22,155–158]. Using gold cores with TiO$_2$ shells also exhibits remarkable photocatalytic activity compared to TiO$_2$ due to enhanced separation of photogenerated charge carriers with the gold core serving as an electron trap [159].

In addition to metals, metal oxides can also be deposited on TiO$_2$ to help improve charge separation and photocatalysis. For example, a Pd-NiO/TiO$_2$ catalyst has been prepared to improve photocatalytic CO$_2$ reduction. Due to the high electron density needed to drive this multielectron reduction, a p–n junction formed by introducing NiO to TiO$_2$ helps to drive hole transport to NiO, whereas the Pd forming a Schottky junction with TiO$_2$ facilitates semiconductor-to-metal electron transfer. These migrations towards NiO and Pd enhance the charge separation and result in high electron density around Pd, which can be used to transform CO_2 efficiently and selectively to CH_4 by reduction [160].

The coupling of the semiconductor to TiO_2 can also be achieved by coupling it with another TiO_2 phase. As mentioned before, mixed-phase TiO_2 of anatase and rutile demonstrates improved photocatalytic performance than by using only pure anatase or rutile. This could be due to the formation of heterojunction when their valence and conduction band edges come into contact [65,161]. Several models were proposed to explain this synergistic effect in mixed phases (Figure 5). First was the model proposed by Bickley et al. based on the positions of the CBs of anatase and rutile in relation to each other [162]. Figure 5a (A) shows the model in which the electron transfer is from anatase to rutile. Separation of charges then happens in anatase and trapping of an electron occurs in the rutile phase. In another "spatial charge-separation model," Hurum and coworkers (Figure 5a(B)) propose that the opposite is happening such that electrons are transferred from the rutile CB to a trapping site in anatase [163].

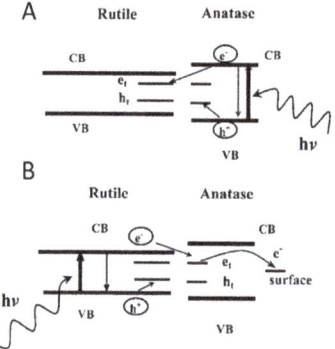

(a) Space-charge separation model

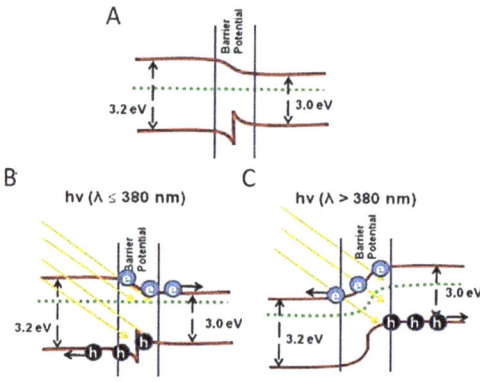

(b) Interfacial model

Figure 5. Models describing the mixed phase synergistic effect due to heterojunction formation: (a) space-charge separation model in which the transfer of electron can occur from (A) anatase to rutile or (B) rutile to anatase. Reprinted (adapted) with permission from Hurum, D.C., et al. Copyright 2003 American Chemical Society [163]; (b) Interfacial model which describes the band bending at the interface of anatase and rutile (A) at equilibrium, (B) when irradiated with light of wavelength ≤380 nm, and (C) when irradiated with >380 nm. Adapted and reprinted with permission from ref. [164]. Copyright 2011 Elsevier B.V.

Meanwhile, the "interfacial model" proposed by Nair et al. looks at the band bending at the interface between anatase and rutile. The electron transfer should occur from the anatase to the rutile upon UV illumination due to the CB energy of the anatase being more negative than that of the rutile. When the illumination is with λ > 380 nm, the rutile is activated, and its CB shifts upward due to accumulated photoinduced electrons enabling the electrons in the rutile to reach the CB of anatase [133]. Thus, the "interfacial model" presents a directional movement of the electrons depending on whether the irradiation is ≤ or >380 nm and upon consideration of the interfacial band bending (Figure 5b) [164]. One can expect that the interfacial nanostructure plays a role in the electron transfer between the components and therefore in the overall photocatalytic performance. Further discussion on the advantages of mixed-phase TiO_2 for photocatalysis can be found in the literature [133].

Fe_2O_3 has also been combined with TiO_2 via photodeposition to enhance charge separation for contaminant decomposition and PEC water splitting. The achieved enhancement of more than 200% in complete mineralization kinetics was ascribed to the transfer of photoelectrons from the TiO_2 to the Fe_2O_3, which in turn favors the rate-determining step of oxygen reduction [165]. Graphitic nanocarbon has also been added to TiO_2 nanomaterials to improve charge separation. By covering short single-wall carbon nanotubes (SWCNT) (~125 nm) around TiO_2 NPs (100 nm) using a hydration-condensation technique, longer lifetimes of photogenerated charge carrier and improved photocatalytic activity for the degradation of an aldehyde was achieved. This was better than nanographene and longer SWCNT hybrid systems. The shorter SWCNT provides greater interfacial contact with each TiO_2 NP, more electron transport channels, and more efficient shuttling of electrons from TiO_2 NP to SWCNT, delaying charge recombination. Improved SWCNT debundling with the short ones also affords these advantages to a larger portion of the composite [166].

The semiconductor junction can also be quite complex, involving materials such as MXenes. For example, Biswal and coworkers designed a Ti_3C_2/N, S-TiO_2/g-C_3N_4 heterojunction to boost the spatial separation of charges and their transport in light of a photocatalytic water-splitting application [166]. This heterostructure was produced by thermal annealing and ultrasonic-assisted impregnation for H_2 production that is up to ~4-fold higher than pristine S-doped titania. The dual heterojunction formed (a n–n heterojunction with a Schottky junction) likely not only enables effective charge carrier separation as CT channels [166] but also reduces the band gap due to the adjustment of the energy bands. $In(OH)_3$-TiO_2 heterostructures were also formed for enhanced photocatalytic H_2 evolution. The band-gap tuning and improved charge separation resulted in up to a >15-fold increase in activity compared to commercial P25 [167].

2.4.2. Enhanced Light Harvesting

The morphology or nanostructure of TiO_2 also improves photocatalytic efficiency due to enhanced light harvesting. As mentioned above, nanoparticle aggregation (~500 nm) can improve light harvesting due to its high scattering effect, resulting in photon reabsorption. This increased visible-range absorption in turn increases the number of excited charges as seen in the increased current density. Such aggregates display unsmooth surfaces, resulting in better molecule adsorption in large surface areas and pore sizes [168].

Nanotextured TiO_2 substrate produced by nanomolding also displays efficient light harvesting. The hierarchical nanopattern of dual-scale nanoscale craters featuring smaller bumps couples both the longer and shorter wavelengths of light resulting in a light trapping effect for efficient light utilization and at a wide angular range [169]. Similarly, cicada-wing-like structures were used as imprints to form nanohole structures of TiO_2 decorated with Ag NPs (10–25 nm) for methylene blue (MB) photocatalytic degradation. The structure did not only exhibit extended absorption to the visible range but also greater light absorption, likely due to the SPR effect from Ag and the nanotexture of TiO_2. This is based on the photocatalytic decomposition rate obtained for the Ag-TiO_2 nanotexture being 2.7 times higher than the nanotextured TiO_2 alone but more than 7 times higher than P25. This shows that even with just nanotextured TiO_2, improved photocatalytic performance can be seen. As discussed in the previous section, the Schottky barrier formed between Ag and TiO$_2$ could also improve the charge separation [170]. Hollow particles of *TiO$_2$ decorated* with Au@Ag core–shell NPs also display enhanced light harvesting due to the combined strengths of the components of having a strong, broadened localized SPR, large specific surface area, and favorable light scattering properties [171].

Orderly arrays of nanostructures, such as TNA, serve as effective light scattering layers according to the Mie scattering theory. The Mie scattering effect displayed by anodized TNA or NPs has been used in solar cells to harvest more sunlight and enhance charge conduction [172]. Mie scattering is important in explaining particle size-dependent Raman enhancement observed with semiconductors [173,174] and is brought about by the plasmon resonance at the surface of the sphere causing signal enhancement that depends on the size as one comes closer to the lowest transverse electric mode of NPs. In addition to the Mie effect, size quantization also affects the Raman intensity obtained on TiO$_2$ NPs [175].

The surface-enhanced Raman (SER) effect on semiconductors has also been well-observed [40,41,43–45], and the influence of nanostructuring on SER scattering, in particular on TiO$_2$, has been investigated. Whereas CT and chemical contribution can provide an enhancement factor (EF) of ~10^3 [46,47], EM enhancement is also afforded in TiO$_2$ nanostructures of a high aspect ratio, such as nanotubes and nanofibers [42,48–50]. Han et al. [49] showed concrete evidence of morphology-dependent EM enhancement using cyt b$_5$ heme as an indirectly-attached SER probe to reduce the chemical contribution to the Raman signals. Using EM field calculations, the particle's aspect ratio was shown to increase the "hot spots" (regions of enhanced EM field) at the TiO$_2$-water interface [49], improving the structure's light-harvesting ability. Hence, other morphologies of higher anisotropy, such as TiO$_2$ nanotubes, were further studied, showing a similar morphology-dependent EM field enhancement [42,50] (Figure 6). TNAs were shown to exhibit different Raman enhancements depending on the tube length, which fits the EM field calculation showing hot spots along the nanotube length [42,50] (Figure 6a). The TNA of high EM field enhancement was shown to perform better as a photocatalyst for visible-light-degradation of an azo dye pollutant immobilized on TNA (Figure 6b). Interestingly, the TNA's optical response (i.e., its EM field enhancement) correlates with the photocatalytic degradation rate occurring on it [176].

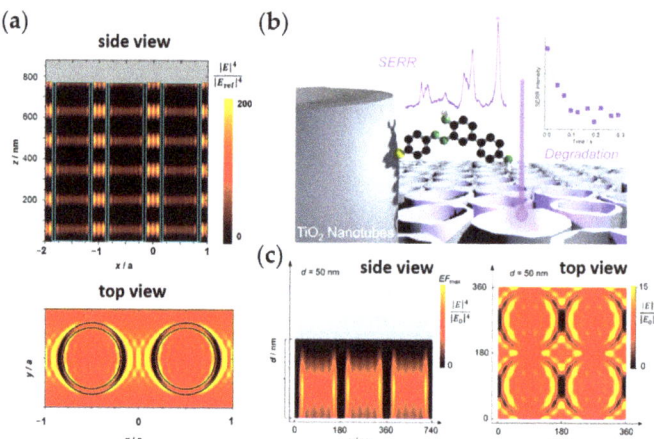

Figure 6. Electromagnetic (EM) field enhancement studies on Ti-based nanotubular arrays: (**a**) calculated EM field enhancement for the high aspect-ratio TiO$_2$ nanotube array (TNA). The side and top view of the calculation show the distribution of the localized field hot spots for 3 and 2 tubes, respectively. The enhancement factor (EF) scale bar is shown at the left of the side view image. Adapted with permission from [50]. Copyright 2018 Wiley-VCH Verlag GmbH and Co. KGaA,

Weinheim. (**b**) Photocatalytic azo-dye degradation on TNA correlates with EM enhancement. The inset shows the surface-enhanced resonance Raman (SERR) spectra of the adsorbed azo dye on TNA of high EM field enhancement and the corresponding exponential fit of the decay rate of the SERR intensity of the dye peak (marked by *). Adapted with permission from ref. [176]. Copyright 2019, the authors. Published by Wiley-VCH Verlag GmbH and Co. KGaA. Open access article. CC BY 4.0 license (https://creativecommons.org/licenses/by/4.0/); (**c**) EM field enhancement calculations for the partially-collapsed nanotube structure of TiN from nitridated TNA. The side and top view of the calculation show the distribution of the localized field hot spots for 3 and 4 tubes, respectively. The EF scale bars are shown at the left of the images. Adapted with permission from [177]. Copyright 2022 by the authors. Licensee MDPI, Basel, Switzerland. This article is an open-access article distributed under the terms and conditions of the Creative Commons Attribution (CC BY) license (https://creativecommons.org/licenses/by/4.0/).

Similar to other TiO_2 nanostructures (see above), incorporating other components to form nanocomposites with TiO_2 nanotubes can further improve not only the charge separation but also the light-harvesting ability. For example, S-doping or the addition of CdS NPs to TiO_2 nanotubes resulted in enhanced visible-light water splitting [178–180]. Ultrafine Pt NPs were also added into TiO_2 nanotubes for the efficient photocatalytic formation of methane from carbon dioxide and water. The nanotubes allowed for a homogeneous distribution of Pt NPs, which accept electrons and become sites for reduction, thereby also allowing efficient separation of charges [133]. Furthermore, even structures obtained from TNA somehow retain the light enhancement afforded by the 2D periodic arrangement of the nanotubes (Figure 6c). For example, nitridation of the TNA resulting in a partially collapsed nanotube structure of TiN also shows wavelength-dependent EM field enhancement and corresponding light enhancement [177].

The 2D periodic arrangement in TNAs enables them to behave as photonic crystals—with photonic lattices reflecting the light of certain wavelengths—bringing about localized EM field enhancement [42,50,181–184]. Interestingly, this photonic crystal-like character has also been observed in inverse-opal (IO) structures, which also achieved SER EF of around 10^4 (though likely due to both chemical and EM contributions) [42,50,181], and which can also be made from TiO_2. IO TiO_2 also shows promising performance as photocatalysts [185–188], with their light harvesting extended to the visible range [187,188] and their ability to generate slow photons [188–190]. The slow photons have been shown to significantly increase the interaction of TiO_2 with light and can work synergistically to amplify the chemical enhancement in the catalyst [186].

2.4.3. Black TiO_2

The photocatalytic efficiency of TiO_2 nanomaterials can be improved by enhancing the separation of their charge carriers and improving their light harvesting and absorption properties (Sections 2.4.1 and 2.4.2). Therefore, having a material that encompasses both is an ultimate surface-engineering achievement in this regard. Black TiO_2 ticks both requirements and reasonably has then become a hot topic in TiO_2 photocatalysis in the last decade or so.

Though previous studies already describe a similar material, as indicated in reviews [191–193], all papers seem to point to the work of Chen et al. [194] for introducing (and coining the term) "black TiO_2" to describe the partially hydrogenated titania nanocrystals which exhibit a reduced bandgap due to a disordered layer at the surface of its crystals. This material exhibited a redshifted absorption onset to near-infrared (compared to the starting TiO_2 nanomaterial), which was not surprising considering its visible color change. That is, due to the hydrogenation process, the crystals changed from white to black (hence the name). Consequently, this also results in a decreased bandgap of ~2.18 eV, making black TiO_2 a good catalyst for visible-light irradiation. Additionally, it also exhibits good stability, making it an ideal catalyst for use under continued irradiation. From calculations, it also presents localized photogenerated charge carriers, indicative of slowed-down recombination, which is beneficial for photocatalysis. This makes the work of Chen et al. the first reported use of black titania for photocatalytic purposes [192].

From then on, many studies have been carried out to synthesize, characterize, and evaluate the photocatalytic performance of black TiO_2 nanomaterials. Different methods have been developed to reduce TiO_2 without the use of high pressure, as was conducted in the work of Chen et al. [194]. These include (electro-)chemical reduction [195,196], solvothermal hydrogenation [197], thermal reduction [198], reduction at the solid phase (reductant + heat) [199], anodization (and annealing) [200], ultrasonication [201] plasma treatment [202], gel combustion [203], or a combination of these [204]. Most of these strategies are similar to the synthesis of TiO_2 nanomaterials presented in Table 1, with a reductant source/ reducing condition (either chemical, thermal, hydrogen, or reducing gases, such as hydrogen, nitrogen, and argon) added. Since black TiO_2 is formed by the reduction of TiO_2, it is also called "reduced TiO_2" [205–207] or "hydrogenated TiO_2" [208] and represented with the formula TiO_{2-x}, the $-x$ indicating the formation of oxygen vacancies [205,206].

What is interesting then is the concept of forming a novel material due to the introduction of surface defects, and yet, as this is also a TiO_2 nanomaterial, it can also exist in different morphologies and structural states, resulting in a plethora of black TiO_2 of various properties and photocatalytic performance. Table 2 gives examples of these materials and their photocatalytic performance in terms of organic compound degradation and hydrogen generation, with the latter being an important solar-driven application of black TiO_2. Chen et al. [193] give a comprehensive review of black TiO_2 nanomaterials, including their properties and examples of application, whereas Naldoni et al. [192] give a good summary of the photocatalytic H_2 generation on black TiO_2. The readers are encouraged to take a look at these reviews.

Table 2. Examples of black (and colored) TiO$_2$ nanomaterials and their photocatalytic performance in terms of degradation of organic compounds or hydrogen evolution.

Material	Degradation/Removal of Organics	H$_2$ Generation	Reference Material/Comparison	Ref.
Black TiO$_2$ nanocrystals/NPs	~7.5× faster MB degradation, solar illumination	0.1 ± 0.02 mmol h^{-1} g^{-1} (2 orders higher (solar simulator or visible IR light))	Degradation: pristine TiO$_2$ nanocrystals H$_2$ generation: most semiconductor photocatalysts	[194]
	Up to k = 0.68 min^{-1} MO degradation, 2.4× faster (simulated sunlight)	Up to 5.2 mmol h^{-1} g^{-1}, 1.7× faster (simulated sunlight)	Pristine P25 degradation: k = 0.28 min^{-1} H$_2$ generation: 5.2 mmol h^{-1} g^{-1}	[199]
	Up to apparent k (k_{app}) = 0.998 h^{-1} or 0.0166 min^{-1} acetaminophen removal, 1.9× faster than P25 and 4.9× faster than sintered P25 (solar illum. AM 1.5G)		P25: k = 0.527 h^{-1} or 0.00878 min^{-1} Sintered P25: 0.203 h^{-1} or 0.00338 min^{-1}	[209]
		Estimate: ~15 μmol h^{-1} g^{-1} ~5–7× higher (AM 1.5 illum.; 100 mW cm^{-2})	Anatase nanopowders; estimate: ~2–3 μmol h^{-1} g^{-1}	[210]
	Anatase; ~1.5× faster, MB degradation finished in 18 min (solar illumination)		Degussa-P25: MB degradation finished in 18 min. (solar illumination)	[203]
Black hydroxylated TiO$_2$ (ultrasonic.)	5.8× (solar illumination) and 7.2× (visible light) faster acid fuchsin decomposition; amorphous state		Original sol TiO$_2$ (non-ultrasonically processed)	[201]
Black TiO$_2$ nanotube array (TNA)	Estimate: 10–15% (3–4×) better photocatalytic degradation (brilliant blue KN-R dye; 175 W Xe lamp)		Pristine TNA	[196]
Ordered mesoporous black TiO$_2$		136.2 μmol h^{-1}, ~2× higher (solar)	Pristine mesoporous TiO$_2$: 76 μmol h^{-1}	[198]
Mesoporous black TiO$_2$ hollow spheres		241 μmol h^{-1} (0.1) g^{-1}, ~2× higher (solar)	Black TiO$_2$ NPs: 118 μmol h^{-1} (0.1) g^{-1} Mesoporous TiO$_2$ hollow spheres: 81 μmol h^{-1} (0.1) g^{-1}	[197]
Defective black TiO$_2$ (dimpled morphology, anodization)	High oxygen vacancy concentration (C$_{Vo}$): up to ~80% RhB degradation (after 4 h), ~1.3× better than low C$_{Vo}$ and ~4× better than TNA.		Low C$_{Vo}$: up to 60% RhB degradation (after 4 h) TNA: ~20% RhB degradation (after 4 h)	[200]

Table 2. Cont.

Material	Degradation/Removal of Organics	H_2 Generation	Reference Material/Comparison	Ref.
Grey TiO_2 nanoparticles (flow furnace)		Estimate: ~75 µmol h^{-1} g^{-1} ~25–37× higher (AM 1.5 illum.; 100 mW cm^{-2})	Anatase nanopowders: estimate: ~2–3 µmol h^{-1} g^{-1}	[210]
Grey TiO_2 nanoparticles (hydrogen. at high P)		Estimate: ~80–85 µmol h^{-1} g^{-1} ~27–42× higher (AM 1.5 illum.; 100 mW cm^{-2})	Anatase nanopowders: estimate: ~2–3 µmol h^{-1} g^{-1}	[210]
Colored TiO_2 (dark blue)	Up to $C/C_0 = 0.14$ (estimated $k = 0.197$ min^{-1}), 1.4× faster MO degradation (300 W, Xe lamp UV-vis light)	Up to max. prod. of 6.5 mmol h^{-1} g^{-1}, 7.2× higher (UV-vis light); ~180 µmol h^{-1} g^{-1} (vis-IR)	Pristine P25 Degradation: $C/C_0 = 0.24$ (estimated $k = 0.143$ min^{-1}) H_2 generation: 0.9 mmol h^{-1} g^{-1}	[195]
Blue TiO_2(B) single-crystal nanorods	$k = 0.0146$ min^{-1}; 97.01% RhB degradation (after 150 min), 6.9× and 2.1× better than TiO_2 NPs and TiO_2 NRs, respectively (vis light); 98.56% deg. RhB (solar light), 99.12% deg. Phenol (solar); reaction constant (k_{rxn}) = 0.0250 (RhB) and 0.0366 (phenol), 8.8× higher than TiO_2 NPs	Up to 149. µmol h^{-1} g^{-1} (AM 1.5 illumination), ~26.6× higher than TiO_2 NPs	Degradation: TiO_2 NPs: $k = 0.0016$ min^{-1}; 14.06% RhB degradation (after 150 min, vis light) TiO_2 NRs: $k = 0.0053$ min^{-1}; 46.44% RhB degradation (after 150 min, vis light) H_2 evolution: TiO_2 NPs: 5.6 µmol h^{-1} g^{-1} (AM 1.5 illumination) TiO_2 NRs: 40.8 µmol h^{-1} g^{-1} (AM 1.5 illumination)	[211]

Some examples included in Table 2 are of different color naming (termed "colored TiO_2"), such as green, grey, and blue TiO_2 [192,195,208,210,211]. This is based on the understanding that the visual colors exhibited by TiO_2 are brought about by intrinsic defects, such as due to the presence of Ti^{3+} and/or oxygen vacancies [192,195,200,208,211] or by doping with impurities [192,202]. Such defects create extra electronic states in the TiO_2 bandgap, i.e., intraband gap states, which alters the optoelectronic properties of TiO_2 [192]. Whether this is also the case for the color of black TiO_2 is still a controversy. While some reports claim that the formation of these color-inducing intrinsic defects in TiO_2 results from the hydrogenation [205], with the color depending on the extent of reduction and reducing condition [208], others propose that the black color is due to the disordered surface [194,202]. The disordered surface is caused by hydrogen and allows hydrogen to swiftly navigate around and induce electronic structural changes [212]. Midgap band states are formed because of the changes in the structure [203] brought about by the excessive lattice disorder. They can form an extended energy state by overlapping with the edge of the conduction band and also possibly combining with the valence band [202].

An effort to further unravel the relationship between the defect nature and photocatalytic activity of reduced TiO_2 was performed by Will et al. [208] by considering that the intrinsic defects created on the surface are pertinent to the photocatalytic process and the location of the defect depends on the structural state and reducing conditions. They found that the introduction of Ti^{3+} at the surface results in a surface with substoichiometry, which activates the surface for photocatalysis. However, too long hydrogenation or too much Ti^{3+} is detrimental to its activity, as these provide additional recombination sites or prevent efficient interfacial CT. Surface roughness and strain were also not important for the activation of photocatalysis.

The photocatalytic activity of black TiO_2 nanomaterial can be further enhanced by forming appropriate heterojunctions with other materials [208], a similar strategy used with TiO_2 nanomaterials. Further, amorphous black TiO_2 can also be synthesized and used for photocatalysis [201], which is important for applications such as for bone implants. Black TiO_2 was shown to exhibit biocompatibility [213], regenerative properties [214], photothermal properties [213–215], and microbicidal action [213,215–217], among others. As such, black TiO_2 shows promise for cancer treatment [214,215], as a bone implant coating, and for disinfection purposes [213,215–217]. Similar to TiO_2, the photo(electro)catalytic disinfection with black TiO_2 nanomaterials is also claimed to occur due to ROS, in particular, superoxide and/or hydroxyl radicals [216,217]. Nevertheless, the biosafety of black TiO_2 needs to be further studied to intensify its application in the biomedical field.

3. Dark Catalysis on Ti-Based Oxides

In addition to photocatalysis, TiO_2, and Ti-based oxides also manifest "dark catalysis". Here, we refer to dark catalysis in the context of the catalytic activity observed in the absence of irradiation. Early works on dark catalysis on TiO_2 seem to have stemmed from the wide use of Ti alloys for biomedical applications. Due to the superb biocompatibility and good mechanical strength of Ti and Ti-based alloys, they are used as bone and dental implants. Thus, an interest in understanding the influence of Ti implants on the inflammatory response led to studies that looked at the $Ti-H_2O_2$ system in the dark. Ambiguities in the results of photocatalytic studies on whether the observed catalytic effect was brought about by light irradiation or by nanoparticle size also contribute to the catalytic effect observed in the dark.

As early as 1989, there was an interest in the influence of implants on the inflammatory response [218,219]. The role of Ti in Fenton-type reactions was examined [219,220] and thought to occur during the inflammatory condition. Ellipsometry studies showed that in the presence of H_2O_2, metals such as Ti readily oxidize to form metal hydroxides or metal oxides such that the body mainly interacts with the oxidized Ti instead of the bare metal [219]. Further, in the dark, TiO_2 is found to catalyze the H_2O_2 decomposition based on observed oxygen evolution, though it is thought to unlikely occur through

·OH radicals [218]. The latter is based on ESR and spectroscopic results showing low ·OH formation for the Ti-H_2O_2 system in the dark [219]. When a Ti(IV)-H_2O_2 complex coordinates to a H_2O_2, a TiOOH matrix can be formed on the surface. This matrix can trap superoxide radicals, making the Ti(III) (reduced from Ti(IV)) likely inaccessible [218,219].

The addition of H_2O_2 could also effectively promote the catalytic activity of TiO_2 [220–227]. In Section 2, H_2O_2 and peroxides play a role in the photocatalysis of TiO_2 (in the presence of UV, water, and oxygen). Once produced, peroxides and H_2O_2 can perform dark catalysis on TiO_2. Such were the findings of Krishnan et al. in their investigation of the changes in the surface of photocatalytic bulk TiO_2 powder in terms of UV exposure, as well as the presence of water vapor and O_2 [228]. Using advanced XPS, changes in Ti 2p, O 1s, and the bridging and terminal OH were investigated. Maintaining the TiO_2 activation state for a certain period in the dark was found to require the presence of water vapor and oxygen. The prolonged oxidative capacity of the TiO_2 powder in the dark was ascribed to the appearance of peroxides and dissolved H_2O_2. Though the highest catalytic activity was observed in the combined presence of UV, water vapor, and oxygen, the nonreversal behavior of the XPS spectra upon UV light removal (Figure 7, Phase 5) points to continued TiO_2 activity and indicates that the presence of H_2O and O_2 is enough to retain the dark catalytic activity for a time period (of around $1/3$ of the duration when all three factors were present) [228]. Though this may seem to be only due to the residual effect from photocatalysis, the fact that prolonged and sustained catalysis even after removal of the light source continued and produced ROS points to the need to understand what is happening during this time. Understanding the continued catalysis in the dark will enable us to further exploit the advantages of this process.

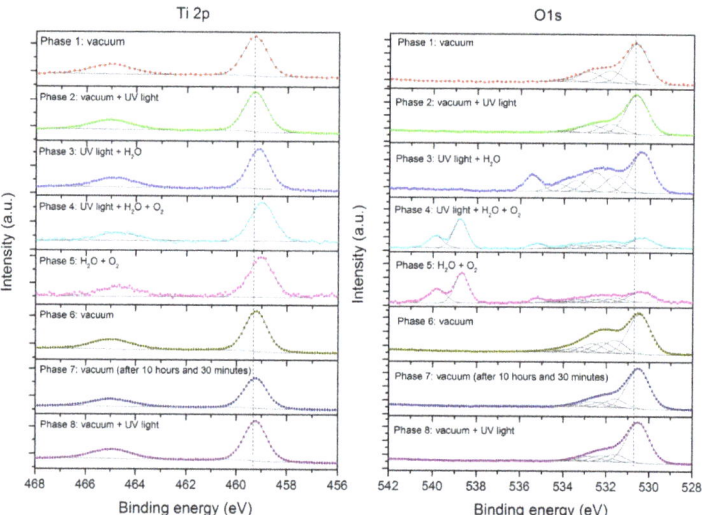

Figure 7. X-ray photoelectron Ti 2p and O 1s spectra of TiO_2 exposed and unexposed step-wise to UV light, water, and oxygen. Phase 5 shows the presence of peroxides and dissolved H_2O_2 indicative of residual catalysis. Reprinted from ref. [228]. Copyright 2017, the author(s). Published by Springer Nature. This is an open-access article distributed under the terms of the Creative Commons CC BY license (https://creativecommons.org/licenses/by/4.0/).

Parallel to inflammation studies for biomedical implants, dye decomposition using TiO_2 for purposes such as water treatment also continued to develop to extend the light beyond the UV range—towards visible light, ambient light, and even in the dark. Hence, from this field, an interest in dark catalysis on TiO_2 has also developed. One of these works is on the bleaching of MB in the presence of TiO_2 and H_2O_2 by Randorn et al. [220]. Even in the absence of light, they observed that catalytic degradation on TiO_2 could occur, which was better on hydrated TiO_2 (a hydrated amorphous TiO_2 with a high surface area) than on Degussa P25. They noted that the mechanism could be different from photocatalysis with photogenerated charge carriers and instead must involve Ti^{3+}/Ti^{4+} in a Fenton-reaction-like superoxide-driven process, whereby H_2O_2 is consumed directly:

$$Ti^{3+}(s) + H_2O_2 \rightarrow \cdot OH + OH^- + Ti^{4+}(s) \tag{1}$$

where (s) indicates that the metal ions are from dangling bonds at the solid surface [220]. However, the catalytic effect in the dark observed in this work cannot be unambiguously distinguished from the surface adsorption effect in bleaching.

Sanchez et al. used a suspension of TiO_2 and H_2O_2 to degrade MB in the dark and found that the TiO_2 surface area and the concentration of H_2O_2 are crucial in catalysis. Using ESR, they found that free radicals are present in the mixture in the dark and attributed the observed catalysis mainly to ·OH and hydroperoxyl radicals (HO_2·) [224]. The presence of the HO_2·, together with other ROS (superoxide and hydroxyl radicals), was also detected by ESR in the work of Wiedmer et al. [226] in which MB degradation was performed nonirradiated on TiO_2 (micro- and) nanoparticles with (3 vol.%) H_2O_2. The $\cdot O_2^-/\cdot OOH$ radicals seem to play a significant role in the dye degradation, as these radicals are present for those that show high dye degradation even if ·OH is more energetically favorable on anatase and is the most reactive among these oxygen-centered radicals. On the other hand, Zhang et al. [225] attribute the improved performance of the TiO_2-H_2O_2 mainly to ·OH formation. In their work, ·OH (E^0 = 2.80 eV) radicals can be formed by using facet- and defect-engineered TiO_2 to heterogeneously activate H_2O_2 (E^0 = 1.78 eV) into a defect-centered mechanism for Fenton-like catalysis. This involves surface Ti^{3+}. The Ti^{3+} donates electrons to the H_2O_2 and generates $\cdot O_2^-/\cdot OOH$ and ·OH in the process [221,225].

Facets also have an influence on (photo-) and dark catalysis on TiO_2. For example, (001) is considered the most reactive facet in anatase, likely due to its very high anisotropic stress. The surface reconstructs to reduce this stress by forming ridged atoms in every fourth unit cell and likely by the creation of ridge atom vacancies [229], which can interact with charge carriers and ROS.

Wei et al. showed that TiO_2 (B) nanosheets and H_2O_2 can degrade dye molecules in the dark, though the process is accelerated in the presence of visible light or heat. They attributed this catalytic activity to the reaction of the nanosheets with H_2O_2 which can generate superoxide radicals [223]. Jose et al. attributed the dark catalytic H_2O_2 decomposition with hydrogen titanate nanotubes to occur primarily by generating and attacking $\cdot O_2^-$ rather than the hydroxyl radical [230]. Using (delaminated) titanate nanosheets, efficient removal of high concentrations of dyes at a wide range of pH can be achieved. The mechanism of this non-light-driven catalytic degradation involves the formation of the yellow complex surface Ti-OOH, the key species to strongly oxidize and degrade organic dyes into smaller molecules. Initially, H_2O_2 adsorbs and is followed by an exchange with Ti-OH groups at the surface [227]. In effect, the active species observed in the said work can be thought of as a Ti-coordinated hydroperoxyl unit.

The enhancement of the TiO_2 catalysis in the dark upon H_2O_2 addition is due to the formation of ROS on the surface, including ·OH and $\cdot O_2^-/\cdot OOH$. Wu et al. investigated the mechanism of this process by using TiO_2 NPs with single-electron-trapped oxygen vacancy (SETOV). SETOVs are common TiO_2 intrinsic defects. In the presence of H_2O_2, TiO_2 with SETOVs can efficiently degrade organic dyes catalytically in the dark [221]. Using XPS and ESR, they found that SETOV mainly activates H_2O_2 in the dark by a direct contribution of electrons, which, in the process, forms both $\cdot O_2^-/\cdot OOH$ and ·OH to enhance the system's catalytic activity. The steps in the mechanism for the dark catalytic ROS generation are given in Table 3.

Table 3. Steps in the mechanism for the catalytic ROS generation in the dark. Activation of H_2O_2 performed on TiO_2 with single-electron-trapped oxygen vacancies (SETOVs) is proposed in ref. [221].

	Reaction *
1	$H_2O_2 + Ti\text{–}OH \rightarrow Ti\text{–}OOH + H_2O$
2	$V_O^{\cdot} + Ti\text{–}OOH \rightarrow Ti\text{–}\cdot OOH$
3	$V_O^{\cdot} + Ti\text{–}\cdot OOH \rightarrow Ti\text{–}OH + \frac{1}{2} O_2$
4	$V_O^{\cdot} + O_2 \rightarrow \cdot O_2^{-}$
5	$V_O^{\cdot} + H_2O_2 \rightarrow \cdot OH + OH^{-}$
6	$H_2O_2 + \cdot OH \rightarrow \cdot OOH + H_2O$

* $V_O^{\cdot}$ pertains to the single electron in SETOVs.

Oxygen vacancies in general are said to play a pertinent role in dark catalysis. Such is the case for the decomposition of N_2O on anatase (001) and (101), which, during the reaction, involves filling the vacancy [192,229]. The concentration of oxygen vacancies can be increased by calcination at a higher temperature. The oxygen vacancies reductively interact with oxygen to form O_2^{-}, which can increase the current density (for Hg^{2+} reactions, for example) on TiO_2 [229,231].

In terms of biomedical applications, there is also growing evidence that ROS formation and its adverse effects are induced in the presence of TiO_2 even in the absence of light. Unexposed anatase NPs induced higher levels of ROS within human hepatoma cells compared to unexposed rutile, with the former causing oxidative DNA damage [118,223]. DNA oxidative damage seems to be only brought about by nanoparticles as with ordinary TiO_2 particles, without irradiation, cell survival was not affected (though the number of DNA strand breaks was also increased) [118,232].

Microbicidal Effect of TiO_2 in the Dark

A similar discussion to Section 2.3 is presented here but for the disinfection with TiO_2 nanostructures in the dark. This is useful for certain TiO_2 applications, such as with implants that will be in the dark after surgery. To prevent inflammation, strategies include ensuring the implant material surface has antimicrobial properties. As titanium naturally grows oxide and may be induced to grow thicker and more stable oxide films (see below), some natural bactericidal effect is also already afforded on Ti and its alloys. This is important since it is almost impossible to achieve a completely bacteria-free environment for surgery as most operating rooms are contaminated quickly and easily [36,37,233]. To highlight the catalytic effect of TiO_2 on antibacterial action for biomedical implants, the discussion here is limited to the bactericidal effect of TiO_2. Therefore, readers interested in strategies to improve antibacterial properties on Ti implants are referred to other reviews which have already summarized such strategies [234–237].

The microbial-killing action of TiO_2 in the dark has been observed and recognized for a long time. Matsunaga et al. [119] observed that even in the absence of irradiation, ~10–12% of the *S. cerevisiae* cells were killed. A decrease in the colony-forming units of *S. sobrinus* with TiO_2 in the dark was also seen by Saito et al. [121]. Other works then followed, mainly on photocatalytic disinfection with TiO_2, in which the bactericidal effect of TiO_2 in the dark was observed and recognized [120,122]. However, these works point to the disinfection effect in the dark that is residual from the bactericidal phototreatment. This effect is similar to and has been pointed out in the work of Krishnan et al. (see Section 3) [228]. They pointed out that the long-lived reactivity of TiO_2 in the dark could explain the observed extended bactericidal effect of TiO_2 in the dark. Indeed, Rincón and Pulgarin [238] also attribute this "residual disinfection effect" to the photoinduced generation of ROS that damaged and continued to kill bacteria in the dark [122,238]. These studies point to the fact that light may be needed for initiation but may not be continuously necessary for various applications [37,228], which is a beneficial finding for TiO_2-based biomedical applications in relation to presurgical irradiated disinfection. In addition to ROS generation, TiO_2 is also claimed to display bactericidal and self-sterilizing effects by altering the material's surface free energy and electrostatic interaction with the microbial cell wall [65]. Moreover, as discussed in the previous section, ROS generation seems to occur not only due to photocatalysis but also in the absence of light.

Black TiO_2, on the other hand, shows an electrochemical (EC) microbicidal effect which could be of the same disinfection rate as the photocatalytic effect [213]. Though radicals were not seen in ESR in the dark on black TiO_2, in comparison to PEC treatment, the EC microbicidal result indicates that light is not necessary for black TiO_2 to display microbicidal action.

4. Ti and Ti-Based Oxides for Biomedical Implants

In the field of bioimplant application, one should also consider other aspects of the material for targeted implant usage. Ti and TiO_2, for example, find applications in dental and bone implants due to their excellent biocompatibility and good mechanical strength. For bone implant application, for example, the material's mechanical properties have an influence on the postimplantation healing of the affected bone area and the performance of the implant. When the material's stiffness (Young's modulus) is too high compared to the bone, the distribution of the postsurgical physiological load on the periprosthetic bone changes such that the implant handles more of the load, and the bone receives insufficient stress that it needs, i.e., "stress-shielding" occurs. This results in bone resorption, loss of density, and eventual atrophy, resulting in aseptic loosening causing implant failure (Figure 8a) [51,239–241]. Studies show that aseptic loosening accounts for at least 20–33% of orthopedic revisions due to implant failure [51,239,242].

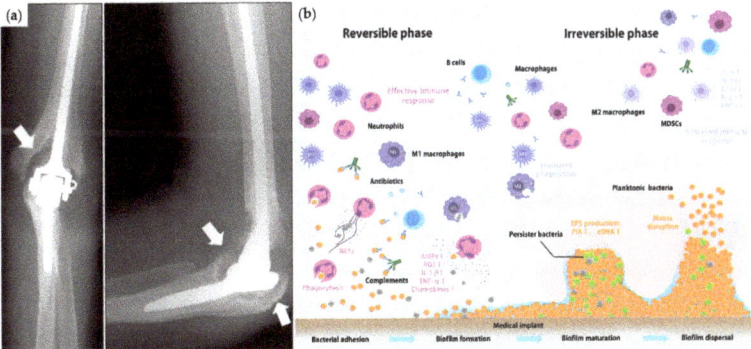

Figure 8. Challenges for bone (and dental) implants: (**a**) aseptic loosening of the implant commonly occurring due to "stress shielding". Reprinted with permission from ref. [241]. Copyright 2019 by the Korean Orthopaedic Association. This is an open-access article distributed under the terms of the Creative Commons Attribution Noncommercial License (https://creativecommons.org/licenses/by-nc/4.0/); (**b**) occurrence of inflammation and biofilm formation. If osseointegration fails when host cells do not strongly attach to the implant, microbes can colonize the surface, leading to biofilm formation and inflammation. Reprinted (adapted) from ref. [38]. Copyright 2022, The Author(s). Published by Springer Nature. This is an open-access article distributed under the terms of the Creative Commons CC BY license (https://creativecommons.org/licenses/by/4.0/).

Titanium and its commonly available alloys have a Young's modulus of 100–150 GPa, which is still higher than that of bone (10–30 GPa) [1,243–245]. As such, efforts to reduce the alloys' stiffness have been investigated. For example, β-type Ti alloys (which can be formed by adding stabilizers, such as Nb, Ta, V, or Mo) [243,246] can have a Young's modulus of ~50–80 GPa and can even reach as low as 40 GPa when subjected to severe cold-working [1,29,30,243,244,247]. Ti displays nontoxicity high strength [245,248], which also has to be considered together with stiffness [1,29,31,244–247] (opposing in nature) when designing alloys for implant application. Together with these two, the corrosion properties should also be considered [1,29,31,243,245,249].

Corrosion response determines how the material behaves when exposed to the physiological electrolyte and environment during and post-implantation [29–31]. Ti and Ti-based alloys were developed mainly to improve the mechanical properties of implant materials, especially for load-bearing purposes by increasing the fatigue properties. Due to the thin, passive TiO_2 film that develops on the surface, Ti and its alloys also display good corrosion properties. This thin TiO_2 film is stable in natural and artificial physiological fluids [243], and elements added for alloying should therefore not disrupt the oxide layer formation. Nb and Zr, for example, can be added as alloying elements to Ti because their (mixed) oxides remain passive and contribute to corrosion resistance. A challenge alongside the oxide formation in alloys is in the case of uneven distribution of the elements in the different phases, unstable formation of the passive oxide film could occur and would lower the material's resistance to corrosion [1,29,30,250]. All these aforementioned properties of Ti and its alloys, therefore, have to be considered for their advancement in use as modern implant materials. Further, when improving processability, such as in additive manufacturing or by adding components to achieve other functionalities, the influence of such modifications on the aforementioned properties has to be considered [1,29,30].

A critical functionality for bone implants is the formation of a robust and lasting attachment between the implant and the bone [251]. Therefore, efforts to increase the success rate of implants also entail improving bone adhesion and growth on the implant surface. This is an advantage for Ti alloys which are known to exhibit osseointegration, allowing for direct anchoring of the implant to the bone even though Ti is considered inert. Nowadays, the understanding of osseointegration considers the implant as a foreign body from which the body tries to defend and shield itself by forming bone tissues to surround the implant [39,243]. Effective implant osseointegration will not only promote healing but also prevent infection by not allowing pathogens and microbes to colonize the implant surface. This so-called *race for the surface* [34–37] determines whether the implant will succeed or fail, especially after surgery and during wound healing [34]. This depends on whether the host cells can attach to the implant irreversibly before bacterial cells do so in the irreversible phase of biofilm formation (Figure 8b) [38]. To further improve the interface between bone and the implant, biocompatible oxides, such as TiO_2, are used to facilitate this bridging. The roughness of the surface contributes to the attachment dynamics displayed by the bacteria and the host cells towards the implant, making nanostructured TiO_2 beneficial for such cases [36]. For further interest, the readers are referred to more extensive reviews on the surface modification for biomedical and antibacterial properties [51,52].

While the implant surface is quite important for its successful osseointegration, the bioinert native TiO_2 layer (2–5 nm) [53], however, does not allow the implant to bind easily and strongly with bone tissues. Further, this layer can be disrupted due to tribological factors (e.g., fretting) in the presence of fluoride (as for dental implants) or caused by oxidative stress brought about by highly aggressive ROS, such as when inflammation occurs due to the implantation process or during implant degradation (see Section 4.1). Because of the debilitating effect of the disruption of the thin native oxide film on Ti-based implants, efforts were done to produce thicker and more stable layers of Ti-based oxides to improve the materials' surface bioactivity [53–55], favoring bone cell adhesion and proliferation and matrix mineralization promotion [31,56]. Direct oxidation of Ti implants by treatment with H_2O_2 or NaOH to induce the formation of a TiO_2 layer can be performed to enhance the bioactive fixation of Ti-based implants [252]. Other efforts also include growing Ti-based nanoporous oxides [33] and nanotubular oxide layers [53] on glass-forming alloys, such as Ti-Y-Al-Co for the former and Ti-Zr-Si(-Nb) for the latter (Figure 9). Similar to other alloys, while the alloying elements are needed for the desired mechanical and corrosion properties for certain biomedical applications, mixed metal oxides could form. With the prospect of growing nanotubular layers, for example, the effect of the alloying elements on the tube dimensions should also be considered. These alloying elements can have different electrochemical oxidation rates and stabilities in the electrolyte [53]. Nb_2O_5, for instance, is more resistant in F^-- induced dissolution compared to TiO_2 [253]. Thus, Nb could slow down (or accelerate) the nanotube growth depending on the anodization electrolyte used, whereas Zr usually increases the nanotube length at the expense of the diameter growth. Such effects could result in two- (or multi-) scale diameters of the nanotubes [53]. Titanium alloys, such as Ti6Al4V, that were pretreated with H_2O_2 can also develop a relatively thick and porous surface layer, which could promote precipitation of hydroxycarbonate apatite to achieve a seamless transition between the peri-implant bone region and the implant materials, improving osseointegration [243].

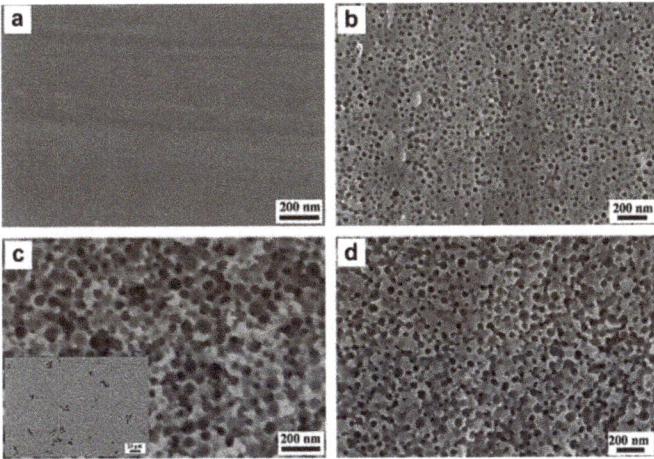

(a) Nanoporous oxides on $Ti_{45}Y_{11}Al_{24}Co_{20}$

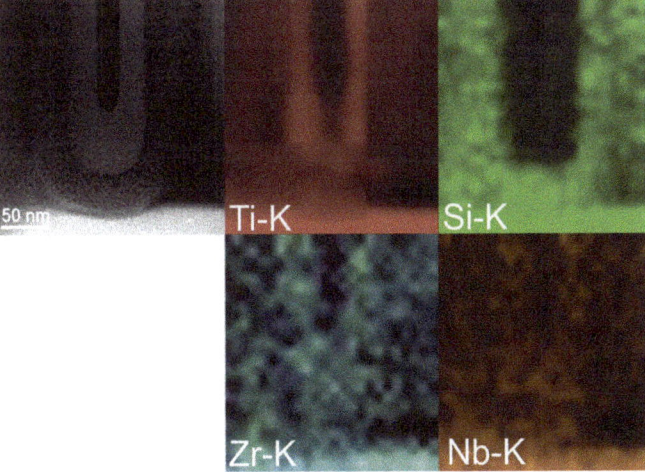

(b) Nanotubular oxides on $Ti_{60}Zr_{10}Si_{15}Nb_{15}$

Figure 9. Nanostructured oxide layers grown on glassy alloys: (**a**) nanoporous oxides on a Ti-Y-Al-Co alloy. The electron microscope images show the nanopores formed after various treatments. Reprinted (adapted) with permission from ref. [33]. Copyright 2009 Elsevier Ltd.; (**b**) nanotubular oxides grown on Ti-Zr-Si-Nb alloy. Electron microscope images show the layered tubular structure and the distribution of the different alloying elements in the oxide. Reprinted (adapted) with permission from ref. [53]. Copyright 2016 Elsevier B.V.

4.1. Safety of Ti-Based Implants and Inflammation

In addition to the safety concern regarding titanium dioxide NPs [254] (also see Section 2.2.1), Ti-based materials, while generally considered safe, have also been increasingly scrutinized regarding their toxicity [114]. Extensive reviews on this, such as Kim et al.'s [114] exist, and when interested, the readers are encouraged to read them. The main concern regarding Ti as an implant material is the possibility of its degradation-induced debris formation and chronic accumulation. This can cause inflammation [118,255], such as perimucositis or peri-implantitis [118]. Further, these debris can also accumulate in the spleen, bone marrow, and liver and may result in systemic diseases and other health issues [114].

The degradation of Ti-based materials due to implantation results in the formation of a thick layer of TiO_2 on the surface of the implant (initially determined from color change [256]), indicative of the occurrence of corrosion processes [243,256,257]. Evidence of material dissolution is also present. Such is the case for the β-phase of Ti6Al4V [258] in which its selective dissolution could originate from the attack of H_2O_2. Other evidences of Ti degradation including oxide growth within, oxide-induced stress corrosion cracking, and the presence of much-concerning periprosthetic debris have also been observed [243]. Alloys such as Ti-Al-V can also cause inflammation by inducing the release of mediators (prostaglandin E2, tumor necrosis factor, etc.) and may affect the periprosthetic tissues to cause osteolysis [114,255,259]. While the degradation of Ti-based materials could happen due to inflammation, implant deterioration could also be due to other factors which may be electrochemical, chemical, biological, and/or mechanical in nature [243].

Inflammation is the immune system's response to detrimental stimuli involving white blood cells (leukocytes). This occurs, for instance, due to the wound or the presence of infectious species or foreign debris. The leukocytes respond by either engulfing the invaders (phagocytosis) or by increasing their O_2 consumption to produce ROS [243,260,261]. Cell-signaling proteins are also produced by leukocytes to recruit more leukocytes, amplifying the process. When phagocytosis could not occur due to the size of the target (e.g., large implant debris), macrophages merge together to produce foreign body giant cells [243,262]. In the case of bones, these foreign body giant cells can be the osteoclasts (bone resorption cells), which can also form phagocytosis and produce ROS [243,263].

At different phases of inflammation, ROS can be produced by specific enzymes, and the biochemical reactions involved are depicted in Figure 10 [243,261]. As H_2O_2 plays a key role in the inflammation process involving the immune system, it has been used extensively for in vitro corrosion studies in simulated inflammatory conditions. Based on the observed effects of inflammation, inflammatory studies are therefore carried out and evaluated by looking at the metal release, phase dissolution, and oxide formation on the material under evaluation [243]. Electrolytes closer to the physiological condition have also been used, whereby it was observed that a synergetic effect could occur with the presence of H_2O_2 and albumin in terms of metal release and material implant dissolution but not oxide layer growth (at least for Ti6Al4V) [114,243].

While inflammation can be useful, such as at the early stage (acute) needed to heal the wound and prevent peri-implant infection (duration ~1 week), inflammation that lasts for weeks or months (chronic) results in health issues and can generate pain, irreversible cell degradation or DNA damage, and implant damage [243]. Periprosthetic inflammation has been found to correlate with the increased level of Ti (whether dissolved or as particles) and could result in bone resorption of the surrounding region (in the case of peri-implantitis) [114,262]. Additionally, Ti exposure has also been related to the occurrence of yellow nail syndrome, wherein the person's nail exhibits discoloration associated with sinus inflammation and coughing, among other symptoms [114,264–266]. In addition to Ti, considering its alloys, the other constituents could also result in health issues pertaining to those elements (Co, Cr, and Ni for instance have higher toxicity) [114,243].

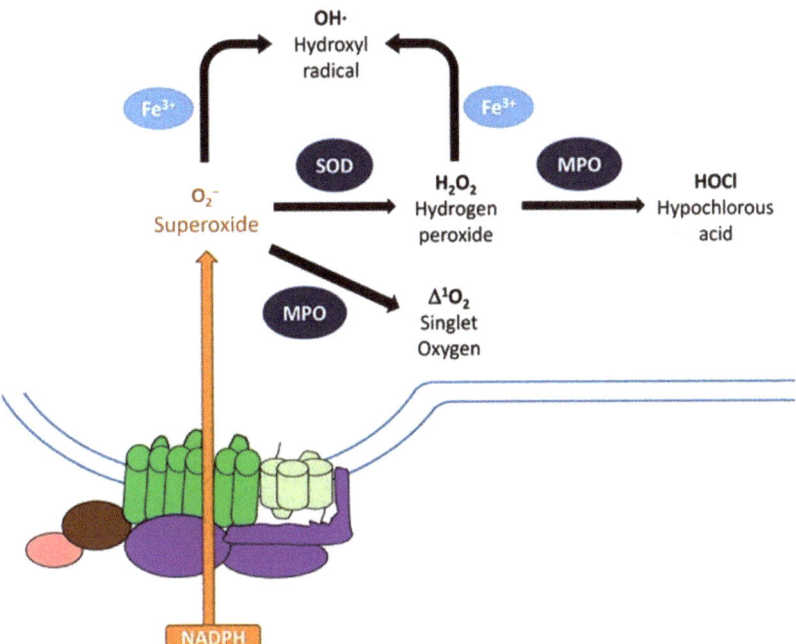

Figure 10. Biochemical reactions involved in reactive oxygen species ($\cdot O_2^-$, $\cdot OH$, H_2O_2, singlet oxygen, and HOCl) formation during inflammation. Reproduced (adapted) with permission from [261]. Copyright 2017 Nguyen, Green, and Mecsas. This is an open-access article distributed under the terms of the Creative Commons Attribution License (CC BY) (https://creativecommons.org/licenses/by/4.0/).

5. Conclusions and Future Perspectives

Catalysis on TiO_2 is mainly used and is more effective in the presence of light of sufficient energy (i.e., with $E \geq E_g$). Nowadays, with many modern forms of TiO_2, including black TiO_2, and advanced structures incorporating them, this can be extended to visible and IR light. This strategy (extending the absorption range or tuning E_g) and other means to improve TiO_2 photocatalysis remain relevant in furthering the applications which benefit from this field. Extensive knowledge of TiO_2 photocatalytic mechanisms can be compared and contrasted to what is thus far understood regarding dark TiO_2 catalysis. Both processes involve ROS generation; however, due to the absence of light needed for charge carrier generation in the case of dark catalysis, it seems that defects are crucial sources of charges to activate ROS formation. This may be the case with black (and colored) TiO_2, whereby surface defects could further promote ROS generation and, consequently, (photo)catalytic activity. Thus, in terms of implant application, looking at both photo- and dark catalysis could give a more holistic overview as implants could be also exposed to light prior to implantation, resulting in the so-called residual disinfection effect.

The residual disinfection observed after the removal of irradiation can also be taken advantage of by developing strategies to address the growing concern about antibacterial resistance. Deepening the understanding of what is happening during residual catalysis could help design materials, processes, and strategies to address such challenges and prevent implant/device failure. For example, while there is a general understanding of the involvement of ROS, intracellular peroxidation, and the disruption and direct attack of TiO_2 NPs themselves in this (photoinduced) residual bactericidal effect of TiO_2 [122], further details on what is happening could be beneficial in obtaining a nuanced understanding to aid designing materials with improved bactericidal action. The specific mechanisms for each different microbial species also need to be figured out. As many of these microbial species evolve continuously, such as by developing into different strains, such mechanisms should also be updated regularly. The fact that the viability of bacteria differs when inside and outside a biofilm should also be considered. Though mechanisms of actions of TiO_2 against bacteria outside and within a biofilm have been proposed [51,65,125,126,129], these understandings need to be further deepened.

In addition to photocatalysis and residual catalysis on TiO_2, it is important to also explore and further establish TiO_2 catalysis in the dark. Whereas over the past years, significant development has been carried out to unravel the dark ROS formation mechanism on TiO_2, such as by looking at the role of SETOVs in dark activation of H_2O_2 on TiO_2, the mechanism is yet to be confirmed (if possible) in non-SETOV-incorporated TiO_2 nanostructures. The role of other intrinsic defects (Ti^{3+} and surface oxygen vacancies) on the dark catalysis of TiO_2 should also be further studied, and a mechanism including these remains to be proposed. The role of extrinsic defects, e.g., for doped TiO_2 structures, in dark catalysis also needs to be studied. This is important, especially when considering that Ti alloys include other elements which could introduce impurities to TiO_2 or form mixed oxides with Ti. Further knowledge of dark catalysis on TiO_2 can also be helpful in advancing the use of TiO_2 for biomedical applications. Implants and devices will be in the dark after surgery and important in vivo processes to ultimately determine implantation failure or success, such as inflammation, tissue regeneration, and possibly antibacterial action around the implant also occurs in the absence of irradiation. Thus, the role of ROS in dark TiO_2 catalysis in view of these processes is crucial in addressing the challenges in Ti-based implant application.

Understanding the dark catalysis on TiO_2 in vitro can shed light on what could be happening with TiO_2/Ti implants in vivo and help in rationally designing materials that could take advantage of ROS formation and/or catalysis for implant application. The sensitivity of ROS generation observed in several studies to factors such as pH and concentration, the presence/absence of oxygen, and other bio-/molecules points to a possibility of a different mechanism happening in the physiological condition and also during inflammation. In vitro mechanistic studies on these implant materials using physiological conditions should therefore be eventually extended to in vivo studies.

Black TiO_2 nanomaterials also introduce new opportunities to widen the application of TiO_2 in the biomedical field. The fact that it can be produced on amorphous TiO_2 and on other Ti alloys without necessitating heat treatment makes it attractive for bioimplants—which could be made from amorphous alloys. Further, in vivo (/in situ) phototreatment could be made possible as the light absorbance of black TiO_2 could be tuned to lie within the biological near-IR window. Its redshifted broadened absorbance allows for photothermal application in relation to implant use, which is beneficial for cancer treatment, photothermal antibacterial disinfection, and so on. However, its biosafety has to be confirmed.

The development of new materials, while it is useful and advances the field, also entails the need to be investigated in terms of their ROS generation mechanism. For example, the incorporation of other components to further improve other desired properties, surface treatments, and the use of alloys forming (supposedly mixed) surface oxides can have an influence on the dark catalytic properties; their mechanisms would also need to be studied. Novel materials for implant application will also result in the formation of new microstructures of corrosion products (e.g., the oxide layer), which will also need to be evaluated in terms of their vulnerability to ROS attack. This is considering the plethora of processing available, the nanostructures that can be formed, the incorporation of bulk and surface modification, and the possibility of protocol changes with the advancements in the medical field, such as in the fight against antibacterial resistance. There are also findings on other Ti alloys that cannot be explained by the current existing understanding, and this points to the possibility that they involve modifications in the mechanism known for pure TiO_2 alone. Additionally, a number of applications using TiO_2 make use of its amorphous state, such as for TiO_2 grown on glassy alloys that are used in dental implants. Apart from using it as a control in crystallinity effect studies, this structural state seemed to have been forgotten, especially in terms of understanding its catalytic activity (if any) and properties and performance in the presence of ROS. This is also relevant when considering black TiO_2, which can be grown as an amorphous material and possibly also on an amorphous material. It is also necessary to investigate the catalytic activity of black TiO_2, both photocatalytic and in the dark, and now also PEC and EC processes (with an increasing number of studies presenting such as comparative results), in terms of ROS generation when relevant. Therefore, the understanding of TiO_2 catalysis needs to be extended to (and maybe modified for) these materials, which will make the already complex question (of understanding dark ROS generation on titanium-oxide-based implants and addressing challenges in the biomedical field in light of TiO_2 catalysis) a tad more complex.

Funding: This research received no external funding.

Data Availability Statement: No new data were created or analyzed in this study. Data sharing is not applicable to this article.

Acknowledgments: The author would like to thank all the scientists and engineers that she had fruitful discussions with regarding catalysis on TiO_2 nanomaterials. Special thanks are given to colleagues she worked closely with in the past and those she continuously exchanges ideas and knowledge with at present: E.P. Enriquez (AdMU-Chem) for introducing her to TiO_2 nanomaterials and DSSCs, P. Hildebrandt (TU Berlin) for the many discussions they had regarding vibrational spectroscopy, I.M. Weidinger, H.K. Ly, and I.H. Öner (TU Dresden) for the interesting works they had on TNAs, and A. Gebert (IFW Dresden) for the on-going discussions regarding the understanding of catalytic processes related with TiO_2 in biomedical applications.

Conflicts of Interest: The author declares no conflict of interest.

References

1. Kaur, M.; Singh, K. Review on Titanium and Titanium Based Alloys as Biomaterials for Orthopaedic Applications. *Mater. Sci. Eng. C* **2019**, *102*, 844–862. [CrossRef] [PubMed]
2. Buddee, S.; Wongnawa, S.; Sirimahachai, U.; Puetpaibool, W. Recyclable UV and Visible Light Photocatalytically Active Amorphous TiO_2 Doped with M (III) Ions (M = Cr and Fe). *Mater. Chem. Phys.* **2011**, *126*, 167–177. [CrossRef]
3. Nowotny, M.K.; Bogdanoff, P.; Dittrich, T.; Fiechter, S.; Fujishima, A.; Tributsch, H. Observations of P-Type Semiconductivity in Titanium Dioxide at Room Temperature. *Mater. Lett.* **2010**, *64*, 928–930. [CrossRef]
4. Nowotny, J.; Bak, T.; Nowotny, M.K.; Sheppard, L.R. Titanium Dioxide for Solar-Hydrogen II. Defect Chemistry. *Int. J. Hydrogen Energy* **2007**, *32*, 2630–2643. [CrossRef]
5. Fujishima, A.; Honda, K. Electrochemical Photolysis of Water at a Semiconductor Electrode. *Nature* **1972**, *238*, 37–38. [CrossRef]
6. De Souza, M.L.; Tristao, D.C.; Corio, P. Vibrational Study of Adsorption of Congo Red onto TiO_2 and the LSPR Effect on Its Photocatalytic Degradation Process. *RSC Adv.* **2014**, *4*, 23351–23358. [CrossRef]
7. Bhat, V.T.; Duspara, P.A.; Seo, S.; Abu Bakar, N.S.B.; Greaney, M.F. Visible Light Promoted Thiol-Ene Reactions Using Titanium Dioxide. *Chem. Commun.* **2015**, *51*, 4383–4385. [CrossRef]
8. Tayade, R.J.; Surolia, P.K.; Kulkarni, R.G.; Raksh, V.; Tayade, R.J.; Surolia, P.K.; Kulkarni, R.G.; Jasra, R.V. Photocatalytic Degradation of Dyes and Organic Contaminants in Water Using Nanocrystalline Anatase and Rutile TiO_2. *Sci. Technol. Adv. Mater.* **2007**, *8*, 455–462. [CrossRef]
9. Bin Mukhlish, M.Z.; Najnin, F.; Rahman, M.M.; Uddin, M.J. Photocatalytic Degradation of Different Dyes Using TiO_2 with High Surface Area: A Kinetic Study. *J. Sci. Res.* **2013**, *5*, 301–314. [CrossRef]
10. Hernandez, S.; Hidalgo, D.; Sacco, A.; Chiodoni, A.; Lamberti, A.; Cauda, V.; Tressoab, E.; Saraccob, G. Comparison of Photocatalytic and Transport Properties of TiO_2 and ZnO Nanostructures for Solar-Driven Water Splitting. *Phys. Chem. Chem. Phys.* **2015**, *17*, 7775–7786. [CrossRef]
11. Samsudin, E.M.; Abd Hamid, S.B. Effect of Band Gap Engineering in Anionic-Doped TiO_2 Photocatalyst. *Appl. Surf. Sci.* **2017**, *391*, 326–336. [CrossRef]
12. Dozzi, M.V.; D'Andrea, C.; Ohtani, B.; Valentini, G.; Selli, E. Fluorine-Doped TiO_2 Materials: Photocatalytic Activity vs Time-Resolved Photoluminescence. *J. Phys. Chem. C* **2013**, *117*, 25586–25595. [CrossRef]
13. Ajmal, A.; Majeed, I.; Malik, R.N.; Idriss, H.; Nadeem, M.A. Principles and Mechanisms of Photocatalytic Dye Degradation on TiO 2 Based Photocatalysts: A Comparative Overview. *RSC Adv.* **2014**, *4*, 37003–37026. [CrossRef]
14. Giovannetti, R.; Amato, C.A.D.; Zannotti, M.; Rommozzi, E.; Gunnella, R.; Minicucci, M.; Di Cicco, A. Visible Light Photoactivity of Polypropylene Coated Nano-TiO_2 for Dyes Degradation in Water. *Sci. Rep.* **2015**, *5*, 17801. [CrossRef]

15. Chowdhury, P.; Moreira, J.; Gomaa, H.; Ray, A.K. Visible-Solar-Light-Driven Photocatalytic Degradation of Phenol with Dye-Sensitized TiO$_2$: Parametric and Kinetic Study. *Ind. Eng. Chem. Res.* **2012**, *51*, 4523–4532. [CrossRef]
16. Shang, M.; Hou, H.; Gao, F.; Wang, L.; Yang, W. Mesoporous Ag@TiO$_2$ Nanofibers and Their Photocatalytic Activity for Hydrogen Evolution. *RSC Adv.* **2017**, *7*, 30051–30059. [CrossRef]
17. Chouirfa, H.; Bouloussa, H.; Migonney, V.; Falentin-Daudré, C. Review of Titanium Surface Modification Techniques and Coatings for Antibacterial Applications. *Acta Biomater.* **2019**, *83*, 37–54. [CrossRef]
18. Zhang, H.; Chen, G.; Bahnemann, D.W. Photoelectrocatalytic Materials for Environmental Applications. *J. Mater. Chem.* **2009**, *19*, 5089–5121. [CrossRef]
19. Nowotny, J.; Bak, T.; Nowotny, M.K.; Sheppard, L.R. Titanium Dioxide for Solar-Hydrogen I. Functional Properties. *Int. J. Hydrogen Energy* **2007**, *32*, 2609–2629. [CrossRef]
20. Han, X.X.; Chen, L.; Kuhlmann, U.; Schulz, C.; Weidinger, I.M.; Hildebrandt, P. Magnetic Titanium Dioxide Nanocomposites for Surface-Enhanced Resonance Raman Spectroscopic Determination and Degradation of Toxic Anilines and Phenols. *Angew. Chem.—Int. Ed.* **2014**, *53*, 2481–2484. [CrossRef]
21. Zhu, P.; Nair, A.S.; Shengjie, P.; Shengyuan, Y.; Ramakrishna, S. Facile Fabrication of TiO$_2$-Graphene Composite with Enhanced Photovoltaic and Photocatalytic Properties by Electrospinning. *ACS Appl. Mater. Interfaces* **2012**, *4*, 581–585. [CrossRef] [PubMed]
22. Schneider, J.; Matsuoka, M.; Takeuchi, M.; Zhang, J.; Horiuchi, Y.; Anpo, M.; Bahnemann, D.W. Understanding TiO$_2$ Photocatalysis: Mechanisms and Materials. *Chem. Rev.* **2014**, *114*, 9919–9986. [CrossRef] [PubMed]
23. Fujishima, A.; Zhang, X.; Tryk, D.A. TiO$_2$ Photocatalysis and Related Surface Phenomena. *Surf. Sci. Rep.* **2008**, *63*, 515–582. [CrossRef]
24. Kohtani, S.; Kawashima, A.; Miyabe, H. Reactivity of Trapped and Accumulated Electrons in Titanium Dioxide Photocatalysis. *Catalysts* **2017**, *7*, 303. [CrossRef]
25. Qian, R.; Zong, H.; Schneider, J.; Zhou, G.; Zhao, T.; Li, Y.; Yang, J.; Bahnemann, D.W.; Pan, J.H. Charge Carrier Trapping, Recombination and Transfer during TiO$_2$ Photocatalysis: An Overview. *Catal. Today* **2019**, *335*, 78–90. [CrossRef]
26. Hoffmann, M.R.; Martin, S.T.; Choi, W.; Bahnemann, D.W. Environmental Applications of Semiconductor Photocatalysis. *Chem. Rev.* **1995**, *95*, 69–96. [CrossRef]
27. Li, G.; Chen, L.; Graham, M.E.; Gray, K.A. A Comparison of Mixed Phase Titania Photocatalysts Prepared by Physical and Chemical Methods: The Importance of the Solid-Solid Interface. *J. Mol. Catal. A Chem.* **2007**, *275*, 30–35. [CrossRef]
28. Pillai, S.C.; Periyat, P.; George, R.; Mccormack, D.E.; Seery, M.K.; Hayden, H.; Colreavy, J.; Corr, D.; Hinder, S.J. Synthesis of High-Temperature Stable Anatase TiO$_2$ Photocatalyst. *J. Phys. Chem. C* **2007**, *111*, 1605–1611. [CrossRef]
29. Hariharan, A.; Goldberg, P.; Gustmann, T.; Maawad, E.; Pilz, S.; Schell, F.; Kunze, T.; Zwahr, C.; Gebert, A. Designing the Microstructural Constituents of an Additively Manufactured near β Ti Alloy for an Enhanced Mechanical and Corrosion Response. *Mater. Des.* **2022**, *217*, 110618. [CrossRef]
30. Gebert, A.; Oswald, S.; Helth, A.; Voss, A.; Gostin, P.F.; Rohnke, M.; Janek, J.; Calin, M.; Eckert, J. Effect of Indium (In) on Corrosion and Passivity of a Beta-Type Ti-Nb Alloy in Ringer's Solution. *Appl. Surf. Sci.* **2015**, *335*, 213–222. [CrossRef]
31. Pilz, S.; Gebert, A.; Voss, A.; Oswald, S.; Göttlicher, M.; Hempel, U.; Eckert, J.; Rohnke, M.; Janek, J.; Calin, M. Metal Release and Cell Biological Compatibility of Beta-Type Ti-40Nb Containing Indium. *J. Biomed. Mater. Res.—Part B Appl. Biomater.* **2018**, *106*, 1686–1697. [CrossRef]
32. Sopha, H.; Krbal, M.; Ng, S.; Prikryl, J.; Zazpe, R.; Yam, F.K.; Macak, J.M. Highly Efficient Photoelectrochemical and Photocatalytic Anodic TiO$_2$ Nanotube Layers with Additional TiO$_2$ Coating. *Appl. Mater. Today* **2017**, *9*, 104–110. [CrossRef]
33. Jayaraj, J.; Park, J.M.; Gostin, P.F.; Fleury, E.; Gebert, A.; Schultz, L. Nano-Porous Surface States of Ti-Y-Al-Co Phase Separated Metallic Glass. *Intermetallics* **2009**, *17*, 1120–1123. [CrossRef]
34. Pham, V.T.H.; Truong, V.K.; Orlowska, A.; Ghanaati, S.; Barbeck, M.; Booms, P.; Fulcher, A.J.; Bhadra, C.M.; Buividas, R.; Baulin, V.; et al. Race for the Surface: Eukaryotic Cells Can Win. *ACS Appl. Mater. Interfaces* **2016**, *8*, 22025–22031. [CrossRef]
35. Mehrjou, B.; Mo, S.; Dehghan-Baniani, D.; Wang, G.; Qasim, A.M.; Chu, P.K. Antibacterial and Cytocompatible Nanoengineered Silk-Based Materials for Orthopedic Implants and Tissue Engineering. *ACS Appl. Mater. Interfaces* **2019**, *11*, 31605–31614. [CrossRef]
36. Gallo, J.; Holinka, M.; Moucha, C. Antibacterial Surface Treatment for Orthopaedic Implants. *Int. J. Mol. Sci.* **2014**, *15*, 13849–13880. [CrossRef]
37. An, Y.H.; Friedman, R.J. Prevention of Sepsis in Total Joint Arthroplasty. *J. Hosp. Infect.* **1996**, *33*, 93–108. [CrossRef]
38. Dong, J.; Wang, W.; Zhou, W.; Zhang, S.; Li, M.; Li, N.; Pan, G.; Zhang, X.; Bai, J.; Zhu, C. Immunomodulatory Biomaterials for Implant-Associated Infections: From Conventional to Advanced Therapeutic Strategies. *Biomater. Res.* **2022**, *26*, 72. [CrossRef]
39. Trindade, R.; Albrektsson, T.; Galli, S.; Prgomet, Z.; Tengvall, P.; Wennerberg, A. Osseointegration and Foreign Body Reaction: Titanium Implants Activate the Immune System and Suppress Bone Resorption during the First 4 Weeks after Implantation. *Clin. Implant Dent. Relat. Res.* **2018**, *20*, 82–91. [CrossRef]
40. Pérez León, C.; Kador, L.; Peng, B.; Thelakkat, M. Characterization of the Adsorption of Ru-Bpy Dyes on Mesoporous TiO$_2$ Films with UV-Vis, Raman, and FTIR Spectroscopies. *J. Phys. Chem. B* **2006**, *110*, 8723–8730. [CrossRef]
41. Goff, A.H.; Joiret, S.; Falaras, P.; Curie, M. Raman Resonance Effect in a Monolayer of Polypyridyl Ruthenium (II) Complex Adsorbed on Nanocrystalline TiO$_2$ via Phosphonated Terpyridyl Ligands. *J. Phys. Chem. B* **1999**, *103*, 9569–9575. [CrossRef]

42. Öner, I.H.; Querebillo, C.J.; David, C.; Gernert, U.; Walter, C.; Driess, M.; Leimkühler, S.; Ly, K.H.; Weidinger, I.M.; Leimk, S.; et al. Hohe Elektromagnetische Feldverstärkung in Nanotubularen TiO_2-Elektroden. *Angew. Chem.* **2018**, *130*, 7344–7348. [CrossRef]
43. Shoute, L.C.T.; Loppnow, G.R. Excited-State Dynamics of Alizarin-Sensitized TiO_2 Nanoparticles from Resonance Raman Spectroscopy. *J. Chem. Phys.* **2002**, *117*, 842–850. [CrossRef]
44. Blackbourn, R.L.; Johnson, C.S.; Hupp, J.T. Surface Intervalence Enhanced Raman Scattering from $Fe(CN)_6$ on Colloidal Titanium Dioxide. A Mode-by-Mode Description of the Franck—Condon Barrier to Interfacial Charge Transfer. *J. Am. Chem. Soc.* **1991**, *113*, 1060–1062. [CrossRef]
45. Finnie, K.S.; Bartlett, J.R.; Woolfrey, J.L. Vibrational Spectroscopic Study of the Coordination of (2,2′-Bipyridyl-4,4′-Dicarboxylic Acid)Ruthenium(II) Complexes to the Surface of Nanocrystalline Titania. *Langmuir* **1998**, *14*, 2744–2749. [CrossRef]
46. Yang, L.; Jiang, X.; Ruan, W.; Zhao, B.; Xu, W.; Lombardi, J.R. Observation of Enhanced Raman Scattering for Molecules Adsorbed on TiO_2 Nanoparticles: Charge-Transfer Contribution. *J. Phys. Chem. C* **2008**, *112*, 20095–20098. [CrossRef]
47. Musumeci, A.; Gosztola, D.; Schiller, T.; Dimitrijevic, N.M.; Mujica, V.; Martin, D.; Rajh, T. SERS of Semiconducting Nanoparticles (TiO_2 Hybrid Composites). *J. Am. Chem. Soc.* **2009**, *131*, 6040–6041. [CrossRef]
48. Maznichenko, D.; Venkatakrishnan, K.; Tan, B. Stimulating Multiple SERS Mechanisms by a Nanofibrous Three-Dimensional Network Structure of Titanium Dioxide (TiO_2). *J. Phys. Chem. C* **2013**, *117*, 578–583. [CrossRef]
49. Han, X.X.; Köhler, C.; Kozuch, J.; Kuhlmann, U.; Paasche, L.; Sivanesan, A.; Weidinger, I.M.; Hildebrandt, P. Potential-Dependent Surface-Enhanced Resonance Raman Spectroscopy at Nanostructured TiO_2: A Case Study on Cytochrome b_5. *Small* **2013**, *9*, 4175–4181. [CrossRef]
50. Öner, I.H.; Querebillo, C.J.; David, C.; Gernert, U.; Walter, C.; Driess, M.; Leimkühler, S.; Ly, K.H.; Weidinger, I.M. High Electromagnetic Field Enhancement of TiO_2 Nanotube Electrodes. *Angew. Chem. Int. Ed.* **2018**, *57*, 7225–7229. [CrossRef]
51. Lu, X.; Wu, Z.; Xu, K.; Wang, X.; Wang, S.; Qiu, H.; Li, X.; Chen, J. Multifunctional Coatings of Titanium Implants Toward Promoting Osseointegration and Preventing Infection: Recent Developments. *Front. Bioeng. Biotechnol.* **2021**, *9*, 783816. [CrossRef]
52. Xue, T.; Attarilar, S.; Liu, S.; Liu, J.; Song, X.; Li, L.; Zhao, B.; Tang, Y. Surface Modification Techniques of Titanium and Its Alloys to Functionally Optimize Their Biomedical Properties: Thematic Review. *Front. Bioeng. Biotechnol.* **2020**, *8*, 603072. [CrossRef]
53. Sopha, H.; Pohl, D.; Damm, C.; Hromadko, L.; Rellinghaus, B.; Gebert, A.; Macak, J.M. Self-Organized Double-Wall Oxide Nanotube Layers on Glass-Forming Ti-Zr-Si(-Nb) Alloys. *Mater. Sci. Eng. C* **2017**, *70*, 258–263. [CrossRef]
54. Yang, B.; Uchida, M.; Kim, H.M.; Zhang, X.; Kokubo, T. Preparation of Bioactive Titanium Metal via Anodic Oxidation Treatment. *Biomaterials* **2004**, *25*, 1003–1010. [CrossRef]
55. Sul, Y.T.; Johansson, C.B.; Petronis, S.; Krozer, A.; Jeong, Y.; Wennerberg, A.; Albrektsson, T. Characteristics of the Surface Oxides on Turned and Electrochemically Oxidized Pure Titanium Implants up to Dielectric Breakdown: The Oxide Thickness, Micropore Configurations, Surface Roughness, Crystal Structure and Chemical Composition. *Biomaterials* **2002**, *23*, 491–501. [CrossRef]
56. Herzer, R.; Gebert, A.; Hempel, U.; Hebenstreit, F.; Oswald, S.; Damm, C.; Schmidt, O.G.; Medina-Sánchez, M. Rolled-Up Metal Oxide Microscaffolds to Study Early Bone Formation at Single Cell Resolution. *Small* **2021**, *17*, 2005527. [CrossRef]
57. Nosaka, Y.; Nosaka, A.Y. Generation and Detection of Reactive Oxygen Species in Photocatalysis. *Chem. Rev.* **2017**, *117*, 11302–11336. [CrossRef]
58. Nosaka, Y.; Nosaka, A. Understanding Hydroxyl Radical (•OH) Generation Processes in Photocatalysis. *ACS Energy Lett.* **2016**, *1*, 356–359. [CrossRef]
59. Jedsukontorn, T.; Meeyoo, V.; Saito, N.; Hunsom, M. Effect of Electron Acceptors H_2O_2 and O_2 on the Generated Reactive Oxygen Species $^1O_2$ and OH· in TiO_2-Catalyzed Photocatalytic Oxidation of Glycerol. *Cuihua Xuebao/Chin. J. Catal.* **2016**, *37*, 1975–1981. [CrossRef]
60. Gao, R.; Stark, J.; Bahnemann, D.W.; Rabani, J. Quantum Yields of Hydroxyl Radicals in Illuminated TiO_2 Nanocrystallite Layers. *J. Photochem. Photobiol. A Chem.* **2002**, *148*, 387–391. [CrossRef]
61. Diesen, V.; Jonsson, M. Formation of H_2O_2 in TiO_2 Photocatalysis of Oxygenated and Deoxygenated Aqueous Systems: A Probe for Photocatalytically Produced Hydroxyl Radicals. *J. Phys. Chem. C* **2014**, *118*, 10083–10087. [CrossRef]
62. Lawless, D.; Serpone, N.; Meisel, D. Role of OH· Radicals and Trapped Holes in Photocatalysis. A Pulse Radiolysis Study. *J. Phys. Chem.* **1991**, *95*, 5166–5170. [CrossRef]
63. Zhang, J.; Nosaka, Y. Mechanism of the OH Radical Generation in Photocatalysis with TiO_2 of Different Crystalline Types. *J. Phys. Chem. C* **2014**, *118*, 10824–10832. [CrossRef]
64. Liao, H.; Reitberger, T. Generation of Free OH_{aq} Radicals by Black Light Illumination of Degussa (Evonik) P25 TiO_2 Aqueous Suspensions. *Catalysts* **2013**, *3*, 418–443. [CrossRef]
65. Kumaravel, V.; Nair, K.M.; Mathew, S.; Bartlett, J.; Kennedy, J.E.; Manning, H.G.; Whelan, B.J.; Leyland, N.S.; Pillai, S.C. Antimicrobial TiO_2 Nanocomposite Coatings for Surfaces, Dental and Orthopaedic Implants. *Chem. Eng. J.* **2021**, *416*, 129071. [CrossRef]
66. Kakuma, Y.; Nosaka, A.Y.; Nosaka, Y. Difference in TiO_2 Photocatalytic Mechanism between Rutile and Anatase Studied by the Detection of Active Oxygen and Surface Species in Water. *Phys. Chem. Chem. Phys.* **2015**, *17*, 18691–18698. [CrossRef]
67. Fu, Z.; Liang, Y.; Wang, S.; Zhong, Z. Structural Phase Transition and Mechanical Properties of TiO_2 under High Pressure. *Phys. Status Solidi Basic Res.* **2013**, *250*, 2206–2214. [CrossRef]
68. Rich, C.C.; Knorr, F.J.; McHale, J.L. Trap State Photoluminescence of Nanocrystalline and Bulk TiO_2: Implications for Carrier Transport. *Mater. Res. Soc. Symp. Proc.* **2010**, *1268*, 117–122. [CrossRef]

69. Kurtz, R.L.; Stock-Bauer, R.; Msdey, T.E.; Román, E.; De Segovia, J.L. Synchrotron Radiation Studies of H_2O Adsorption on TiO_2(110). *Surf. Sci.* **1989**, *218*, 178–200. [CrossRef]
70. Tôrres, A.R.; Azevedo, E.B.; Resende, N.S.; Dezotti, M. A Comparison between Bulk and Supported TiO_2 Photocatalysts in the Degradation of Formic Acid. *Braz. J. Chem. Eng.* **2007**, *24*, 185–192. [CrossRef]
71. Ma, H.; Lenz, K.A.; Gao, X.; Li, S.; Wallis, L.K. Comparative Toxicity of a Food Additive TiO_2, a Bulk TiO_2, and a Nano-Sized P25 to a Model Organism the Nematode C. Elegans. *Environ. Sci. Pollut. Res.* **2019**, *26*, 3556–3568. [CrossRef]
72. Macak, J.M.; Zlamal, M.; Krysa, J.; Schmuki, P. Self-Organized TiO_2 Nanotube Layers as Highly Efficient Photocatalysts. *Small* **2007**, *3*, 300–304. [CrossRef]
73. Bai, H.; Liu, L.; Liu, Z.; Sun, D.D. Hierarchical 3D Dendritic TiO_2 Nanospheres Building with Ultralong 1D Nanoribbon/Wires for High Performance Concurrent Photocatalytic Membrane Water Purification. *Water Res.* **2013**, *47*, 4126–4138. [CrossRef]
74. En Du, Y.; Niu, X.; Bai, Y.; Qi, H.; Guo, Y.; Chen, Y.; Wang, P.; Yang, X.; Feng, Q. Synthesis of Anatase TiO_2 Nanocrystals with Defined Morphologies from Exfoliated Nanoribbons: Photocatalytic Performance and Application in Dye-Sensitized Solar Cell. *ChemistrySelect* **2019**, *4*, 4443–4457. [CrossRef]
75. Wang, X.; Xia, R.; Muhire, E.; Jiang, S.; Huo, X.; Gao, M. Highly Enhanced Photocatalytic Performance of TiO_2 Nanosheets through Constructing TiO_2/TiO_2 Quantum Dots Homojunction. *Appl. Surf. Sci.* **2018**, *459*, 9–15. [CrossRef]
76. Kang, L.; Liu, X.Y.; Wang, A.; Li, L.; Ren, Y.; Li, X.; Pan, X.; Li, Y.; Zong, X.; Liu, H.; et al. Photo–Thermo Catalytic Oxidation over a TiO_2-WO_3-Supported Platinum Catalyst. *Angew. Chem.* **2020**, *132*, 13009–13016. [CrossRef]
77. Panniello, A.; Curri, M.L.; Diso, D.; Licciulli, A.; Locaputo, V.; Agostiano, A.; Comparelli, R.; Mascolo, G. Nanocrystalline TiO_2 Based Films onto Fibers for Photocatalytic Degradation of Organic Dye in Aqueous Solution. *Appl. Catal. B Environ.* **2012**, *121–122*, 190–197. [CrossRef]
78. Ali, T.; Tripathi, P.; Azam, A.; Raza, W.; Ahmed, A.S.; Ahmed, A.; Muneer, M. Photocatalytic Performance of Fe-Doped TiO_2 Nanoparticles under Visible-Light Irradiation. *Mater. Res. Express* **2017**, *4*, 015022. [CrossRef]
79. Zhang, H.; Miao, G.; Ma, X.; Wang, B.; Zheng, H. Enhancing the Photocatalytic Activity of Nanocrystalline TiO_2 by Co-Doping with Fluorine and Yttrium. *Mater. Res. Bull.* **2014**, *55*, 26–32. [CrossRef]
80. Hu, D.; Li, R.; Li, M.; Pei, J.; Guo, F.; Zhang, S. Photocatalytic Efficiencies of WO_3/TiO_2 Nanoparticles for Exhaust Decomposition under UV and Visible Light Irradiation. *Mater. Res. Express* **2018**, *5*, 095029. [CrossRef]
81. Naufal, B.; Ullattil, S.G.; Periyat, P. A Dual Function Nanocrystalline TiO_2 Platform for Solar Photocatalysis and Self Cleaning Application. *Sol. Energy* **2017**, *155*, 1380–1388. [CrossRef]
82. Dong, G.; Wang, Y.; Lei, H.; Tian, G.; Qi, S. Hierarchical Mesoporous Titania Nanoshell Encapsulated on Polyimide Nano Fiber as Flexible, Highly Reactive, Energy Saving and Recyclable Photocatalyst for Water Purification. *J. Clean. Prod.* **2020**, *253*, 120021. [CrossRef]
83. Xie, J.; Wen, W.; Jin, Q.; Xiang, X.B.; Wu, J.M. TiO_2 Nanotrees for the Photocatalytic and Photoelectrocatalytic Phenol Degradation. *New J. Chem.* **2019**, *43*, 11050–11056. [CrossRef]
84. Li, Y.; Zhang, L.; Wu, W.; Dai, P.; Yu, X.; Wu, M.; Li, G. Hydrothermal Growth of TiO_2 Nanowire Membranes Sensitized with CdS Quantum Dots for the Enhancement of Photocatalytic Performance. *Nanoscale Res. Lett.* **2014**, *9*, 270. [CrossRef]
85. Luan, S.; Qu, D.; An, L.; Jiang, W.; Gao, X.; Hua, S.; Miao, X.; Wen, Y.; Sun, Z. Enhancing Photocatalytic Performance by Constructing Ultrafine TiO_2 Nanorods/g-C3N4 Nanosheets Heterojunction for Water Treatment. *Sci. Bull.* **2018**, *63*, 683–690. [CrossRef]
86. Zhang, H.; Yu, M. Photocatalytic Activity of TiO_2 Nanofibers: The Surface Crystalline Phase Matters. *Nanomaterials* **2019**, *9*, 535. [CrossRef]
87. Teodorescu-Soare, C.T.; Catrinescu, C.; Dobromir, M.; Stoian, G.; Arvinte, A.; Luca, D. Growth and Characterization of TiO_2 Nanotube Arrays under Dynamic Anodization. Photocatalytic Activity. *J. Electroanal. Chem.* **2018**, *823*, 388–396. [CrossRef]
88. Zhao, Y.; Wang, C.; Hu, J.; Li, J.; Wang, Y. Photocatalytic Performance of TiO_2 Nanotube Structure Based on TiN Coating Doped with Ag and Cu. *Ceram. Int.* **2021**, *47*, 7233–7240. [CrossRef]
89. Ariyanti, D.; Mo'Ungatonga, S.; Li, Y.; Gao, W. Formation of TiO_2 Based Nanoribbons and the Effect of Post-Annealing on Its Photocatalytic Activity. *IOP Conf. Ser. Mater. Sci. Eng.* **2018**, *348*, 012002. [CrossRef]
90. Shaban, M.; Ashraf, A.M.; Abukhadra, M.R. TiO_2 Nanoribbons/Carbon Nanotubes Composite with Enhanced Photocatalytic Activity; Fabrication, Characterization, and Application. *Sci. Rep.* **2018**, *8*, 781. [CrossRef]
91. Wan, Y.; Wang, J.; Wang, X.; Xu, H.; Yuan, S.; Zhang, Q.; Zhang, M. Preparation of Inverse Opal Titanium Dioxide for Photocatalytic Performance Research. *Opt. Mater.* **2019**, *96*, 109287. [CrossRef]
92. Albu, S.P.; Kim, D.; Schmuki, P. Growth of Aligned TiO_2 Bamboo-Type Nanotubes and Highly Ordered Nanolace. *Angew. Chem.—Int. Ed.* **2008**, *47*, 1916–1919. [CrossRef]
93. Chahrour, K.M.; Yam, F.K.; Eid, A.M.; Nazeer, A.A. Enhanced Photoelectrochemical Properties of Hierarchical Black TiO_{2-x} Nanolaces for Cr (VI) Photocatalytic Reduction. *Int. J. Hydrogen Energy* **2020**, *45*, 22674–22690. [CrossRef]
94. Harris, J.; Silk, R.; Smith, M.; Dong, Y.; Chen, W.T.; Waterhouse, G.I.N. Hierarchical TiO_2 Nanoflower Photocatalysts with Remarkable Activity for Aqueous Methylene Blue Photo-Oxidation. *ACS Omega* **2020**, *5*, 18919–18934. [CrossRef]
95. Wen, W.; Hai, J.; Yao, J.; Gu, Y.J.; Kobayashi, H.; Tian, H.; Sun, T.; Chen, Q.; Yang, P.; Geng, C.; et al. Univariate Lattice Parameter Modulation of Single-Crystal-like Anatase TiO_2 Hierarchical Nanowire Arrays to Improve Photoactivity. *Chem. Mater.* **2021**, *33*, 1489–1497. [CrossRef]

96. Yu, Y.; Wen, W.; Qian, X.Y.; Liu, J.B.; Wu, J.M. UV and Visible Light Photocatalytic Activity of Au/TiO$_2$ Nanoforests with Anatase/Rutile Phase Junctions and Controlled Au Locations. *Sci. Rep.* **2017**, *7*, 41253. [CrossRef]
97. Zhu, X.; Wen, G.; Liu, H.; Han, S.; Chen, S.; Kong, Q.; Feng, W. One-Step Hydrothermal Synthesis and Characterization of Cu-Doped TiO$_2$ Nanoparticles/Nanobucks/Nanorods with Enhanced Photocatalytic Performance under Simulated Solar Light. *J. Mater. Sci. Mater. Electron.* **2019**, *30*, 13826–13834. [CrossRef]
98. Hosseinnia, A.; Keyanpour-Rad, M.; Pazouki, M. Photo-Catalytic Degradation of Organic Dyes with Different Chromophores by Synthesized Nanosize TiO$_2$ Particles. *World Appl. Sci. J.* **2010**, *8*, 1327–1332.
99. Shrivastava, V.S. Photocatalytic Degradation of Methylene Blue Dye and Chromium Metal from Wastewater Using Nanocrystalline TiO$_2$ Semiconductor. *Arch. Appl. Sci. Res.* **2012**, *4*, 1244–1254.
100. Joshi, K.M.; Shrivastava, V.S. Degradation of Alizarine Red-S (A Textiles Dye) by Photocatalysis Using ZnO and TiO$_2$ as Photocatalyst. *Int. J. Environ. Sci.* **2011**, *2*, 8–21.
101. Chen, X.; Mao, S.S. Titanium Dioxide Nanomaterials: Synthesis, Properties, Modifications and Applications. *Chem. Rev.* **2007**, *107*, 2891–2959. [CrossRef] [PubMed]
102. Torimoto, T.; Ito, S.; Kuwabata, S.; Yoneyama, H. Effects of Adsorbents Used as Supports for Titanium Dioxide Loading on Photocatalytic Degradation of Propyzamide. *Environ. Sci. Technol.* **1996**, *30*, 1275–1281. [CrossRef]
103. Fox, M.A.; Doan, K.E.; Dulay, M.T. The Effect of the "Inert" Support on Relative Photocatalytic Activity in the Oxidative Decomposition of Alcohols on Irradiated Titanium Dioxide Composites. *Res. Chem. Intermed.* **1994**, *20*, 711–721. [CrossRef]
104. Rachel, A.; Subrahmanyam, M.; Boule, P. Comparison of Photocatalytic Efficiencies of TiO$_2$ in Suspended and Immobilised Form for the Photocatalytic Degradation of Nitrobenzenesulfonic Acids. *Appl. Catal. B Environ.* **2002**, *37*, 301–308. [CrossRef]
105. Yu, J.C.; Yu, J.; Zhao, J. Enhanced Photocatalytic Activity of Mesoporous and Ordinary TiO$_2$ thin Films by Sulfuric Acid Treatment. *Appl. Catal. B Environ.* **2002**, *36*, 31–43. [CrossRef]
106. Wang, J.A.; Limas-Ballesteros, R.; López, T.; Moreno, A.; Gómez, R.; Novaro, O.; Bokhimi, X. Quantitative Determination of Titanium Lattice Defects and Solid-State Reaction Mechanism in Iron-Doped TiO$_2$ Photocatalysts. *J. Phys. Chem. B* **2001**, *105*, 9692–9698. [CrossRef]
107. Yan, J.; Wu, G.; Guan, N.; Li, L.; Li, Z.; Cao, X. Understanding the Effect of Surface/Bulk Defects on the Photocatalytic Activity of TiO$_2$: Anatase versus Rutile. *Phys. Chem. Chem. Phys.* **2013**, *15*, 10978–10988. [CrossRef]
108. Bai, J.; Zhou, B. Titanium Dioxide Nanomaterials for Sensor Applications. *Chem. Rev.* **2014**, *114*, 10131–10176. [CrossRef]
109. Ophus, E.M.; Rode, L.; Gylseth, B.; Nicholson, D.G.; Saeed, K. Analysis of Titanium Pigments in Human Lung Tissue. *Scand. J. Work. Environ. Health* **1979**, *5*, 290–296. [CrossRef]
110. Grande, F.; Tucci, P. Titanium Dioxide Nanoparticles: A Risk for Human Health? *Mini-Rev. Med. Chem.* **2016**, *16*, 762–769. [CrossRef]
111. Lu, N.; Chen, Z.; Song, J.; Weng, Y.; Yang, G.; Liu, Q.; Yang, K.; Lu, X.; Liu, Y. Size Effect of TiO$_2$ Nanoparticles as Food Additive and Potential Toxicity. *Food Biophys.* **2022**, *17*, 75–83. [CrossRef]
112. Auffan, M.; Rose, J.; Bottero, J.Y.; Lowry, G.V.; Jolivet, J.P.; Wiesner, M.R. Towards a Definition of Inorganic Nanoparticles from an Environmental, Health and Safety Perspective. *Nat. Nanotechnol.* **2009**, *4*, 634–641. [CrossRef]
113. Nel, A.; Xia, T.; Mädler, L.; Li, N. Toxic Potential of Materials at the Nanolevel. *Science* **2006**, *311*, 622–627. [CrossRef]
114. Kim, K.T.; Eo, M.Y.; Nguyen, T.T.H.; Kim, S.M. General Review of Titanium Toxicity. *Int. J. Implant Dent.* **2019**, *5*, 10. [CrossRef]
115. Wang, J.J.; Sanderson, B.J.S.; Wang, H. Cyto-and Genotoxicity of Ultrafine TiO$_2$ Particles in Cultured Human Lymphoblastoid Cells. *Mutat. Res.—Genet. Toxicol. Environ. Mutagen.* **2007**, *628*, 99–106. [CrossRef]
116. Uchino, T.; Tokunaga, H.; Ando, M.; Utsumi, H. Quantitative Determination of OH Radical Generation and Its Cytotoxicity Induced by TiO$_2$-UVA Treatment. *Toxicol. Vitr.* **2002**, *16*, 629–635. [CrossRef]
117. Dodd, N.J.F.; Jha, A.N. Titanium Dioxide Induced Cell Damage: A Proposed Role of the Carboxyl Radical. *Mutat. Res.—Fundam. Mol. Mech. Mutagen.* **2009**, *660*, 79–82. [CrossRef]
118. Skocaj, M.; Filipic, M.; Petkovic, J.; Novak, S. Titanium Dioxide in Our Everyday Life; Is It Safe? *Radiol. Oncol.* **2011**, *45*, 227–247. [CrossRef]
119. Matsunaga, T.; Tomoda, R.; Nakajima, T.; Wake, H. Photoelectrochemical Sterilization of Microbial Cells by Semiconductor Powders. *FEMS Microbiol. Lett.* **1985**, *29*, 211–214. [CrossRef]
120. Maness, P.C.; Smolinski, S.; Blake, D.M.; Huang, Z.; Wolfrum, E.J.; Jacoby, W.A. Bactericidal Activity of Photocatalytic TiO$_2$ Reaction: Toward an Understanding of Its Killing Mechanism. *Appl. Environ. Microbiol.* **1999**, *65*, 4094–4098. [CrossRef]
121. Saito, T.; Iwase, T.; Horie, J.; Morioka, T. Mode of Photocatalytic Bactericidal Action of Powdered Semiconductor TiO$_2$ on Mutans Streptococci. *J. Photochem. Photobiol. B Biol.* **1992**, *14*, 369–379. [CrossRef] [PubMed]
122. Robertson, P.K.J.; Robertson, J.M.C.; Bahnemann, D.W. Removal of Microorganisms and Their Chemical Metabolites from Water Using Semiconductor Photocatalysis. *J. Hazard. Mater.* **2012**, *211–212*, 161–171. [CrossRef] [PubMed]
123. Wei, C.; Lin, W.Y.; Zainal, Z.; Williams, N.E.; Zhu, K.; Kruzlc, A.P.; Smith, R.L.; Rajeshwar, K. Bactericidal Activity of TiO$_2$ Photocatalyst in Aqueous Media: Toward a Solar-Assisted Water Disinfection System. *Environ. Sci. Technol.* **1994**, *28*, 934–938. [CrossRef] [PubMed]
124. Singh, J.; Hegde, P.B.; Avasthi, S.; Sen, P. Scalable Hybrid Antibacterial Surfaces: TiO$_2$ Nanoparticles with Black Silicon. *ACS Omega* **2022**, *7*, 7816–7824. [CrossRef]

125. Kalelkar, P.P.; Riddick, M.; García, A.J. Biomaterial-Based Antimicrobial Therapies for the Treatment of Bacterial Infections. *Nat. Rev. Mater.* **2022**, *7*, 39–54. [CrossRef]
126. Costerton, J.W.; Stewart, P.S.; Greenberg, E.P. Bacterial Biofilms: A Common Cause of Persistent Infections. *Science* **1999**, *284*, 1318–1322. [CrossRef]
127. Nagay, B.E.; Dini, C.; Cordeiro, J.M.; Ricomini-Filho, A.P.; De Avila, E.D.; Rangel, E.C.; Da Cruz, N.C.; Barão, V.A.R. Visible-Light-Induced Photocatalytic and Antibacterial Activity of TiO_2 Codoped with Nitrogen and Bismuth: New Perspectives to Control Implant-Biofilm-Related Diseases. *ACS Appl. Mater. Interfaces* **2019**, *11*, 18186–18202. [CrossRef]
128. Han, X.; Zhang, G.; Chai, M.; Zhang, X. Light-Assisted Therapy for Biofilm Infected Micro-Arc Oxidation TiO_2 Coating on Bone Implants. *Biomed. Mater.* **2021**, *16*, 025018. [CrossRef]
129. Wang, R.; Shi, M.; Xu, F.; Qiu, Y.; Zhang, P.; Shen, K.; Zhao, Q.; Yu, J.; Zhang, Y. Graphdiyne-Modified TiO_2 Nanofibers with Osteoinductive and Enhanced Photocatalytic Antibacterial Activities to Prevent Implant Infection. *Nat. Commun.* **2020**, *11*, 4465. [CrossRef]
130. Horváth, E.; Rossi, L.; Mercier, C.; Lehmann, C.; Sienkiewicz, A.; Forró, L. Photocatalytic Nanowires-Based Air Filter: Towards Reusable Protective Masks. *Adv. Funct. Mater.* **2020**, *30*, 2004615. [CrossRef]
131. Khaiboullina, S.; Uppal, T.; Dhabarde, N.; Subramanian, V.R.; Verma, S.C. Inactivation of Human Coronavirus by Titania Nanoparticle Coatings and Uvc Radiation: Throwing Light on Sars-CoV-2. *Viruses* **2021**, *13*, 19. [CrossRef]
132. Yoshizawa, N.; Ishihara, R.; Omiya, D.; Ishitsuka, M.; Hirano, S.; Suzuki, T. Application of a Photocatalyst as an Inactivator of Bovine Coronavirus. *Viruses* **2020**, *12*, 1372. [CrossRef]
133. He, H.; Liu, C.; Dubois, K.D.; Jin, T.; Louis, M.E.; Li, G. Enhanced Charge Separation in Nanostructured TiO_2 Materials for Photocatalytic and Photovoltaic Applications. *Ind. Eng. Chem. Res.* **2012**, *51*, 11841–11849. [CrossRef]
134. Kočí, K.; Obalová, L.; Matějová, L.; Plachá, D.; Lacný, Z.; Jirkovský, J.; Šolcová, O. Effect of TiO_2 Particle Size on the Photocatalytic Reduction of CO_2. *Appl. Catal. B Environ.* **2009**, *89*, 494–502. [CrossRef]
135. Guo, Q.; Zhou, C.; Ma, Z.; Yang, X. Fundamentals of TiO_2 Photocatalysis: Concepts, Mechanisms, and Challenges. *Adv. Mater.* **2019**, *31*, 1901997. [CrossRef]
136. Satoh, N.; Nakashima, T.; Kamikura, K.; Yamamoto, K. Quantum Size Effect in TiO_2 Nanoparticles Prepared by Finely Controlled Metal Assembly on Dendrimer Templates. *Nat. Nanotechnol.* **2008**, *3*, 106–111. [CrossRef]
137. Li, W.; Ni, C.; Lin, H.; Huang, C.P.; Shah, S.I. Size Dependence of Thermal Stability of TiO_2 Nanoparticles. *J. Appl. Phys.* **2004**, *96*, 6663–6668. [CrossRef]
138. Jansson, I.; Suárez, S.; Garcia-Garcia, F.J.; Sánchez, B. Zeolite-TiO_2 Hybrid Composites for Pollutant Degradation in Gas Phase. *Appl. Catal. B Environ.* **2015**, *178*, 100–107. [CrossRef]
139. Wang, C.Y.; Böttcher, C.; Bahnemann, D.W.; Dohrmann, J.K. A Comparative Study of Nanometer Sized Fe(III)-Doped TiO_2 Photocatalysts: Synthesis, Characterization and Activity. *J. Mater. Chem.* **2003**, *13*, 2322–2329. [CrossRef]
140. Zhu, K.; Neale, N.R.; Miedaner, A.; Frank, A.J. Enhanced Charge-Collection Efficiencies and Light Scattering in Dye-Sensitized Solar Cells Using Oriented TiO_2 Nanotubes Arrays. *Nano Lett.* **2007**, *7*, 69–74. [CrossRef]
141. De Jongh, P.E.; Vanmaekelbergh, D. Trap-Limited Electronic Transport in Assemblies of Nanometer-Size TiO_2 Particles. *Phys. Rev. Lett.* **1996**, *77*, 3427–3430. [CrossRef] [PubMed]
142. Nelson, J.; Haque, S.A.; Klug, D.R.; Durrant, J.R. Trap-Limited Recombination in Dye-Sensitized Nanocrystalline Metal Oxide Electrodes. *Phys. Rev. B* **2001**, *63*, 205321-1–205321-29. [CrossRef]
143. Cao, F.; Oskam, G.; Meyer, G.J.; Searson, P.C. Electron Transport in Porous Nanocrystalline TiO_2 Photoelectrochemical Cells. *J. Phys. Chem.* **1996**, *100*, 17021–17027. [CrossRef]
144. Dloczik, L.; Ileperuma, O.; Lauermann, I.; Peter, L.M.; Ponomarev, E.A.; Redmond, G.; Shaw, N.J.; Uhlendorf, I. Dynamic Response of Dye-Sensitized Nanocrystalline Solar Cells: Characterization by Intensity-Modulated Photocurrent Spectroscopy. *J. Phys. Chem. B* **1997**, *5647*, 10281–10289. [CrossRef]
145. Park, N.G.; Frank, A.J. Evaluation of the Charge-Collection Efficiency of Dye-Sensitized Nanocrystalline TiO_2 Solar Cells. *J. Phys. Chem. B* **1999**, *103*, 782–791. [CrossRef]
146. Paulose, M.; Shankar, K.; Yoriya, S.; Prakasam, H.E.; Varghese, O.K.; Mor, G.K.; Latempa, T.A.; Fitzgerald, A.; Grimes, C.A. Anodic Growth of Highly Ordered TiO_2 Nanotube Arrays to 134 μm in Length. *J. Phys. Chem. B* **2006**, *110*, 16179–16184. [CrossRef]
147. Ohsaki, Y.; Masaki, N.; Kitamura, T.; Wada, Y.; Okamoto, T.; Sekino, T.; Niihara, K.; Yanagida, S. Dye-Sensitized TiO_2 Nanotube Solar Cells: Fabrication and Electronic Characterization. *Phys. Chem. Chem. Phys.* **2005**, *7*, 4157–4163. [CrossRef]
148. Mor, G.K.; Shankar, K.; Paulose, M.; Varghese, O.K.; Grimes, C.A. Use of Highly-Ordered TiO_2 Nanotube Arrays in Dye-Sensitized Solar Cells. *Nano Lett.* **2006**, *6*, 215–218. [CrossRef]
149. Bahnemann, D.W. Current Challenges in Photo Catalysis: Improved Photocatalysts and Appropriate Photoreactor Engineering. *Res. Chem. Intermed.* **2000**, *26*, 207–220. [CrossRef]
150. Qin, Y.; Deng, L.; Wei, S.; Bai, H.; Gao, W.; Jiao, W.; Yu, T. An Effective Strategy for Improving Charge Separation Efficiency and Photocatalytic Degradation Performance Using a Facilely Synthesized Oxidative TiO_2 Catalyst. *Dalt. Trans.* **2022**, *51*, 6899–6907. [CrossRef]
151. Furube, A.; Asahi, T.; Masuhara, H.; Yamashita, H.; Anpo, M. Direct Observation of a Picosecond Charge Separation Process in Photoexcited Platinum-Loaded TiO_2 Particles by Femtosecond Diffuse Reflectance Spectroscopy. *Chem. Phys. Lett.* **2001**, *336*, 424–430. [CrossRef]

152. Yang, L.; Gao, P.; Lu, J.; Guo, W.; Zhuang, Z.; Wang, Q.; Li, W.; Feng, Z. Mechanism Analysis of Au, Ru Noble Metal Clusters Modified on TiO_2(101) to Intensify Overall Photocatalytic Water Splitting. *RSC Adv.* **2020**, *10*, 20654–20664. [CrossRef]
153. Kmetykó, Á.; Szániel, Á.; Tsakiroglou, C.; Dombi, A.; Hernádi, K. Enhanced Photocatalytic H_2 Generation on Noble Metal Modified TiO_2 Catalysts Excited with Visible Light Irradiation. *React. Kinet. Mech. Catal.* **2016**, *117*, 379–390. [CrossRef]
154. Chiarello, G.L.; Aguirre, M.H.; Selli, E. Hydrogen Production by Photocatalytic Steam Reforming of Methanol on Noble Metal-Modified TiO_2. *J. Catal.* **2010**, *273*, 182–190. [CrossRef]
155. Nie, J.; Schneider, J.; Sieland, F.; Zhou, L.; Xia, S.; Bahnemann, D.W. New Insights into the Surface Plasmon Resonance (SPR) Driven Photocatalytic H_2 Production of Au-TiO_2. *RSC Adv.* **2018**, *8*, 25881–25887. [CrossRef]
156. Singh, Y.; Raghuwanshi, S.K. Titanium Dioxide (TiO_2) Coated Optical Fiber-Based SPR Sensor in near-Infrared Region with Bimetallic Structure for Enhanced Sensitivity. *Optik* **2021**, *226*, 165842. [CrossRef]
157. Lin, Z.; Wang, X.; Liu, J.; Tian, Z.; Dai, L.; He, B.; Han, C.; Wu, Y.; Zeng, Z.; Hu, Z. On the Role of Localized Surface Plasmon Resonance in UV-Vis Light Irradiated Au/TiO_2 Photocatalysis Systems: Pros and Cons. *Nanoscale* **2015**, *7*, 4114–4123. [CrossRef]
158. Tian, Y.; Tatsuma, T. Mechanisms and Applications of Plasmon-Induced Charge Separation at TiO_2 Films Loaded with Gold Nanoparticles. *J. Am. Chem. Soc.* **2005**, *127*, 7632–7637. [CrossRef]
159. Wu, L.; Ma, S.; Chen, P.; Li, X. The Mechanism of Enhanced Charge Separation and Photocatalytic Activity for Au@TiO_2 Core-Shell Nanocomposite. *Int. J. Environ. Anal. Chem.* **2023**, *103*, 201–211. [CrossRef]
160. Lan, D.; Pang, F.; Ge, J. Enhanced Charge Separation in NiO and PdCo-Modified TiO_2 Photocatalysts for Efficient and Selective Photoreduction of CO_2. *ACS Appl. Energy Mater.* **2021**, *4*, 6324–6332. [CrossRef]
161. Cao, F.; Xiong, J.; Wu, F.; Liu, Q.; Shi, Z.; Yu, Y.; Wang, X.; Li, L. Enhanced Photoelectrochemical Performance from Rationally Designed Anatase/Rutile TiO_2 Heterostructures. *ACS Appl. Mater. Interfaces* **2016**, *8*, 12239–12245. [CrossRef] [PubMed]
162. Bickley, R.I.; Gonzalez-Carreno, T.; Lees, J.S.; Palmisano, L.; Tilley, R.J.D. A Structural Investigation of Titanium Dioxide Photocatalysts. *J. Solid State Chem.* **1991**, *92*, 178–190. [CrossRef]
163. Hurum, D.C.; Agrios, A.G.; Gray, K.A.; Rajh, T.; Thurnauer, M.C. Explaining the Enhanced Photocatalytic Activity of Degussa P25 Mixed-Phase TiO_2 Using EPR. *J. Phys. Chem. B* **2003**, *107*, 4545–4549. [CrossRef]
164. Nair, R.G.; Paul, S.; Samdarshi, S.K. High UV/Visible Light Activity of Mixed Phase Titania: A Generic Mechanism. *Sol. Energy Mater. Sol. Cells* **2011**, *95*, 1901–1907. [CrossRef]
165. Moniz, S.J.A.; Shevlin, S.A.; An, X.; Guo, Z.X.; Tang, J. Fe_2O_3-TiO_2 Nanocomposites for Enhanced Charge Separation and Photocatalytic Activity. *Chem.—A Eur. J.* **2014**, *20*, 15571–15579. [CrossRef]
166. Al Mayyahi, A.; Everhart, B.M.; Shrestha, T.B.; Back, T.C.; Amama, P.B. Enhanced Charge Separation in TiO_2/Nanocarbon Hybrid Photocatalysts through Coupling with Short Carbon Nanotubes. *RSC Adv.* **2021**, *11*, 11702–11713. [CrossRef]
167. Du, X.; Hu, J.; Xie, J.; Hao, A.; Lu, Z.; Cao, Y. Simultaneously Tailor Band Structure and Accelerate Charge Separation by Constructing Novel $In(OH)_3$-TiO_2 Heterojunction for Enhanced Photocatalytic Water Reduction. *Appl. Surf. Sci.* **2022**, *593*, 153305. [CrossRef]
168. Ge, Z.; Wang, C.; Chen, Z.; Wang, T.; Chen, T.; Shi, R.; Yu, S.; Liu, J. Investigation of the TiO_2 Nanoparticles Aggregation with High Light Harvesting for High-Efficiency Dye-Sensitized Solar Cells. *Mater. Res. Bull.* **2021**, *135*, 111148. [CrossRef]
169. Ram, S.K.; Rizzoli, R.; Desta, D.; Jeppesen, B.R.; Bellettato, M.; Samatov, I.; Tsao, Y.C.; Johannsen, S.R.; Neuvonen, P.T.; Pedersen, T.G.; et al. Directly Patterned TiO_2 Nanostructures for Efficient Light Harvesting in Thin Film Solar Cells. *J. Phys. D Appl. Phys.* **2015**, *48*, 365101. [CrossRef]
170. Zada, I.; Zhang, W.; Zheng, W.; Zhu, Y.; Zhang, Z.; Zhang, J.; Imtiaz, M.; Abbas, W.; Zhang, D. The Highly Efficient Photocatalytic and Light Harvesting Property of Ag-TiO_2 with Negative Nano-Holes Structure Inspired from Cicada Wings. *Sci. Rep.* **2017**, *7*, 17277. [CrossRef]
171. Yun, J.; Hwang, S.H.; Jang, J. Fabrication of Au@Ag Core/Shell Nanoparticles Decorated TiO_2 Hollow Structure for Efficient Light-Harvesting in Dye-Sensitized Solar Cells. *ACS Appl. Mater. Interfaces* **2015**, *7*, 2055–2063. [CrossRef]
172. Yang, H.Y.; Rho, W.Y.; Lee, S.K.; Kim, S.H.; Hahn, Y.B. TiO_2 Nanoparticles/Nanotubes for Efficient Light Harvesting in Perovskite Solar Cells. *Nanomaterials* **2019**, *9*, 326. [CrossRef]
173. Lombardi, J.R.; Birke, R.L. Theory of Surface-Enhanced Raman Scattering in Semiconductors. *J. Phys. Chem. C* **2014**, *118*, 11120–11130. [CrossRef]
174. Hayashi, S.; Koh, R.; Ichiyama, Y.; Yamamoto, K. Evidence for Surface-Enhanced Raman Scattering on Nonmetallic Surfaces: Copper Phthalocyanine Molecules on GaP Small Particles. *Phys. Rev. Lett.* **1988**, *60*, 1085–1089. [CrossRef]
175. Xue, X.; Ji, W.; Mao, Z.; Mao, H.; Wang, Y.; Wang, X.; Ruan, W.; Zhao, B.; Lombardi, J.R. Raman Investigation of Nanosized TiO_2: Effect of Crystallite Size and Quantum Confinement. *J. Phys. Chem. C* **2012**, *116*, 8792–8797. [CrossRef]
176. Querebillo, C.J.; Öner, H.I.; Hildebrandt, P.; Ly, K.H.; Weidinger, I.M. Accelerated Photo-Induced Degradation of Benzidine-p-Aminothiophenolate Immobilized at Light-Enhancing TiO_2 Nanotube Electrodes. *Chem. Eur. J.* **2019**, *25*, 16048–16053. [CrossRef]
177. Öner, I.H.; David, C.; Querebillo, C.J.; Weidinger, I.M.; Ly, K.H. Electromagnetic Field Enhancement of Nanostructured TiN Electrodes Probed with Surface-Enhanced Raman Spectroscopy. *Sensors* **2022**, *22*, 487. [CrossRef]
178. Zhang, X.; Lei, L.; Zhang, J.; Chen, Q.; Bao, J.; Fang, B. A Novel CdS/S-TiO_2 Nanotubes Photocatalyst with High Visible Light Activity. *Sep. Purif. Technol.* **2009**, *66*, 417–421. [CrossRef]
179. Shin, S.W.; Lee, J.Y.; Ahn, K.S.; Kang, S.H.; Kim, J.H. Visible Light Absorbing TiO_2 Nanotube Arrays by Sulfur Treatment for Photoelectrochemical Water Splitting. *J. Phys. Chem. C* **2015**, *119*, 13375–13383. [CrossRef]

180. Shen, J.; Meng, Y.; Xin, G. CdS/TiO$_2$ Nanotubes Hybrid as Visible Light Driven Photocatalyst for Water Splitting. *Rare Met.* **2011**, *30*, 280–283. [CrossRef]
181. Qi, D.; Lu, L.; Wang, L.; Zhang, J. Improved SERS Sensitivity on Plasmon-Free TiO$_2$ Photonic Microarray by Enhancing Light-Matter Coupling. *J. Am. Chem. Soc.* **2014**, *136*, 9886–9889. [CrossRef] [PubMed]
182. Joannopoulos, J.D.; Pierre, R.; Villeneuve, S.F. Photonic Crystals:Putting a New Twist on Light. *Nature* **1997**, *386*, 7. [CrossRef]
183. Al-Haddad, A.; Wang, Z.; Xu, R.; Qi, H.; Vellacheri, R.; Kaiser, U.; Lei, Y. Dimensional Dependence of the Optical Absorption Band Edge of TiO$_2$ Nanotube Arrays beyond the Quantum Effect. *J. Phys. Chem. C* **2015**, *119*, 16331–16337. [CrossRef]
184. Yip, C.T.; Huang, H.; Zhou, L.; Xie, K.; Wang, Y.; Feng, T.; Li, J.; Tam, W.Y. Direct and Seamless Coupling of TiO$_2$ Nanotube Photonic Crystal to Dye-Sensitized Solar Cell: A Single-Step Approach. *Adv. Mater.* **2011**, *23*, 5624–5628. [CrossRef]
185. Gesesse, G.D.; Li, C.; Paineau, E.; Habibi, Y.; Remita, H.; Colbeau-Justin, C.; Ghazzal, M.N. Enhanced Photogenerated Charge Carriers and Photocatalytic Activity of Biotemplated Mesoporous TiO$_2$ Films with a Chiral Nematic Structure. *Chem. Mater.* **2019**, *31*, 4851–4863. [CrossRef]
186. Chen, J.I.L.; Loso, E.; Ebrahim, N.; Ozin, G.A. Synergy of Slow Photon and Chemically Amplified Photochemistry in Platinum Nanocluster-Loaded Inverse Titania Opals. *J. Am. Chem. Soc.* **2008**, *130*, 5420–5421. [CrossRef]
187. Zhang, X.; John, S. Enhanced Photocatalysis by Light-Trapping Optimization in Inverse Opals. *J. Mater. Chem. A* **2020**, *8*, 18974–18986. [CrossRef]
188. Huo, J.; Yuan, C.; Wang, Y. Nanocomposites of Three-Dimensionally Ordered Porous TiO$_2$ Decorated with Pt and Reduced Graphene Oxide for the Visible-Light Photocatalytic Degradation of Waterborne Pollutants. *ACS Appl. Nano Mater.* **2019**, *2*, 2713–2724. [CrossRef]
189. Chen, J.I.L.; Von Freymann, G.; Choi, S.Y.; Kitaev, V.; Ozin, G.A. Slow Photons in the Fast Lane in Chemistry. *J. Mater. Chem.* **2008**, *18*, 369–373. [CrossRef]
190. Sordello, F.; Duca, C.; Maurino, V.; Minero, C. Photocatalytic Metamaterials: TiO$_2$ Inverse Opals. *Chem. Commun.* **2011**, *47*, 6147–6149. [CrossRef]
191. Rajaraman, T.S.; Parikh, S.P.; Gandhi, V.G. Black TiO$_2$: A Review of Its Properties and Conflicting Trends. *Chem. Eng. J.* **2020**, *389*, 123918. [CrossRef]
192. Naldoni, A.; Altomare, M.; Zoppellaro, G.; Liu, N.; Kment, Š.; Zbořil, R.; Schmuki, P. Photocatalysis with Reduced TiO$_2$: From Black TiO$_2$ to Cocatalyst-Free Hydrogen Production. *ACS Catal.* **2019**, *9*, 345–364. [CrossRef]
193. Chen, X.; Liu, L.; Huang, F. Black Titanium Dioxide (TiO$_2$) Nanomaterials. *Chem. Soc. Rev.* **2015**, *44*, 1861–1885. [CrossRef]
194. Chen, X.; Liu, L.; Yu, P.Y.; Mao, S.S. Increasing Solar Absorption for Photocatalysis with Black Hydrogenated Titanium Dioxide Nanocrystals. *Science* **2011**, *331*, 746–750. [CrossRef]
195. Tan, H.; Zhao, Z.; Niu, M.; Mao, C.; Cao, D.; Cheng, D.; Feng, P.; Sun, Z. A Facile and Versatile Method for Preparation of Colored TiO$_2$ with Enhanced Solar-Driven Photocatalytic Activity. *Nanoscale* **2014**, *6*, 10216–10223. [CrossRef]
196. Zhu, L.; Ma, H.; Han, H.; Fu, Y.; Ma, C.; Yu, Z.; Dong, X. Black TiO$_2$ Nanotube Arrays Fabricated by Electrochemical Self-Doping and Their Photoelectrochemical Performance. *RSC Adv.* **2018**, *8*, 18992–19000. [CrossRef]
197. Hu, W.; Zhou, W.; Zhang, K.; Zhang, X.; Wang, L.; Jiang, B.; Tian, G.; Zhao, D.; Fu, H. Facile Strategy for Controllable Synthesis of Stable Mesoporous Black TiO$_2$ Hollow Spheres with Efficient Solar-Driven Photocatalytic Hydrogen Evolution. *J. Mater. Chem. A* **2016**, *4*, 7495–7502. [CrossRef]
198. Zhou, W.; Li, W.; Wang, J.Q.; Qu, Y.; Yang, Y.; Xie, Y.; Zhang, K.; Wang, L.; Fu, H.; Zhao, D. Ordered Mesoporous Black TiO$_2$ as Highly Efficient Hydrogen Evolution Photocatalyst. *J. Am. Chem. Soc.* **2014**, *136*, 9280–9283. [CrossRef]
199. Zhu, G.; Yin, H.; Yang, C.; Cui, H.; Wang, Z.; Xu, J.; Lin, T.; Huang, F. Black Titania for Superior Photocatalytic Hydrogen Production and Photoelectrochemical Water Splitting. *ChemCatChem* **2015**, *7*, 2614–2619. [CrossRef]
200. Dong, J.; Han, J.; Liu, Y.; Nakajima, A.; Matsushita, S.; Wei, S.; Gao, W. Defective Black TiO$_2$ Synthesized via Anodization for Visible-Light Photocatalysis. *ACS Appl. Mater. Interfaces* **2014**, *6*, 1385–1388. [CrossRef]
201. Fan, C.; Chen, C.; Wang, J.; Fu, X.; Ren, Z.; Qian, G.; Wang, Z. Black Hydroxylated Titanium Dioxide Prepared via Ultrasonication with Enhanced Photocatalytic Activity. *Sci. Rep.* **2015**, *5*, 11712. [CrossRef] [PubMed]
202. Islam, S.Z.; Reed, A.; Nagpure, S.; Wanninayake, N.; Browning, J.F.; Strzalka, J.; Kim, D.Y.; Rankin, S.E. Hydrogen Incorporation by Plasma Treatment Gives Mesoporous Black TiO$_2$ Thin Films with Visible Photoelectrochemical Water Oxidation Activity. *Microporous Mesoporous Mater.* **2018**, *261*, 35–43. [CrossRef]
203. Ullattil, S.G.; Periyat, P. A "one Pot" Gel Combustion Strategy towards Ti^{3+} Self-Doped "Black" Anatase TiO$_{2-x}$ Solar Photocatalyst. *J. Mater. Chem. A* **2016**, *4*, 5854–5858. [CrossRef]
204. Sun, L.; Xie, J.; Li, Q.; Wang, F.; Xi, X.; Li, L.; Wu, J.; Shao, R.; Chen, Z. Facile Synthesis of Thin Black TiO$_{2-x}$ Nanosheets with Enhanced Lithium-Storage Capacity and Visible Light Photocatalytic Hydrogen Production. *J. Solid State Electrochem.* **2019**, *23*, 803–810. [CrossRef]
205. Kang, Q.; Cao, J.; Zhang, Y.; Liu, L.; Xu, H.; Ye, J. Reduced TiO$_2$ Nanotube Arrays for Photoelectrochemical Water Splitting. *J. Mater. Chem. A* **2013**, *1*, 5766–5774. [CrossRef]
206. Zhang, M.; Pei, Q.; Chen, W.; Liu, L.; He, T.; Chen, P. Room Temperature Synthesis of Reduced TiO$_2$ and Its Application as a Support for Catalytic Hydrogenation. *RSC Adv.* **2017**, *7*, 4306–4311. [CrossRef]
207. He, M.; Ji, J.; Liu, B.; Huang, H. Reduced TiO$_2$ with Tunable Oxygen Vacancies for Catalytic Oxidation of Formaldehyde at Room Temperature. *Appl. Surf. Sci.* **2019**, *473*, 934–942. [CrossRef]

208. Will, J.; Wierzbicka, E.; Wu, M.; Götz, K.; Yokosawa, T.; Liu, N.; Tesler, A.B.; Stiller, M.; Unruh, T.; Altomare, M.; et al. Hydrogenated Anatase TiO$_2$ Single Crystals: Defects Formation and Structural Changes as Microscopic Origin of Co-Catalyst Free Photocatalytic H$_2$ evolution Activity. *J. Mater. Chem. A* **2021**, *9*, 24932–24942. [CrossRef]
209. Katal, R.; Salehi, M.; Davood Abadi Farahani, M.H.; Masudy-Panah, S.; Ong, S.L.; Hu, J. Preparation of a New Type of Black TiO$_2$ under a Vacuum Atmosphere for Sunlight Photocatalysis. *ACS Appl. Mater. Interfaces* **2018**, *10*, 35316–35326. [CrossRef]
210. Liu, N.; Zhou, X.; Nguyen, N.T.; Peters, K.; Zoller, F.; Hwang, I.; Schneider, C.; Miehlich, M.E.; Freitag, D.; Meyer, K.; et al. Black Magic in Gray Titania: Noble-Metal-Free Photocatalytic H$_2$ Evolution from Hydrogenated Anatase. *ChemSusChem* **2017**, *10*, 62–67. [CrossRef]
211. Zhang, Y.; Xing, Z.; Liu, X.; Li, Z.; Wu, X.; Jiang, J.; Li, M.; Zhu, Q.; Zhou, W. Ti^{3+} Self-Doped Blue TiO$_2$(B) Single-Crystalline Nanorods for Efficient Solar-Driven Photocatalytic Performance. *ACS Appl. Mater. Interfaces* **2016**, *8*, 26851–26859. [CrossRef]
212. Chen, X.; Liu, L.; Liu, Z.; Marcus, M.A.; Wang, W.C.; Oyler, N.A.; Grass, M.E.; Mao, B.; Glans, P.A.; Yu, P.Y.; et al. Properties of Disorder-Engineered Black Titanium Dioxide Nanoparticles through Hydrogenation. *Sci. Rep.* **2013**, *3*, 1510. [CrossRef]
213. Yang, F.; Zhang, Z.; Li, Y.; Xiao, C.; Zhang, H.; Li, W.; Zhan, L.; Liang, G.; Chang, Y.; Ning, C.; et al. In Situ Construction of Black Titanium Oxide with a Multilevel Structure on a Titanium Alloy for Photothermal Antibacterial Therapy. *ACS Biomater. Sci. Eng.* **2022**, *8*, 2419–2427. [CrossRef]
214. Zhang, W.; Gu, J.; Li, K.; Zhao, J.; Ma, H.; Wu, C.; Zhang, C.; Xie, Y.; Yang, F.; Zheng, X. A Hydrogenated Black TiO$_2$ Coating with Excellent Effects for Photothermal Therapy of Bone Tumor and Bone Regeneration. *Mater. Sci. Eng. C* **2019**, *102*, 458–470. [CrossRef]
215. Janczarek, M.; Endo-Kimura, M.; Wang, K.; Wei, Z.; Akanda, M.M.A.; Markowska-Szczupak, A.; Ohtani, B.; Kowalska, E. Is Black Titania a Promising Photocatalyst? *Catalysts* **2022**, *12*, 1320. [CrossRef]
216. Zhang, M.; Wu, N.; Yang, J.; Zhang, Z. Photoelectrochemical Antibacterial Platform Based on Rationally Designed Black TiO$_{2-x}$ Nanowires for Efficient Inactivation against Bacteria. *ACS Appl. Bio Mater.* **2022**, *5*, 1341–1347. [CrossRef]
217. Campbell, L.; Nguyen, S.H.; Webb, H.K.; Eldridge, D.S. Photocatalytic Disinfection of *S. aureus* Using Black TiO$_{2-x}$ under Visible Light. *Catal. Sci. Technol.* **2022**, *13*, 62–71. [CrossRef]
218. Tengvall, P.; Lundström, I.; Sjöqvist, L.; Elwing, H.; Bjursten, L.M. Titanium-Hydrogen Peroxide Interaction: Model Studies of the Influence of the Inflammatory Response on Titanium Implants. *Biomaterials* **1989**, *10*, 166–175. [CrossRef]
219. Tengvall, P.; Elwing, H.; Sjöqvist, L.; Lundström, I.; Bjursten, L.M. Interaction between Hydrogen Peroxide and Titanium: A Possible Role in the Biocompatibility of Titanium. *Biomaterials* **1989**, *10*, 118–120. [CrossRef]
220. Randorn, C.; Wongnawa, S.; Boonsin, P. Bleaching of Methylene Blue by Hydrated Titanium Dioxide. *ScienceAsia* **2004**, *30*, 149–156. [CrossRef]
221. Wu, Z.; Guo, K.; Cao, S.; Yao, W.; Piao, L. Synergetic Catalysis Enhancement between H$_2$O$_2$ and TiO$_2$ with Single-Electron-Trapped Oxygen Vacancy. *Nano Res.* **2020**, *13*, 551–556. [CrossRef]
222. Zou, J.; Gao, J.; Xie, F. An Amorphous TiO$_2$ Sol Sensitized with H$_2$O$_2$ with the Enhancement of Photocatalytic Activity. *J. Alloys Compd.* **2010**, *497*, 420–427. [CrossRef]
223. Wei, Z.; Liu, D.; Wei, W.; Chen, X.; Han, Q.; Yao, W.; Ma, X.; Zhu, Y. Ultrathin TiO$_2$(B) Nanosheets as the Inductive Agent for Transfrering H$_2$O$_2$ into Superoxide Radicals. *ACS Appl. Mater. Interfaces* **2017**, *9*, 15533–15540. [CrossRef] [PubMed]
224. Sánchez, L.D.; Taxt-Lamolle, S.F.M.; Hole, E.O.; Krivokapić, A.; Sagstuen, E.; Haugen, H.J. TiO$_2$ Suspension Exposed to H$_2$O$_2$ in Ambient Light or Darkness: Degradation of Methylene Blue and EPR Evidence for Radical Oxygen Species. *Appl. Catal. B Environ.* **2013**, *142–143*, 662–667. [CrossRef]
225. Zhang, A.Y.; Lin, T.; He, Y.Y.; Mou, Y.X. Heterogeneous Activation of H$_2$O$_2$ by Defect-Engineered TiO$_{2-X}$ Single Crystals for Refractory Pollutants Degradation: A Fenton-like Mechanism. *J. Hazard. Mater.* **2016**, *311*, 81–90. [CrossRef]
226. Wiedmer, D.; Sagstuen, E.; Welch, K.; Haugen, H.J.; Tiainen, H. Oxidative Power of Aqueous Non-Irradiated TiO$_2$-H$_2$O$_2$ Suspensions: Methylene Blue Degradation and the Role of Reactive Oxygen Species. *Appl. Catal. B Environ.* **2016**, *198*, 9–15. [CrossRef]
227. Zhou, C.; Luo, J.; Chen, Q.; Jiang, Y.; Dong, X.; Cui, F. Titanate Nanosheets as Highly Efficient Non-Light-Driven Catalysts for Degradation of Organic Dyes. *Chem. Commun.* **2015**, *51*, 10847–10849. [CrossRef]
228. Krishnan, P.; Liu, M.; Itty, P.A.; Liu, Z.; Rheinheimer, V.; Zhang, M.H.; Monteiro, P.J.M.; Yu, L.E. Characterization of Photocatalytic TiO$_2$ Powder under Varied Environments Using near Ambient Pressure X-Ray Photoelectron Spectroscopy. *Sci. Rep.* **2017**, *7*, 43298. [CrossRef]
229. Vorontsov, A.V.; Valdés, H.; Smirniotis, P.G.; Paz, Y. Recent Advancements in the Understanding of the Surface Chemistry in TiO$_2$ Photocatalysis. *Surfaces* **2020**, *3*, 72–92. [CrossRef]
230. Jose, M.; Haridas, M.P.; Shukla, S. Predicting Dye-Adsorption Capacity of Hydrogen Titanate Nanotubes via One-Step Dye-Removal Method of Novel Chemically-Activated Catalytic Process Conducted in Dark. *J. Environ. Chem. Eng.* **2014**, *2*, 1980–1988. [CrossRef]
231. Xiong, F.; Yin, L.-L.; Wang, Z.; Jin, Y.; Sun, G.; Gong, X.-Q.; Huang, W. Surface Reconstruction-Induced Site-Specific Charge Separation and Photocatalytic Reaction on Anatase TiO$_2$ (001) Surface. *J. Phys. Chem. C* **2017**, *121*, 9991–9999. [CrossRef]
232. Petković, J.; Küzma, T.; Rade, K.; Novak, S.; Filipič, M. Pre-Irradiation of Anatase TiO$_2$ Particles with UV Enhances Their Cytotoxic and Genotoxic Potential in Human Hepatoma HepG2 Cells. *J. Hazard. Mater.* **2011**, *196*, 145–152. [CrossRef]

233. Humphreys, H. Surgical Site Infection, Ultraclean Ventilated Operating Theatres and Prosthetic Joint Surgery: Where Now? *J. Hosp. Infect.* **2012**, *81*, 71–72. [CrossRef]
234. Van Hengel, I.A.J.; Tierolf, M.W.A.M.; Fratila-apachitei, L.E.; Apachitei, I.; Zadpoor, A.A. Antibacterial Titanium Implants Biofunctionalized by Plasma Electrolytic Oxidation with Silver, Zinc, and Copper: A Systematic Review. *Int. J. Mol. Sci.* **2021**, *22*, 3800. [CrossRef]
235. Ferraris, S.; Spriano, S. Antibacterial Titanium Surfaces for Medical Implants. *Mater. Sci. Eng. C* **2016**, *61*, 965–978. [CrossRef]
236. Yu, J.; Zhou, M.; Zhang, L.; Wei, H. Antibacterial Adhesion Strategy for Dental Titanium Implant Surfaces: From Mechanisms to Application. *J. Funct. Biomater.* **2022**, *13*, 169. [CrossRef]
237. Akshaya, S.; Rowlo, P.K.; Dukle, A.; Nathanael, A.J. Antibacterial Coatings for Titanium Implants: Recent Trends and Future Perspectives. *Antibiotics* **2022**, *11*, 1719. [CrossRef]
238. Rincón, A.G.; Pulgarin, C. Absence of E. Coli Regrowth after Fe^{3+} and TiO_2 Solar Photoassisted Disinfection of Water in CPC Solar Photoreactor. *Catal. Today* **2007**, *124*, 204–214. [CrossRef]
239. Delanois, R.E.; Mistry, J.B.; Gwam, C.U.; Mohamed, N.S.; Choksi, U.S.; Mont, M.A. Current Epidemiology of Revision Total Knee Arthroplasty in the United States. *J. Arthroplast.* **2017**, *32*, 2663–2668. [CrossRef]
240. Savio, D.; Bagno, A. When the Total Hip Replacement Fails: A Review on the Stress-Shielding Effect. *Processes* **2022**, *10*, 612. [CrossRef]
241. Kwak, J.M.; Koh, K.H.; Jeon, I.H. Total Elbow Arthroplasty: Clinical Outcomes, Complications, and Revision Surgery. *CiOS Clin. Orthop. Surg.* **2019**, *11*, 369–379. [CrossRef] [PubMed]
242. Shen, X.; Shukla, P. A Review of Titanium Based Orthopaedic Implants (Part-I): Physical Characteristics, Problems and the Need for Surface Modification. *Int. J. Peen. Sci. Technol.* **2020**, *1*, 301–332.
243. Prestat, M.; Thierry, D. Corrosion of Titanium under Simulated Inflammation Conditions: Clinical Context and in Vitro Investigations. *Acta Biomater.* **2021**, *136*, 72–87. [CrossRef] [PubMed]
244. Zhang, Q.H.; Cossey, A.; Tong, J. Stress Shielding in Periprosthetic Bone Following a Total Knee Replacement: Effects of Implant Material, Design and Alignment. *Med. Eng. Phys.* **2016**, *38*, 1481–1488. [CrossRef] [PubMed]
245. Eliaz, N. Corrosion of Metallic Biomaterials: A Review. *Materials* **2019**, *12*, 407. [CrossRef]
246. Chen, Q.; Thouas, G.A. Metallic Implant Biomaterials. *Mater. Sci. Eng. R Reports* **2015**, *87*, 1–57. [CrossRef]
247. Niinomi, M.; Nakai, M. Titanium-Based Biomaterials for Preventing Stress Shielding between Implant Devices and Bone. *Int. J. Biomater.* **2011**, *2011*, 836587. [CrossRef]
248. Rack, H.J.; Qazi, J.I. Titanium Alloys for Biomedical Applications. *Mater. Sci. Eng. C* **2006**, *26*, 1269–1277. [CrossRef]
249. Hansen, D.C. Metal Corrosion in the Human Body: The Ultimate Bio-Corrosion Scenario. *Electrochem. Soc. Interface* **2008**, *17*, 31–34. [CrossRef]
250. Kulkarni, M.; Mazare, A.; Gongadze, E.; Perutkova; Kralj-Iglic, V.; Milošev, I.; Schmuki, P.; Iglič, A.; Mozetič, M. Titanium Nanostructures for Biomedical Applications. *Nanotechnology* **2015**, *26*, 062002. [CrossRef]
251. Ventre, M.; Coppola, V.; Iannone, M.; Netti, P.A.; Tekko, I.; Larrañeta, E.; Rodgers, A.M.; Scott, C.J.; Kissenpfennig, A.; Donnelly, R.F.; et al. Nanotechnologies for Tissue Engineering and Regeneration. In *Nanotechnologies in Preventive and Regenerative Medicine*; Elsevier: Amsterdam, The Netherlands, 2018; pp. 93–206.
252. Wu, J.M. *Nanostructured TiO_2 Layers on Ti for Bone Bonding*; Elsevier: Amsterdam, The Netherlands, 2020; ISBN 9780081029992.
253. Li, Y.; Xu, J. Is Niobium More Corrosion-Resistant than Commercially Pure Titanium in Fluoride-Containing Artificial Saliva? *Electrochim. Acta* **2017**, *233*, 151–166. [CrossRef]
254. Reeves, J.F.; Davies, S.J.; Dodd, N.J.F.; Jha, A.N. Hydroxyl Radicals (•OH) Are Associated with Titanium Dioxide (TiO_2) Nanoparticle-Induced Cytotoxicity and Oxidative DNA Damage in Fish Cells. *Mutat. Res.—Fundam. Mol. Mech. Mutagen.* **2008**, *640*, 113–122. [CrossRef]
255. Eger, M.; Hiram-Bab, S.; Liron, T.; Sterer, N.; Carmi, Y.; Kohavi, D.; Gabet, Y. Mechanism and Prevention of Titanium Particle-Induced Inflammation and Osteolysis. *Front. Immunol.* **2018**, *9*, 2963. [CrossRef]
256. Pan, J.; Thierry, D.; Leygraf, C. Electrochemical and XPS Studies of Titanium for Biomaterial Applications with Respect to the Effect of Hydrogen Peroxide. *J. Biomed. Mater. Res.* **1994**, *28*, 113–122. [CrossRef]
257. Sundgren, J.-E.; Bodö, P.; Lundström, I. Auger Electron Spectroscopic Studies of the Interface between Human Tissue and Implants of Titanium and Stainless Steel. *J. Colloid Interface Sci.* **1986**, *110*, 9–20. [CrossRef]
258. Gilbert, J.L.; Mali, S.; Urban, R.M.; Silverton, C.D.; Jacobs, J.J. In Vivo Oxide-Induced Stress Corrosion Cracking of Ti-6Al-4V in a Neck-Stem Modular Taper: Emergent Behavior in a New Mechanism of in Vivo Corrosion. *J. Biomed. Mater. Res.—Part B Appl. Biomater.* **2012**, *100 B*, 584–594. [CrossRef]
259. Haynes, D.R.; Rogers, S.D.; Hay, S.; Pearcy, M.J.; Howie, D.W. The Differences in Toxicity and Release of Bone-Resorbing Mediators Induced by Titanium and Cobalt-Chromium-Alloy Wear Particles. *J. Bone Jt. Surg.* **1993**, *75*, 825–834. [CrossRef]
260. Liu, Y.; Gilbert, J.L. The Effect of Simulated Inflammatory Conditions and Fenton Chemistry on the Electrochemistry of CoCrMo Alloy. *J. Biomed. Mater. Res.—Part B Appl. Biomater.* **2018**, *106*, 209–220. [CrossRef]
261. Nguyen, G.T.; Green, E.R.; Mecsas, J. Neutrophils to the ROScue: Mechanisms of NADPH Oxidase Activation and Bacterial Resistance. *Front. Cell. Infect. Microbiol.* **2017**, *7*, 373. [CrossRef]
262. Sheikh, Z.; Brooks, P.J.; Barzilay, O.; Fine, N.; Glogauer, M. Macrophages, Foreign Body Giant Cells and Their Response to Implantable Biomaterials. *Materials* **2015**, *8*, 5671–5701. [CrossRef]

263. Väänänen, H.K.; Zhao, H.; Mulari, M.; Halleen, J.M. The Cell Biology of Osteoclast Function. *J. Cell Sci.* **2000**, *113*, 377–381. [CrossRef] [PubMed]
264. Berglund, F.; Carlmark, B. Titanium, Sinusitis, and the Yellow Nail Syndrome. *Biol. Trace Elem. Res.* **2011**, *143*, 1–7. [CrossRef] [PubMed]
265. Decker, A.; Daly, D.; Scher, R.K. Role of Titanium in the Development of Yellow Nail Syndrome. *Ski. Appendage Disord.* **2015**, *1*, 28–30. [CrossRef] [PubMed]
266. Ataya, A.; Kline, K.P.; Cope, J.; Alnuaimat, H. Titanium Exposure and Yellow Nail Syndrome. *Respir. Med. Case Rep.* **2015**, *16*, 146–147. [CrossRef]

Disclaimer/Publisher's Note: The statements, opinions and data contained in all publications are solely those of the individual author(s) and contributor(s) and not of MDPI and/or the editor(s). MDPI and/or the editor(s) disclaim responsibility for any injury to people or property resulting from any ideas, methods, instructions or products referred to in the content.

Article

Ultra-Violet-Assisted Scalable Method to Fabricate Oxygen-Vacancy-Rich Titanium-Dioxide Semiconductor Film for Water Decontamination under Natural Sunlight Irradiation

Mohammed Alyami

Physics Department, College of Science and Humanities in Al-Kharj, Prince Sattam Bin Abdulaziz University, Al-Kharj 11942, Saudi Arabia; m.alyami@psau.edu.sa

Abstract: This work reports the fabrication of titanium dioxide (TiO_2) nanoparticle (NPs) films using a scalable drop-casting method followed by ultra-violet (UV) irradiation for creating defective oxygen vacancies on the surface of a fabricated TiO_2 semiconductor film using an UV lamp with a wavelength oof 255 nm for 3 h. The success of the use of the proposed scalable strategy to fabricate oxygen-vacancy-rich TiO_2 films was assessed through UV–Vis spectroscopy, X-ray photoelectron spectroscopy (XPS), X-ray diffraction (XRD), and scanning electron microscopy (SEM). The Ti 2p XPS spectra acquired from the UV-treated sample showed the presence of additional Ti^{3+} ions compared with the untreated sample, which contained only Ti^{4+} ions. The band gap of the untreated TiO_2 film was reduced from 3.2 to 2.95 eV after UV exposure due to the created oxygen vacancies, as evident from the presence of Ti^{3+} ions. Radiation exposure has no significant influence on sample morphology and peak pattern, as revealed by the SEM and XRD analyses, respectively. Furthermore, the photocatalytic activity of the fabricated TiO_2 films for methylene-blue-dye removal was found to be 99% for the UV-treated TiO_2 films and compared with untreated TiO_2 film, which demonstrated only 77% at the same operating conditions under natural-sunlight irradiation. The proposed UV-radiation method of oxygen vacancy has the potential to promote the wider application of photo-catalytic TiO_2 semiconductor films under visible-light irradiation for solving many environmental and energy-crisis challenges for many industrial and technological applications.

Keywords: sunlight; oxygen vacancy; TiO_2; photocatalytic activity

Citation: Alyami, M. Ultra-Violet-Assisted Scalable Method to Fabricate Oxygen-Vacancy-Rich Titanium-Dioxide Semiconductor Film for Water Decontamination under Natural Sunlight Irradiation. *Nanomaterials* **2023**, *13*, 703. https://doi.org/10.3390/nano13040703

Academic Editors: Wei Zhou and Eric Waclawik

Received: 4 January 2023
Revised: 4 February 2023
Accepted: 10 February 2023
Published: 12 February 2023

Copyright: © 2023 by the author. Licensee MDPI, Basel, Switzerland. This article is an open access article distributed under the terms and conditions of the Creative Commons Attribution (CC BY) license (https://creativecommons.org/licenses/by/4.0/).

1. Introduction

The environmental hazard posed by gaseous pollution has recently raised serious concerns [1–6]. Specifically, several harmful health-related and environmental issues emanate from NO_x-based pollutants, while various measures to control these pollutants revolve around the fabrication of photo-catalyst with desirable features and the ability to oxidize harmful pollutants for environmental remediation [7–10]. Photocatalyst-coated surface-based reactors have proven to be more practical for long-term operation than photocatalytic powder-based reactors [11–14]. As a promising photo-electrode and photocatalyst, titanium dioxide (TiO_2) has enjoyed wider applicability in photocatalytic hydrogen generation, solar cells, and the remediation of organic contaminants among other photo-catalytic applications [15–17]. Furthermore, TiO_2 is recognized as a low-cost, highly effective and photo-catalyst of interest as a result of its promising thermal and chemical stabilities, desirable electronic features, and environmental benignity, among others [18–21]. Pristine TiO_2 semiconductors are characterized by a wide band gap that can only utilize the UV part of the light spectrum with wavelengths shorter than 385 nm, which is just 5% of the sunlight energy capacity. The extension of spectrum usability to visible regions requires further and more extensive research [22–25]. Additionally, the rapid recombination of photo-generated holes and electrons further restricts the practical applicability of these semiconductors [26,27]. Previous theoretical and experimental efforts to extend the separation period of photo-generated carriers and to reach a narrower band gap included the formation of defects, metal and non-metal doping, hydrogenation, noble-metal deposition, defect engineering, sensitization, and hetero-junction formation, among others [28,29]. Oxygen vacancy potentially modulates the semiconductor band gap and influences the band properties and structure [30–32]. Such defects as vacancies also improve charge-carrier-separation efficiency through electronic-conductivity improvement [33]. Similarly, oxygen vacancies conveniently modify the electronic structure in the reaction site's vicinity and eventually facilitate intermediate adsorption, while the photo-catalytic activity is improved [34–37]. Oxygen vacancies are widely employed in defect formation in photo-catalysts for enhancing the photocatalytic activity of TiO_2 semiconductors. The experimental characterization and theoretical computations in the literature revealed the creation of a mid-gap state below the conduction band due to the oxygen vacancies incorporated in the semiconducting materials, as well as the generation of Ti^{3+} centers. These centers prevent rapid electron-hole recombination while the energy gap is lowered by the mid-gap states created by the oxygen vacancies. Therefore, the photocatalytic activity of TiO_2 semiconductors can be effectively enhanced through oxygen-vacancy creation. Immense efforts were deployed in order to achieve the synthesis of oxygen-vacancy-mediated TiO_2 semiconductors for addressing several photo-catalytic challenges [4,38,39]. Current methods and techniques for incorporating and controlling oxygen vacancies in TiO_2 semiconductors include thermal treatment with hydrogen, oxygen-depletion-based thermal treatment, self-doping, particle bombardment using high energy, and UV irradiation [4,33,38,39]. This work proposes the UV-irradiation method of oxygen-vacancy creation, which is characterized by simplicity, effectiveness, and versatility compared to other methods, such as the hydrothermal technique. The created oxygen vacancies were assessed using various spectroscopic techniques, including XPS and UV–Vis, among others, to reveal the relation between photocatalytic performance and the presence of oxygen vacancy.

2. Experimental Section

2.1. Sample Preparation

The TiO_2 nanoparticles (<25 nm, 99.5%, Sigma Aldrich, USA) were mixed with ethanol to obtain a concentration of 5 mg/L using a magnetic stirrer. The stirring lasted 30 min at 500 rpm, while the drop-casting technique was employed for applying TiO_2-nanoparticles dispersion on a glass substrate. The glass substrate was subsequently placed on a hot plate at a temperature of 80 degrees Celsius for ethanol removal. This eventually produced a solid film of TiO_2 nanoparticles on the glass substrate. Two different substrates were synthesized using aforementioned experimental conditions, while one substrate with solid TiO_2 nanoparticles was exposed to UV radiation of 255 nm at a distance of 15 cm from a 6-watt UV lamp (UVP UVGL, Analytik Jena, Germany) while the second substrate was not treated with UV exposure. The schematic diagram illustrates the processes and an experimental procedure is presented in Figure 1.

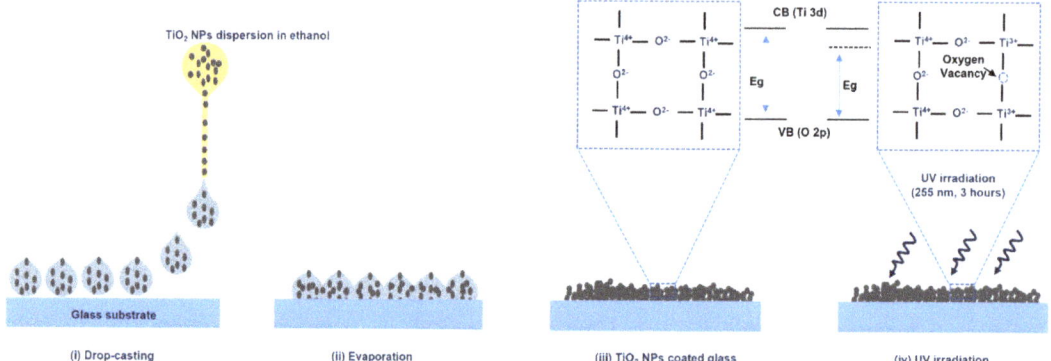

Figure 1. Fabrication strategy aimed at the fabrication of untreated and UV-treated TiO$_2$ films.

2.2. Characterization

Using an X-ray diffractometer (A Shimadzu XRD-6000, Kyoto, Japan) operating in the range $10° \leq 2\theta \leq 80°$, the crystal structure of the as-prepared TiO$_2$ film was examined. The morphology of the as-prepared TiO$_2$ film was examined by using scanning electron microscopy (FESEM, Quanta FEG250, FEI, Hillsboro, OR, USA). Furthermore X-ray photo-electron spectroscopy (XPS; Thermo scientific K-alpha XPS spectrometer, Thermo Fisher Scientific, Waltham, USA) was used to distinguish and characterize the chemical makeup of the as-prepared TiO$_2$ films. A monochromic Al ka source with a characteristic energy of 1486.6 eV was utilized. A spectrophotometer was used to log the UV–Vis absorption spectra of the as-prepared TiO$_2$ films.

2.3. Photocatalytic Activity

By making use of a constructed immobilized photocatalytic reactor [33] fitted with a magnetic stirrer working at 500 r.p.m., the photocatalytic performance of the as-prepared TiO$_2$ films ~ 3 cm^{-2} was investigated for an aqueous solution of methylene-blue dye (~5 ppm) under solar radiation. The setup was left wholly in the dark for about 60 min before each experiment to allow the adsorption–desorption equilibrium to be reached. The as-prepared sample was then placed in sunlight in Al Kharj City, Saudi Arabia, at noon. From the methylene-blue-dye absorption spectra over 300–750 nm, the ratio of methylene-blue-dye concentration at t min (Ct) to methylene-blue-dye concentration at 0 min (C0) was calculated at regular time increments of ~15 min.

3. Results and Discussion

The bandgap and optical absorption of the readymade TiO$_2$ film in its primal and UV-irradiated state were investigated via UV-Vis diffuse-reflectance spectroscopy. From this analysis, it was observed that the UV-treated TiO$_2$ films had a higher absorption in the visible region than the untreated TiO$_2$ films (Figure 2a). In terms of the bandgap energy, the UV-treated TiO$_2$ films demonstrated a narrower bandgap of approximately 2.95 eV, compared to the 3.2 eV observed in the untreated TiO$_2$ films (Figure 2b). Valence-band energy can be obtained from XPS valance-band spectra through extrapolation to the binding-energy axis. The energy of a valence band is the energy of the band of electron orbitals with which electrons jump out (and move into the conduction band) when excited. The determination of the valence-band energy from the acquired XPS valence-band spectra is presented in Figure 2c,d. The valence-band energy of the TiO$_2$ that was not exposed to UV radiation is presented in Figure 2c, while that of the TiO$_2$ sample subjected to UV radiation for possible oxygen-vacancy creation is presented in Figure 2d. The valence-band energy for both the UV-treated and the untreated TiO$_2$ film had a similar value because the top of the valence band was governed by the O 2p states, while the bottom of the conduction band was controlled by the Ti 3d state. Therefore, a defect (in Ti^{3+}) that formed just below the conduction band was responsible for the variation in the conduction-band energy for the UV-treated and untreated TiO_2 films. Hence, the UV-treated sample was characterized by a reduced/narrow energy gap after the UV exposure.

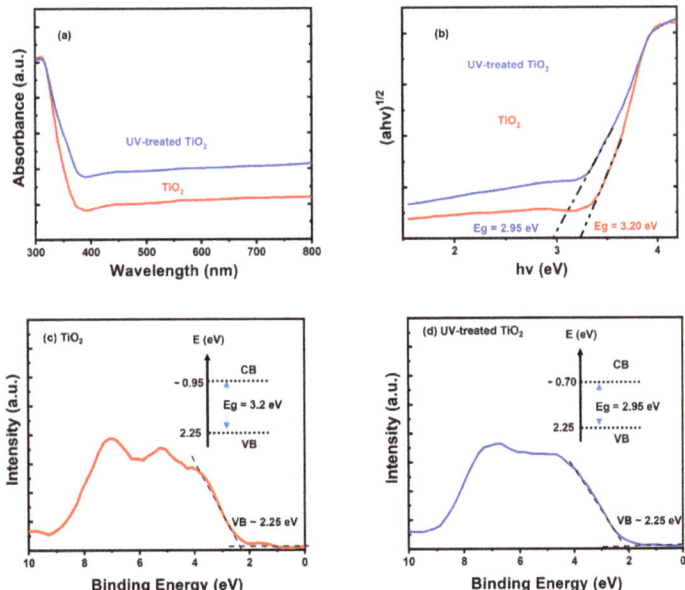

Figure 2. Optical characterization: (**a**) UVVis absorbance spectra, (**b**) estimation of the band-gap energies from Tauc's plots, (**c**,**d**) XPS valence-band spectra of untreated and UV-treated TiO_2 films, along with their electronic band structures, as indicated.

Photocatalytic Performance

To experimentally verify the results obtained from the optical characterization analysis, the rate of the photodegradation (Ct/C0) of the methylene-blue dye by the TiO_2 films was measured under the sunlight irradiation, as shown in Figure 3a. An outstanding photodegradation rate of approximately 99% was reached in 60 min using the UV-treated TiO_2 film, which considerably surpassed the 77% photodegradation rate yielded by the untreated TiO_2 film under the same operating conditions.

Using the experimental results, the photocatalytic mechanism of the readymade UV-treated TiO_2 film with oxygen vacancies was formulated. Figure 3b is a schematic diagram of the photocatalytic degradation of the methylene-blue dye on the UV-treated TiO_2 film. Electron-hole pairs were produced by the photoexcitation of electrons from the valence band (VB) into the conduction-band (CB)/oxygen-vacancy energy level upon the exposure of the UV-treated TiO_2 film to UV-visible light irradiation. The photo-generated electrons were easily trapped by the oxygen vacancies, resulting in a low recombination rate with holes. Consequently, the electrons lived longer and reduced the amount of oxygen assimilated from the surface of the UV-treated TiO_2 film and created superoxide radicals ($^{\bullet}O_2^-$), which are potent oxidizers of methylene-blue-dye molecules [40,41]. Meanwhile, photo-generated holes also spread out to the surface of the UV-treated TiO_2 film and further oxidized any surface-assimilated methylene-blue dye. Consequently, the UV-treated TiO_2 film with oxygen vacancies demonstrated superior visible-light photocatalytic performance. Generally, the oxygen vacancy defects on the surface of the photocatalyst enhance the separation efficiency of electron-hole pairs and ensure impeccable photocatalytic efficiency [17].

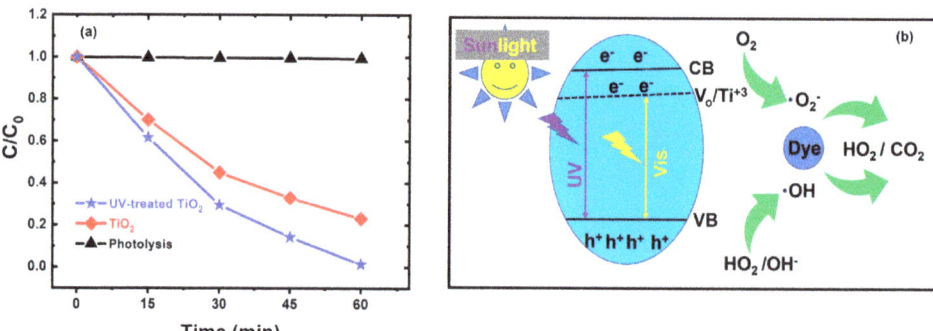

Figure 3. (**a**) Photocatalytic performance of the untreated and UV-treated TiO$_2$ films. (**b**) Schematic illustration of the photocatalytic degradation of the methylene-blue dye on the UV-treated TiO$_2$ film.

The XPS survey spectra acquired from the two prepared TiO$_2$ samples are presented in Figure 4. The spectra were analyzed using Avantage software (version 5.932, Thermo Scientific, Waltham, MA, USA). The elemental identification and chemical states of the prepared samples, along with the binding energies, are shown in Figure 4a,b. Three different peaks were observed in the spectra of the untreated TiO$_2$ semiconductor presented in Figure 4a, which correspond to carbon (C 1s) at a binding energy of 284.80 eV, titanium (Ti 2p3/2) at a binding energy of 458.10 eV, and oxygen (O 1s) band at a binding energy of ~531 eV. However, when the samples were treated with an ultra-violet (UV) beam for oxygen-vacancy creation, similar constituent element peaks and binding-energy positions were observed and are presented in Figure 4b. The carbon signal (C 1s) was observed at a binding energy of 284.53 eV, with a positive binding-energy shift of 0.66 eV due to exposure to the UV beam. The shift in the carbon-containing compound signal was due to the surface-cleaning potential of the UV beam. Furthermore, the titanium Ti 2p state was observed at the 458.94 eV [40] binding-energy peak with a positive binding-energy shift of 0.82 eV due to the possible formation of some new states of titanium after UV exposure. The new states that appeared after the UV treatment were vividly clear in the high-resolution spectra of the Ti 2p signal. The signal corresponding to oxygen O 1s appeared at 530.13 eV for the UV-treated sample, as shown in Figure 4b. The appearance of the new chemical state of oxygen further resulted in a positive binding -energy shift of 0.63 eV as compared to the untreated sample. Table 1 presents the survey -spectra parameters for the sample exposed to UV, as well as the untreated samples. The binding energy and the atomic percentage of the samples before and after UV exposure are also presented in Table 1. The peak areas of each element, except the C 1s band, is increased after exposure to UV light. The atomic percentage of the Ti 2p and O 1s increased after the UV exposure, while that of the carbon C 1s decreased after UV exposure. This indicates the contaminant -removing potentials of UV light.

Table 1. Survey -spectra parameters for untreated and UV-treated samples.

Sample	Name	Peak BE (eV)	Atomic%
untreated	Ti 2p	458.20	23.38
-	O 1s	529.47	45.14
-	C 1s	284.53	31.47
UV -treated	Ti 2p	458.94	24.48
-	O 1s	530.13	50.06
-	C 1s	285.14	25.46

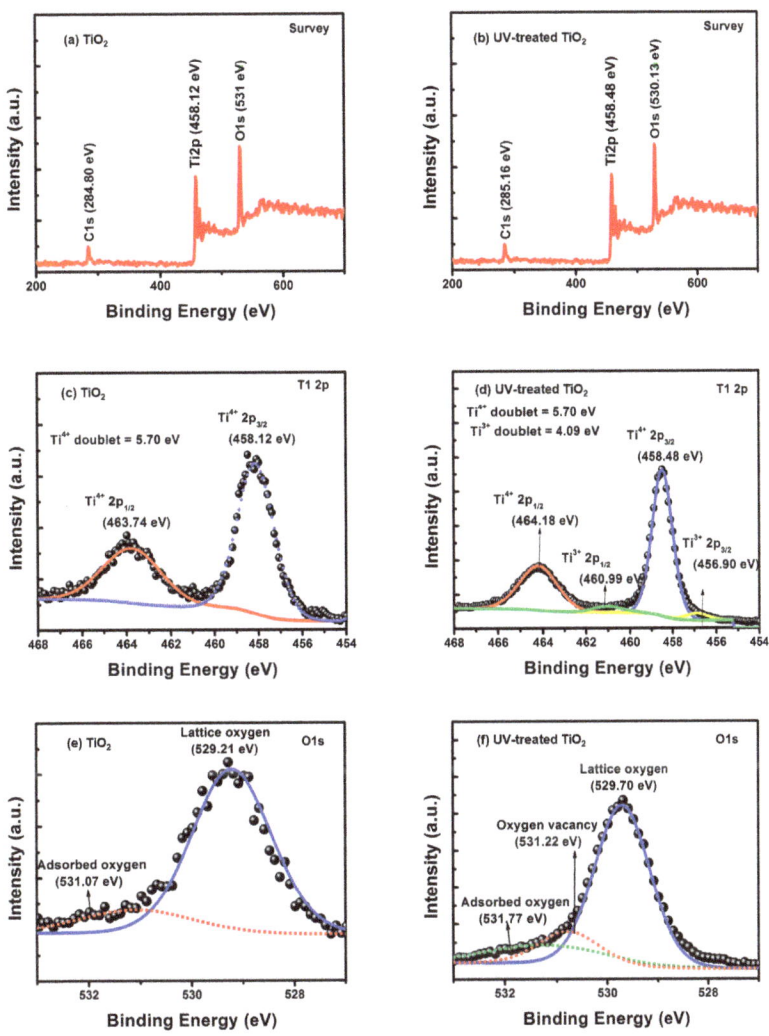

Figure 4. The XPS survey (**a**,**b**) and high-resolution XPS spectra of Ti 2p (**c**,**d**) and O 1s (**e**,**f**) of (**a**) untreated (**b**) UV-treated TiO_2 films, respectively.

The high-resolution spectra of the Ti 2p doublet for the untreated and UV-treated TiO_2 film samples are shown in Figure 4c,d. Two peaks were identified in the spectra of the untreated sample, as shown in Figure 4c, and they were attributed to Ti^{4+} $2p_{3/2}$ and Ti^{4+} $2p_{1/2}$ components. The components were located at binding energies of 458.12 and 463.74 eV, respectively [41]. The energy difference obtained for the doublet components for the untreated TiO_2 sample was 5.7 eV, which shows the presence of an anatase phase in the TiO_2 sample [42]. The spectra for the UV-treated samples are presented in Figure 4d, with four different peaks at different binding energies. The dominant Ti^{4+} $2p_{3/2}$ and Ti^{4+} $2p_{1/2}$ lines of titanium present in the untreated sample were maintained with binding energies of 458.48 and 464.18 eV, respectively. This shows that the Ti^{4+} $2p_{3/2}$ and Ti^{4+} $2p_{1/2}$ states shifted positively due to the change in the surface-charging effect. It is worth mentioning that the anatase phase was maintained after UV exposure, as can be observed from the doublet value of 5.7 eV. Additional Ti^{3+} $2p_{3/2}$ and Ti^{3+} $2p_{1/2}$ states were exhibited at binding energies of 456.90 and 460.99 eV, respectively, and the energy difference was 4.09 eV. Initially, the presence of a Ti^{3+} signal indicates the presence of oxygen vacancies after Ti^{4+} ions undergo a reduction process. The XPS spectra of the oxygen O 1s signal for the untreated and UV-treated samples of TiO_2 semiconductor films are presented in Figure 4e,f. For the untreated-sample spectra presented in Figure 4e, two peaks were exhibited at 529.21 and 531.07 eV [42], which correspond to lattice oxygen and surface-chemisorbed hydroxyl group (any other surface-oxide species are also probable), respectively. For comparison, in the high–resolution-spectra UV-treated samples shown in Figure 4f, three different components are exhibited.

The peaks were located at 529.70, 531.77, and 531.22 eV, which correspond to lattice oxygen, surface-chemisorbed hydroxyl groups, and oxygen vacancies, respectively. The lattice-oxygen peak shifted positively by 0.49 eV, while the surface-chemisorbed hydroxyl group showed a positive shift of 0.70 eV, followed by the occurrence of an oxygen-vacancy peak. The observed binding-energy shift can be attributed to oxygen-vacancy formation, which facilitates electron transfer to Ti and O atoms. The electronic properties of the prepared samples were assessed using the binding-energy difference (BED) approach, which measures the binding-energy difference between O 1s and Ti $2p_{3/2}$ core levels [43]. Using this approach for the data presented in Figure 4c,e, BDE = BE (O 1s)-BE (Ti $2p_{3/2}$) = 529.21 eV −458.12 eV = 71.1 eV. The obtained BED of 71.1 eV shows that the employed peaks were from the Ti^{4+} states in TiO_2. For the dominant components shown in Figure 4d,f, the BED of 71.2 eV also confirms the Ti^{4+} states in the TiO_2. Using the minor components shown in Figure 4d,f, the BED of 74.32 eV confirms the Ti^{3+} states in TiO_2.

The X-ray diffraction (XRD) patterns of the UV-treated and untreated TiO_2 films are presented in Figure 5. The high level of similarity in the diffraction pattern suggests the minimal influence of UV radiation on the crystalline structure of the pure untreated TiO_2 film.

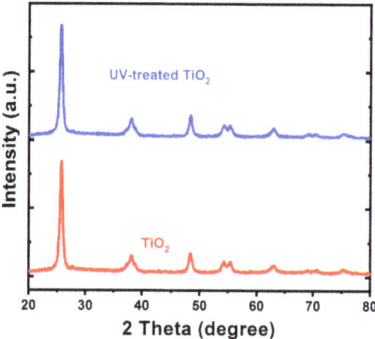

Figure 5. The XRD patterns of the untreated and UV-treated TiO_2 films, as indicated.

The characterization of the UV-treated and untreated TiO_2 films using scanning electron microscopy (SEM) is presented in Figure 6. The morphology of the TiO_2 film after the UV exposure was insignificantly affected by the UV radiation.

Figure 6. The SEM images of the untreated and UV-treated TiO$_2$ films at different magnifications, as indicated.

4. Conclusions

In this work, a solid film of TiO$_2$ nanoparticles was synthesized, and, further, UV radiation was employed for oxygen-vacancy creation to enhance the photocatalytic activity of the semiconductor under visible-light irradiation. Both the UV-treated and untreated samples were characterized using UV–Vis spectroscopy, XPS, XRD, and SEM. The survey XPS spectrum was acquired to show the presence of three constituent elements, carbon, titanium, and oxygen, with their respective chemical states indicated by the C 1s, Ti 2p, and O 1s lines. The high-resolution O 1s spectrum obtained from a TiO$_2$ sample not exposed to UV contained two peaks, which were attributed to lattice oxygen and surface-chemisorbed hydroxyl groups. The binding energies of these two peaks shifted to higher positive values after the sample was exposed to UV radiation, confirming the presence of oxygen vacancies in the UV-radiated sample. The presence of an additional peak ascribed to oxygen vacancies in the spectra of the sample exposed to UV radiation further confirmed the versatility of the proposed UV-irradiation method for oxygen-vacancy creation. The presence of a Ti^{3+} oxidation state in the UV-treated sample due to the reduction of Ti^{4+} offers additional confirmation of the successful creation of oxygen vacancies. The band gap of the untreated TiO$_2$ film was reduced from 3.2 to 2.95 eV after the UV exposure due to the oxygen vacancies created, which was made evident by the presence of Ti^{3+} ions. Radiation exposure has no significant influence on sample morphology or peak pattern, as revealed by the SEM and XRD analyses, respectively. During the methylene-blue dye removal, the UV-treated sample showed 99% capacity, while the untreated sample attained 77% capacity with the same operating conditions under natural-sunlight irradiation. The simplicity, scalability, and versatility of the proposed UV-radiation method of oxygen-vacancy creation can enhance and promote the photocatalytic activity of TiO$_2$ semiconductor films for various desirable photocatalytic applications under solar-light irradiation.

Funding: The authors extend their appreciation to the Deputyship Research and Innovation, Ministry of Education, in Saudi Arabia, for funding this research work through project number IF-PSAU-2021/01/18731.

Institutional Review Board Statement: Not applicable.

Informed Consent Statement: Not applicable.

Data Availability Statement: The data presented in this study are available on request from the corresponding author.

Acknowledgments: The authors would like to acknowledge the support provided by the Deanship of Scientific Research (DSR) at Prince Sattam Bin Abdulaziz University (PSAU)for funding this research work through project number IF-PSAU-2021/01/18731.

Conflicts of Interest: The authors declare no conflict of interest.

References

1. Zhou, M.; Wang, Y.; Zhang, Y.; Sun, L.; Hao, W.; Cao, E.; Yang, Z. Effects of oxygen vacancy on the magnetic properties of Mn(II)-doped anatase TiO_2. *Chem. Phys. Lett.* **2020**, *754*, 137738. [CrossRef]
2. Yuan, F.; Yang, R.; Li, C.; Tan, Y.; Zhang, X.; Zheng, S.; Sun, Z. Enhanced visible-light degradation performance toward gaseous formaldehyde using oxygen vacancy-rich TiO_2-x/TiO_2 supported by natural diatomite. *Build. Environ.* **2022**, *219*, 09216. [CrossRef]
3. Li, J.; Li, X.; Wu, G.; Guo, J.; Yin, X.; Mu, M. Construction of 2D Co-TCPP MOF decorated on B-TiO_2-X nanosheets: Oxygen vacancy and 2D-2D heterojunctions for enhancing visible light-driven photocatalytic degradation of bisphenol A. *J. Environ. Chem. Eng.* **2021**, *9*, 106723. [CrossRef]
4. Zhang, M.; Wang, C.; Li, H.; Wang, J.; Li, M.; Chen, X. Enhanced performance of lithium ion batteries from self-doped TiO_2 nanotube anodes via an adjustable electrochemical process. *Electrochim. Acta* **2019**, *326*, 134972. [CrossRef]
5. Patel, C.; Singh, R.; Dubey, M.; Pandey, S.K.; Upadhyay, S.N. Large, and Uniform Single Crystals of MoS_2 Monolayers for ppb-Level NO_2 Sensing. *Appl. Nano Mater.* **2022**, *5*, 9415–9426. [CrossRef]
6. Patel, C.; Mandal, B.; Jadhav, R.G.; Ghosh, T.; Dubey, M.; Das, A.K.; Htay, M.T.; Atuchin, V.V.; Mukherjee, S, S, N Co-Doped Carbon Dot-Functionalized WO_3 Nanostructures for NO_2 and H_2S Detection. *Appl. Nano Mater.* **2022**, *5*, 2492–2500. [CrossRef]
7. Hu, X.; Li, C.; Song, J.; Zheng, S.; Sun, Z. Multidimensional assembly of oxygen vacancy-rich amorphous TiO_2-BiOBr-sepiolite composite for rapid elimination of formaldehyde and oxytetracycline under visible light. *J. Colloid Interface Sci.* **2020**, *574*, 61–73. [CrossRef] [PubMed]
8. Wang, Y.; Gao, P.; Li, B.; Yin, Z.; Feng, L.; Liu, Y.; Du, Z.; Zhang, L. Enhanced photocatalytic performance of visible-light-driven $CuOx/TiO_2$-x for degradation of gaseous formaldehyde: Roles of oxygen vacancies and nano copper oxides. *Chemosphere* **2022**, *291*, 133007. [CrossRef]
9. Zhang, L.; Bai, X.; Zhao, G.; Shen, X.; Liu, Y.; Bao, X.; Luo, J.; Yu, L.; Zhang, N. A visible light illumination assistant $Li-O_2$ battery based on an oxygen vacancy doped TiO_2 catalyst. *Electrochim. Acta* **2022**, *405*, 139794. [CrossRef]
10. Li, Y.; Wei, X. Deep in-gap states induced by double-oxygen-vacancy clusters in hydrogenated TiO_2. *Phys. B Condens. Matter* **2021**, *600*, 412631. [CrossRef]
11. Miao, Z.; Wang, G.; Zhang, X.; Dong, X. Oxygen vacancies modified TiO_2/Ti_3C_2 derived from MXenes for enhanced photocatalytic degradation of organic pollutants: The crucial role of oxygen vacancy to schottky junction. *Appl. Surf. Sci.* **2020**, *528*, 146929. [CrossRef]
12. Koley, S. Engineering Si doping in anatase and rutile TiO_2 with oxygen vacancy for efficient optical application. *Phys. B Condens. Matter* **2021**, *602*, 412502. [CrossRef]
13. Wang, X.; Lian, M.; Yang, X.; Lu, P.; Zhou, J.; Gao, J.; Liu, C.; Liu, W.; Miao, L. Enhanced activity for catalytic combustion of ethylene by the Pt nanoparticles confined in TiO_2 nanotube with surface oxygen vacancy. *Ceram. Int.* **2022**, *48*, 3933–3940. [CrossRef]
14. Huang, Y.; Li, K.; Zhou, J.; Guan, J.; Zhu, F.; Wang, K.; Liu, M.; Chen, W.; Li, N. Nitrogen-stabilized oxygen vacancies in TiO_2 for site-selective loading of Pt and CoOx cocatalysts toward enhanced photoreduction of CO_2 to CH_4. *Chem. Eng. J.* **2022**, *439*, 135744. [CrossRef]
15. Zhang, H.; Wang, Y.; Zhong, W.; Long, B.; Wu, Y.; Xie, Z. Tailoring the room temperature ferromagnetism in TiO_2 nanoparticles by modulating the concentration of surface oxygen vacancy via La incorporating. *Ceram. Int.* **2019**, *45*, 12949–12956. [CrossRef]
16. Chen, P.; Zheng, H.; Jiang, H.; Liu, J.; Tu, X.; Zhang, W.; Phillips, B.; Fang, L.; Zou, J.-P. Oxygen-vacancy-rich phenanthroline/TiO_2 nanocomposites: An integrated adsorption, detection and photocatalytic material for complex pollutants remediation. *Chin. Chem. Lett.* **2022**, *33*, 907–911. [CrossRef]
17. Gu, Z.; Cui, Z.; Wang, Z.; Chen, T.; Sun, P.; Wen, D. Reductant-free synthesis of oxygen vacancies-mediated TiO_2 nanocrystals with enhanced photocatalytic NO removal performance: An experimental and DFT study. *Appl. Surf. Sci.* **2021**, *544*, 148923. [CrossRef]
18. Xu, M.; Xu, F.; Zhu, K.; Xu, X.; Deng, P.; Wu, W.; Ye, W.; Sun, Z.; Gao, P. Atomic layer deposition technique refining oxygen vacancies in TiO_2 passivation layer for photoelectrochemical ammonia synthesis. *Compos. Commun.* **2022**, *29*, 101037. [CrossRef]
19. Wang, L.; Cai, Y.; Liu, B.; Dong, L. A facile synthesis of brown anatase TiO_2 rich in oxygen vacancies and its visible light photocatalytic property. *Solid State Ion.* **2021**, *361*, 115564. [CrossRef]
20. Zhang, W.; Zhang, W.; Xue, J.; Shen, Q.; Jia, S.; Gao, J.; Liu, X.; Jia, H. Black single-crystal TiO_2 nanosheet array films with oxygen vacancy on {001} facets for boosting photocatalytic CO_2 reduction. *J. Alloys Compd.* **2021**, *870*, 159400. [CrossRef]
21. Gu, Z.; Cui, Z.; Wang, Z.; Qin, K.S.; Asakura, Y.; Hasegawa, T.; Hongo, K.; Maezono, R.; Yin, S. Intrinsic carbon-doping induced synthesis of oxygen vacancies-mediated TiO_2 nanocrystals: Enhanced photocatalytic NO removal performance and mechanism. *J. Catal.* **2021**, *393*, 179–189. [CrossRef]

22. Ma, X.; Wu, J.; Xu, L.; Zhao, B.; Chen, F. Modulation of Pt species on oxygen vacancies enriched TiO_2 via UV illumination for photocatalytic performance optimization. *Colloids Surf. A Physicochem. Eng. Asp.* **2019**, *586*, 124243. [CrossRef]
23. Zhang, Q.; Chen, D.; Song, Q.; Zhou, C.; Li, D.; Tian, D.; Jiang, D. Holey defected TiO_2 nanosheets with oxygen vacancies for efficient photocatalytic hydrogen production from water splitting. *Surf. Interfaces* **2021**, *23*, 100979. [CrossRef]
24. Chu, B.; Ou, X.; Wei, L.; Liu, H.; Chen, K.; Qin, Q.; Meng, L.; Fan, M.; Li, B.; Dong, L. Insight into the effect of oxygen vacancies and OH groups on anatase TiO_2 for CO oxidation: A combined FT-IR and density functional theory study. *Mol. Catal.* **2021**, *511*, 111755. [CrossRef]
25. Zhang, C.; Wang, L.; Etim, U.J.; Song, Y.; Gazit, O.M.; Zhong, Z. Oxygen Vacancies in Cu/TiO_2 Boost Strong Metal-Support Interaction and CO_2 Hydrogenation to Methanol. *J. Catal.* **2022**, *413*, 284–296. [CrossRef]
26. Li, S.; Liu, C.; Chen, P.; Lv, W.; Liu, G. In-situ stabilizing surface oxygen vacancies of TiO_2 nanowire array photoelectrode by N-doped carbon dots for enhanced photoelectrocatalytic activities under visible light. *J. Catal.* **2020**, *382*, 212–227. [CrossRef]
27. Sedaghati, N.; Habibi-Yangjeh, A.; Asadzadeh-Khaneghah, S.; Ghosh, S. Photocatalytic performance of oxygen vacancy rich-TiO_2 combined with $Bi_4O_5Br_2$ nanoparticles on degradation of several water pollutants. *Adv. Powder Technol.* **2021**, *32*, 304–316. [CrossRef]
28. Drozd, V.S.; Zybina, N.A.; Abramova, K.E.; Parfenov, M.Y.; Kumar, U.; Valdés, H.; Smirniotis, P.G.; Vorontsov, A.V. Oxygen vacancies in nano-sized TiO_2 anatase nanoparticles. *Solid State Ion.* **2019**, *339*, 115009. [CrossRef]
29. Han, G.; Kim, J.Y.; Kim, K.J.; Lee, H.; Kim, Y.M. Controlling surface oxygen vacancies in Fe-doped TiO_2 anatase nanoparticles for superior photocatalytic activities. *Appl. Surf. Sci.* **2020**, *507*, 144916. [CrossRef]
30. Park, S.; Baek, S.; Kim, D.W.; Lee, S. Oxygen-vacancy-modified brookite TiO_2 nanorods as visible-light-responsive photocatalysts. *Mater. Lett.* **2018**, *232*, 146–149. [CrossRef]
31. Qi, F.; An, W.; Wang, H.; Hu, J.; Guo, H.; Liu, L.; Cui, W. Combing oxygen vacancies on TiO_2 nanorod arrays with g-C3N4 nanosheets for enhancing photoelectrochemical degradation of phenol. *Mater. Sci. Semicond. Process.* **2020**, *109*, 104954. [CrossRef]
32. Ji, Z.; Wu, J.; Jia, T.; Peng, C.; Xiao, Y.; Liu, Z.; Liu, Q.; Fan, Y.; Han, J.; Hao, L. In-situ growth of TiO_2 phase junction nanorods with Ti^{3+} and oxygen vacancies to enhance photocatalytic activity. *Mater. Res. Bull.* **2021**, *140*, 111291. [CrossRef]
33. Dagdeviren, O.E.; Glass, D.; Sapienza, R.; Cortés, E.; Maier, S.A.; Parkin, I.P.; Grütter, P.; Quesada-Cabrera, R. The Effect of Photoinduced Surface Oxygen Vacancies on the Charge Carrier Dynamics in TiO_2 Films. *Nano Lett.* **2021**, *21*, 8348–8354. [CrossRef] [PubMed]
34. Wang, X.; Zhang, L.; Bu, Y.; Sun, W. Interplay between invasive single atom Pt and native oxygen vacancy in anatase TiO_2 (1 0 1) surface: A theoretical study. *Appl. Surf. Sci.* **2021**, *540*, 148357. [CrossRef]
35. Owolabi, T.O.; Qahtan, T.F.; Abidemi, O.R.; Saleh, T.A. Bismuth oxychloride photocatalytic wide band gap adjustment through oxygen vacancy regulation using a hybrid intelligent computational method. *Mater. Chem. Phys.* **2022**, *290*, 126524. [CrossRef]
36. Amani, T.A.S.; Alansi, M.; Qahtan, T.F.; Al Abass, N.; AlGhamdi, J.M.; Al-Qunaibit, M. Fast Scalable Synthetic Methodology to Prepare Nanoflower-Shaped Bi/BiOCl Br1-x Heterojunction for Efficient Immobilized Photocatalytic Reactors under Visible Light Irradiation. *Adv. Sustain. Syst.* **2021**, *6*, 2100267.
37. Alansi, A.M.; Qahtan, T.F.; Al Abass, N.; Al-qunaibit, M.; Saleh, T.A. In-situ sunlight-driven tuning of photo-induced electron-hole generation and separation rates in bismuth oxychlorobromide for highly efficient water decontamination under visible light irradiation. *J. Colloid Interface Sci.* **2022**, *614*, 58–65. [CrossRef]
38. Wang, X.; Yao, Y.; Gao, W.; Zhan, Z. High-rate and high conductivity mesoporous TiO_2 nano hollow spheres: Synergetic effect of structure and oxygen vacancies. *Ceram. Int.* **2021**, *47*, 13572–13581. [CrossRef]
39. Wang, L.; Wang, S.; Li, M.; Yang, X.; Li, F.; Xu, L.; Zou, Y. Constructing oxygen vacancies and linker defects in MIL-125 @TiO_2 for efficient photocatalytic nitrogen fixation. *J. Alloys Compd.* **2022**, *909*, 164751. [CrossRef]
40. Huang, H.; Yan, H.; Duan, M.; Ji, J.; Liu, X.; Jiang, H.; Liu, B.; Sajid, S.; Cui, P.; Li, Y.; et al. TiO_2 surface oxygen vacancy passivation towards mitigated interfacial lattice distortion and efficient perovskite solar cell. *Appl. Surf. Sci.* **2021**, *544*, 148583. [CrossRef]
41. Raj, C.C.; Srimurugan, V.; Flamina, A.; Prasanth, R. Tuning the carrier density of TiO_2 nanotube arrays by controlling the oxygen vacancies for improved areal capacitance in supercapacitor applications. *Mater. Chem. Phys.* **2020**, *248*, 12292. [CrossRef]
42. He, M.; Cao, Y.; Ji, J.; Li, K.; Huang, H. Superior catalytic performance of Pd-loaded oxygen-vacancy-rich TiO_2 for formaldehyde oxidation at room temperature. *J. Catal.* **2021**, *396*, 122–135. [CrossRef]
43. Atuchin, V.V.; Kesler, V.G.; Pervukhina, N.V.; Zhang, Z. Ti 2p and O 1s core levels and chemical bonding in titanium-bearing oxides. *J. Electron Spectros. Relat. Phenom.* **2006**, *152*, 18–24. [CrossRef]

Disclaimer/Publisher's Note: The statements, opinions and data contained in all publications are solely those of the individual author(s) and contributor(s) and not of MDPI and/or the editor(s). MDPI and/or the editor(s) disclaim responsibility for any injury to people or property resulting from any ideas, methods, instructions or products referred to in the content.

Communication

B-Doped g-C₃N₄/Black TiO₂ Z-Scheme Nanocomposites for Enhanced Visible-Light-Driven Photocatalytic Performance

Yuwei Wang *, Kelin Xu, Liquan Fan *, Yongwang Jiang, Ying Yue and Hongge Jia

Heilongjiang Provincial Key Laboratory of Polymeric Composite Materials, College of Materials Science and Engineering, Qiqihar University, Qiqihar 161006, China
* Correspondence: ywwang@qqhru.edu.cn (Y.W.); 02275@qqhru.edu.cn (L.F.)

Abstract: Black TiO₂ with abundant oxygen vacancies (OVs)/B-doped graphitic carbon nitride (g-C₃N₄) Z-scheme heterojunction nanocomposites are successfully prepared by the one-pot strategy. The OVs can improve not only photogenerated carrier separation, but also the sorption and activation of antibiotic compounds (tetracycline hydrochloride, TC). The prepared heterojunction photocatalysts with a narrow bandgap of ~2.13 eV exhibit excellent photocatalytic activity for the degradation of tetracycline hydrochloride (65%) under visible light irradiation within 30 min, which is several times higher than that of the pristine one. The outstanding photocatalytic property can be ascribed to abundant OVs and B element-dope reducing the bandgap and extending the photo-response to the visible light region, the Z-scheme formation of heterojunctions preventing the recombination of photogenerated electrons and holes, and promoting their effective separation.

Keywords: photocatalysis; black TiO₂; B-doped g-C₃N₄; Z-scheme heterojunction; oxygen vacancy defect

1. Introduction

In recent years, the main aspects of environmental problems have been the energy crisis and pollution, with water pollution receiving special attention [1–3]. Water pollution is principally caused by heavy-metal ion contaminants and organic pollutants such as hormones, dyes, aromatics, pesticides, and perfluorinated organic compounds (PFOCs). Organic pollutants in wastewater, for example, have high toxicity, carcinogenicity, and refractory degradation, posing a significant threat to human health. As a consequence, it is critical to develop efficient technologies for breaking down organic pollutants from water [4]. Photocatalysis, one of the advanced oxidation methods for producing highly oxidizing free radicals, has been identified as a sustainable and ecologically friendly method for the degradation of pollutants. Photocatalytic oxidation has been acknowledged as a significant and successful candidate for eliminating poisonous and harmful contaminants in aqueous environments [5–9].

Numerous photocatalytic materials with superior band structures, visible light adsorption, charge separation, and transport have been created to date. Due to its huge band gap, the original TiO₂—being a mature semiconductor—can only be used to purify wastewater using ultraviolet light, regardless of the fact that TiO₂ has much lower biotoxicity than the majority of semiconductors [6,10–12]. Fortunately, Chen et al. discovered black TiO₂ nanomaterials through a surface hydrogenation strategy, which narrowed the bandgap and extended photo-absorption from ultraviolet to visible light and/or near-infrared [13–16]. The outstanding solar-driven photocatalytic performance represented a breakthrough for wide-spectrum response TiO₂ materials. In recent years, graphitic carbon nitride (g-C₃N₄), with a narrow band gap, excellent stability, and fast charge transfer, has been considered a potential visible light photocatalyst since the groundbreaking work reported by Wang et al. in 2009 [17–20]. However, the quick electron-hole recombination, low quantum efficiency, insufficient specific surface area, and other issues continue to restrict the photocatalytic activity of g-C$_3$N$_4$ [21]. Recently, many groups have documented the use of P- or S-doped g-C$_3$N$_4$ to enhance photocatalytic activity [22–24]. Wang et al. also discovered that even a small amount of boron doping could significantly increase photocatalytic activity [25].

Here, we proposed a one-pot synthesis of a B-doped g-C$_3$N$_4$/black-TiO$_2$ (BCBT) heterojunction nanocomposite photocatalyst using NaBH$_4$ as a solid reducing agent. This catalyst showed significantly higher photocatalytic degradation activity of high-toxic tetracycline hydrochloride (TC) when exposed to both visible light and simulated sunlight. The superiority of this Z-scheme BCBT heterojunction structure is demonstrated by the remarkable photocatalytic activity. More importantly, the photocatalytic degradation mechanism of the heterojunction is further revealed, which provides guidance for the design of a photocatalyst.

2. Experimental

2.1. Materials

Melamine, potassium borohydride (KBH$_4$), and tetracycline hydrochloride (TC) were purchased from Chinese Medicine Group Chemical Reagent Co., Ltd. (Shanghai, China). Degussa P25 (P25, with 85% anatase and 15% rutile) was purchased from Sigma Aldrich (St. Louis, MO, USA). All chemicals were of analytically pure grade and used without further purification.

2.2. Fabrication of Black TiO$_2$/B-Doped g-C$_3$N$_4$ Heterojunction

Further, 2.5 g Melamine, 2.5 g P25, and 1.0 g KBH$_4$ were ground thoroughly for 15 min. Then, the mixture was calcined in a N$_2$ flow at 520 °C for 3 h under normal pressure conditions with a constant heating rate of 5 °C min^{-1}. The obtained composite was washed with deionized water and ethanol three times, and then dried in an oven at 80 °C overnight. Then, the resulting yellow product of BCBT was collected and ground into powder for further use, as detailed in Scheme 1.

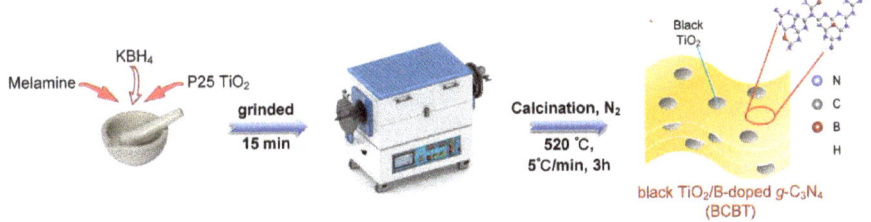

Scheme 1. Schematic illustration for synthesis of the black TiO$_2$/B-doped g-C$_3$N$_4$ (BCBT).

2.3. Characterizations

The structure and phase of materials were determined using a SmartLAb SE X-ray diffractometer (XRD, Rigaku, Tokyo, Japan) with Cu-Kα radiation source at an operating voltage of 40 kV and an operating current of 180 mA. ESCALAB Xi+ X-ray photoelectron spectrometer (XPS, Thermo Fisher, MA, USA) with Al-Kα radiation as the excitation source was used to examine the elements on the surface of the samples. On a Regulus 8220 scanning electron microscope (SEM, Hitachi, Tokyo, Japan) and a JEM-2100 transmission electron microscope (TEM, JEOL, Tokyo, Japan), the microscopic morphologies of the samples were examined. The UV-vis diffuse reflectance spectrum (DRS) was recorded in the range of 200~800 nm using a UV 2600 UV-vis spectrophotometer (Shimadzu, Tokyo, Japan) using BaSO$_4$ as a reference standard. The photoluminescence (PL) spectra of samples were measured using an LS 55 fluorescence spectrometer (Perkin Elmer, MA, USA) with an excitation wavelength of 350 nm.

2.4. Photocatalytic Degradation of Organic Pollutants

Using a Xenon arc lamp (PLS-SXE300+, PerfectLight, Beijing, China) with a cut-off filter ($\lambda > 420$ nm) and tetracycline hydrochloride (TC) as a contaminant, the photocatalytic degradation characteristics of BCBT were investigated. Then, 20 mg of the photocatalyst was added to 100 mL of TC solution with an initial concentration of 10 mg L^{-1}. The solution was stirred for 20 min in the dark. The solution (5 mL) was filtered every 20 min. TC residuals were detected using a UV spectrophotometer. Pure distilled water was served as a reference sample.

2.5. Photoelectrochemical Properties

The photocurrent test, electrochemical impedance spectroscopy (EIS), and the Mott–Schottky plots of the samples were performed on a CHI-660E electrochemical workstation (Chenhua, Shanghai, China). To initiate the photoelectrochemical tests, a Xenon arc lamp (300 W, Beijing Aulight) with a cut-off filter ($\lambda > 420$ nm) was used as the light source. We started by dissolving 20 mg of material in ethanol. With an art airbrush, the dispersion was then uniformly sprayed on an FTO glass. Finally, the BCBT-coated FTO glass was calcined at 350 °C for 2 h in a N_2 environment. The three-electrode electrochemical station included an aqueous Na_2SO_4 solution as the electrolyte, a platinum plate as the counter electrode, FTO glass as the photoanode, and Ag/AgCl as the reference electrode. To de-aerate the solution, the electrolyte was purged with N_2 gas before use.

3. Results and Discussion

As shown in Figure 1a, the strong (200) peak at 32.2° and the (100) peak at 15.9° for the BCN (B-doped C_3N_4), respectively, belonged to the inter-layer and in-plane crystal facets of g-C_3N_4 (ICDD 01-078-1691). The diffraction patterns of the as-prepared b-TiO_2 were well matched with that of the anatase TiO_2 (ICDD 00-004-0477), showing that there was no impurity phase introduced after reduction by $NaBH_4$. All of the diffraction peaks for the BCBT were identical to those for g-C_3N_4 and b-TiO_2. These results indicate that there are no additional impurity peaks, proving that the B-doped g-C_3N_4/black-TiO_2 (BCBT) composite samples were successfully synthesized via the one-pot process. The FT-IR spectra of the BT (black-TiO_2), BCN, and BCBT further validated the existence of BCN and BT (Figure 1b). The typical peaks of g-C_3N_4 can be observed in BCBT at about 3200 cm^{-1} (C-H) and 1250–1650 cm^{-1} (C-N), which are consistent with the BCN [26]. Additionally, the characteristic peak of Ti-O is found at 500–1000 cm^{-1} [14]. All these findings demonstrate the presence of BCN and BT in the BCBT.

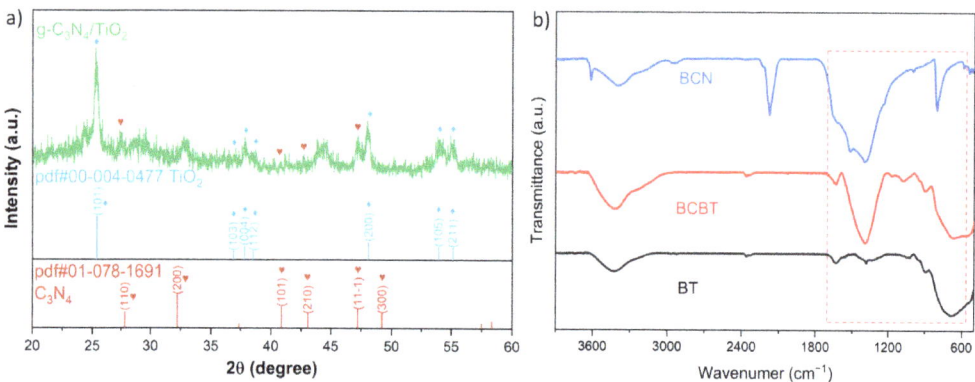

Figure 1. The XRD patterns of BCBT photocatalysts (TiO_2 in blue, C_3N_4 in red) (**a**) and FT-IR spectra of all prepared photocatalysts (**b**).

SEM and TEM were used to characterize the morphology of the obtained samples. In Figure 2a, the thin-layered BCN is associated with the b-TiO$_2$ microspheres, which also demonstrates that b-TiO$_2$ exhibits microspheres with sizes of about 50 nm. Figure 2a shows that the thin-layered BCN is deposited on the surface of b-TiO$_2$ among the BCBT. EDS analysis confirms the existence of Ti, O, B, and C (Figure 2b–e). The TEM image of the BCBT in Figure 2f, which depicts the BCN nanoflakes loaded onto the surfaces of the b-TiO$_2$ nanoparticles, further demonstrates this point and is in line with the findings of the aforementioned SEM studies. These findings demonstrate that b-TiO$_2$ was successfully attached to the BCN surfaces.

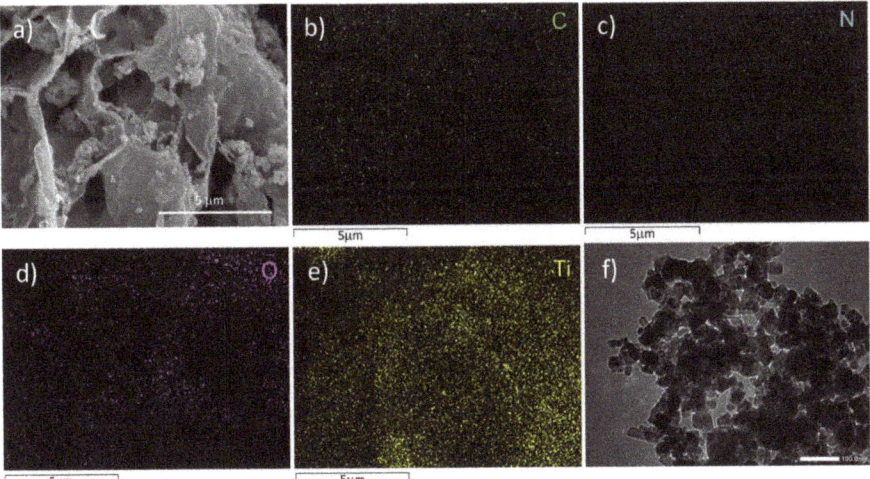

Figure 2. SEM (**a**) and TEM (**f**) images of BCBT; (**b**–**e**) EDX elemental mappings of C, N, O, and Ti.

We examined the change of surface chemical bonding of BCBT induced by NaBH$_4$ treatment with XPS. The XPS survey spectrum reveals the presence of Ti, B, C, N, and O elements (Figure 3a). A tiny change for Ti 2p can be observed in Figure 3b. This may imply that oxygen-bound electrons bound to titanium and oxygen ions turn in oxygen vacancies, which serve as electron traps [27]. Figure 3c displays the high-resolution B 1s peaks of BCN and BCBT. The BCN peak at 191.7 eV represents the typical B-N bond [28,29]. The BCBT peak at 190.4 eV, with a lower binding energy than BCN, shows that some boron atoms are less electropositive than BCN. These demonstrate the efficient charge transfer in the BCBT between b-TiO$_2$ and BCN [30]. A peak near 532.5 eV in the O 1s (Figure 3d) can be ascribed to adsorbed water, which is consistent with a robust interaction between O vacancy sites and water vapor. This peak area clearly grew during the NaBH$_4$ reduction process, which is consistent with the electron transfer to the nearby oxygen vacancies, as shown in the Ti 2p spectrum [31–34].

The light absorption ability is one of the crucial factors in determining photocatalytic performance. The light absorption properties of the as-prepared samples were characterized by the UV-vis diffuse reflectance spectra (UV-vis DRS). The absorption edges of BT and BCN, as seen in Figure 4a, are at wavelengths of around 400 and 460 nm, respectively, while the two photocatalysts all broaden the range of visible light absorption following NaBH$_4$ reduction. One-pot solid synthesis further enhances the light-harvesting abilities of BCBT, which is attributed to the effective charge transfer between the BCN nanoflakes and BT nanoparticles. An additional broad absorption peak with a wavelength of roughly 400~800 nm is observed in the BCBT hybridized photocatalyst. The O vacancies and doped B elements both promote the activation of BCBT's e$^-$-h$^+$ couples when exposed to visible light, increasing BCBT's sensitivity to light. Figure 4b displays the Kubelka–Munk conversion curves for BT, BCN, and BCBT. Band gaps for BT, BCN, and BCBT are estimated to be ~1.98 eV, 2.32 eV, and 2.13 eV, respectively. According to these results, the BCBT, which has a narrower intrinsic bandgap than that of the BCN, is more active in regions of visible light. As a result, it explains why the subsequent photocatalytic activity was improved. The substantial absorption in the visible light range of BCBT is caused by the existence of oxygen vacancies and doped B elements [33,35]. Combining the characterization findings, it can be concluded that the addition of O vacancy sites and doped B elements increases the catalyst's ability to absorb visible light, which is obviously conducive to the photocatalytic performance of defective BCBT.

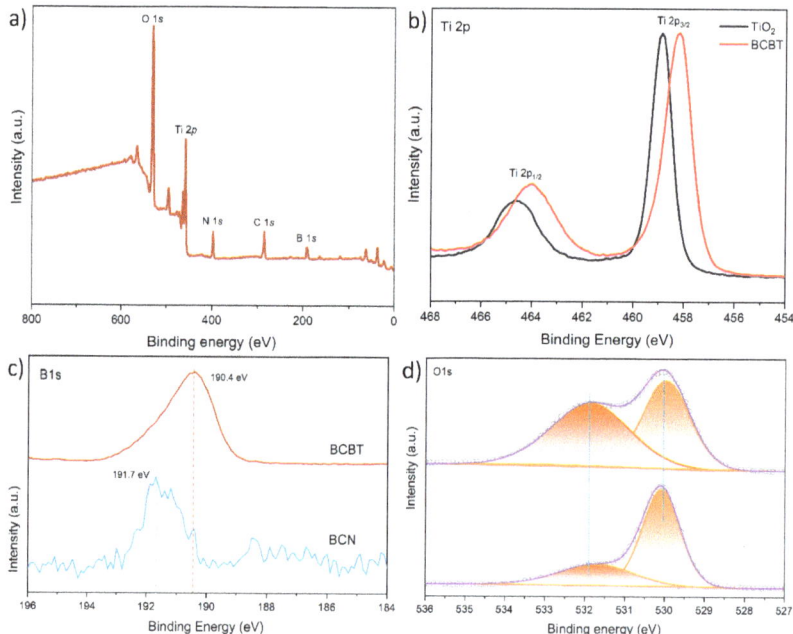

Figure 3. The XPS spectra of BT, BCB, and BCBT: (**a**) survey spectrum of BCT; (**b**) Ti 2p and (**d**) O 1 s of BT and BCBT; and (**c**) B 1 s spectra of BCN and BCBT.

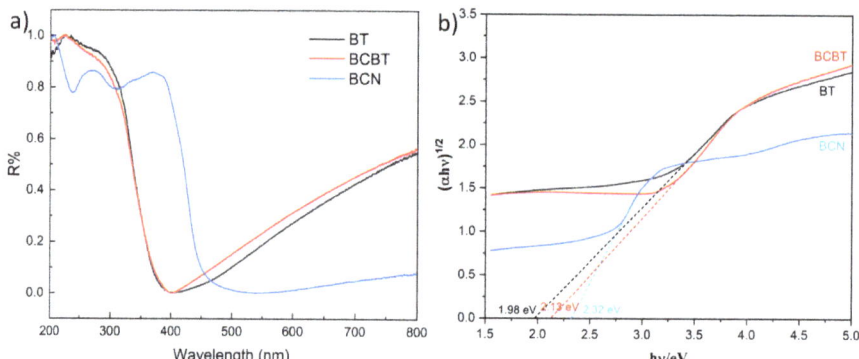

Figure 4. (**a**) UV-vis diffuse reflectance spectra of the as-prepared BT, BCN, and BCBT. (**b**) Relationship of $(\alpha h\nu)^{1/2}$ vs. E (ev).

The research results of photocatalytic activity of several samples for TC degradation are shown in Figure 5a. The reaction conditions are given in Section 2.2. Figure 5a displays the photocatalytic degradation rate of BCN at 27% after 30 min. The photocatalytic activity of BCBT was improved greatly, and degradation efficiency was up to 65% within 30 min. These may be due to the fact that the addition of OVs increases the light absorption range and creates a BCN/BT heterojunction that encourages photogenerated charge separation (PL and EIS spectra). By developing a kinetic model of the reaction, the kinetic behavior of the photocatalytic degradation reaction may be investigated further below.

$$-dC/dt = kC/(1 + kC) \qquad (1)$$

when $kC \ll 1$, it can be simplified to pseudo-first-order dynamics.

$$-dC/dt = kC \qquad (2)$$

where k stands for the pseudo-first-order kinetic constant. The kinetic constants for each molecule are shown in Figure 5b. The simplified pseudo-first-order kinetic formula of L-H demonstrates a remarkable linear relationship between the residual concentration of TC in various samples. The first-order kinetic constants for the fitted kinetic curves of BCN and BCBT in Figure 5b are ~0.0065 and 0.0271 min^{-1}, respectively. The kinetic constants of BCBT are 4.17 times higher than that of BCN, indicating that BCBT's photocatalytic activity greatly increased. Consequently, the BCBT photocatalyst has potential use in wastewater treatment due to its high efficiency, stability, and applicability of antibiotic photodegradation. Additionally, several photocatalysts for the photodegradation of TC published recently are presented in Table 1 and contrasted with the results in this work. The BCBT produced in this work showed superior photodegradation activity with a shorter reaction time when exposed to visible light irradiation when compared to other photocatalysts. This further demonstrates the capability of B-doped g-C_3N_4/black-TiO_2 heterojunction photocatalysts for the photocatalytic degradation of TC.

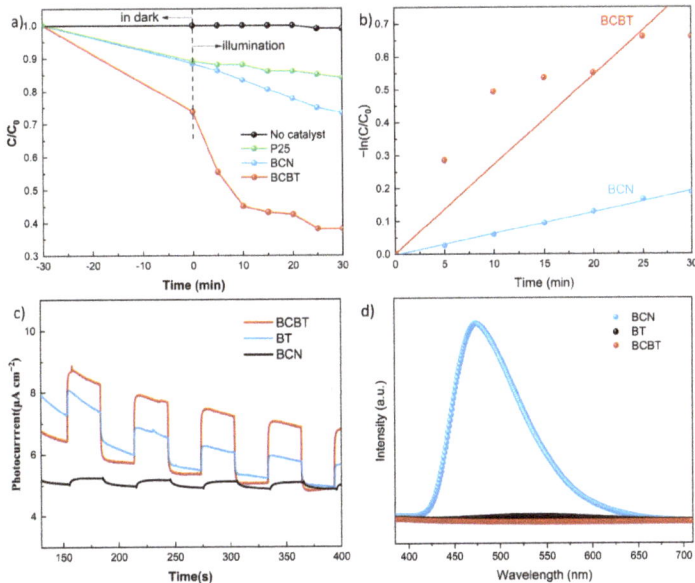

Figure 5. Photocatalytic degradation efficiencies on the degradation of TC under simulated solar light irradiation (**a**), and kinetic linear simulation curves (**b**), photocurrent curves (**c**), and PL spectra (**d**) of BT, BCN, and BCBT, respectively.

Table 1. The comparison of photocatalytic degradation activities of different photocatalysts for TC.

Photocatalyst	Light Source	Dosage of Catalyst (g L^{-1})	TC Concentration (mg L^{-1})	Reaction Time (min)	Rate (min^{-1})	Reference (year)
C nanodots/WO$_3$	150 W XL (λ > 420 nm)	0.5	20	150	0.0067	[36] (2017)
Ag/a-NiMoO$_4$ nanorods	150 W XL (λ > 400 nm)	1.429	20	180	0.0093	[37] (2019)
C-TiO$_2$ nanocomposites	visible-light	0.2	10	160	0.0126	[38] (2019)
BiOCl/TiO$_2$ C nanofibers	300 W XL (λ > 420 nm)	0.5	20	180	0.0085	[39] (2020)
ZnSnO$_3$/g-C$_3$N$_4$	300 W XL (λ > 420 nm)	0.25	10	120	0.0131	[40] (2020)
C-doped 0.5-UNST	300 W XL (λ > 420 nm)	0.5	20	120	0.0134	[41] (2021)
BCBT	300 W XL (λ > 420 nm)	0.2	10	30 (60%)	0.0271	This work

Transient photocurrent responses, which can be utilized to assess charge-transfer properties and photocatalyst stability, were studied using chronoamperometry. As shown in Figure 5c, BCBT has a higher photocurrent density and electron-hole separation efficiency than BT and BCN due to the presence of O vacancy and the heterojunction formation. Additionally, all composite photocurrent responses for both samples are continuous, demonstrating high stability. Figure 5d depicts the PL spectra of BT, BCN, and BCBT. The results reveal that BCBT has the lowest PL response when compared to the other samples, demonstrating that the photogenerated electron-hole pairs efficiently separate after one-pot solid reduction. Charge separation and transfer can be effectively enhanced by decreasing the recombination rate of the photogenerated carriers, directly boosting photocatalytic performance. Electrochemical impedance spectroscopy (EIS) was also used to analyze the migration of the charge carriers. Evaluating the kinetics at the interface requires a knowledge of the as-synthesis electron transfer resistance, which has been expressed as the diameter of the Nyquist circles. The BCBT sample has the median Nyquist circle diameter in contrast to the BT and BCN samples, as shown in Figure S1. Clearly, combining BCN with BT-rich O vacancies promotes the separation of photogenerated charge carriers.

Figure 6a shows the degradation and recovery rates after three cycles of using the BCBT photocatalyst. After three cycles, the 30 min photocatalytic degradation efficiency and recovery rate are still 58.2%, which implies that the sample has high stability, implying the potential applications in fields of environment.

The Mott–Schottky (MS) plots for BT, BCN, and BCBT demonstrated that they were typical n-type semiconductors with relatively positive slopes, as shown in Figure 6b. Calculated from x-intercepts of the linear region, the flat-band potentials of BT, BCN, and BCBT were shown to be −1.01 V, 0.66 V, and 0.38 V vs. SCE. As a result, BT and BCN had conduction band potentials (ECB) of −0.38 V and −0.03 V vs. NHE, respectively. Comparing the ECB of BCBT composite to those of BT and BCN, it appears that there was a significant positive movement. The conduction band potential was believed to have shifted positively as a result of the electrical interactions between BT and BCN, leading to a low conduction band position and a higher observable absorption power for the BCBT composite.

The proposed photocatalytic mechanism over the BCBT photocatalyst is shown in Figure 6c. Both the BCN and BT produced photoinduced carriers when exposed to visible light. The photoinduced electrons were then transported from the CB of the B-doped g-C$_3$N$_4$ to the VB of the black TiO$_2$ to create Z-scheme photocatalysts [42–44]. In addition, the photogenerated electrons produced by the BT reduced oxygen to form O_2^- (E_0 ($O_2/O_2^{·-}$) = −0.33 eV) [45]. The photogenerated holes produced by the VB of BCN were sufficiently positive to cause the oxidation of OH$^-$ to OH (E_0(OH$^-$/·OH = +1.89 eV) [46].

Then, the TC interacted with RSs (reactive species: O^{2-}, ·OH, and h^+) to promote the degradation process. In addition, the absorption of visible light increases when in situ black TiO_2 is combined with B-doped g-C_3N_4. The Ovs level and the introduction of B components considerably increase the BCBT photocatalysts' ability to absorb light, giving them exceptional photo-absorption properties.

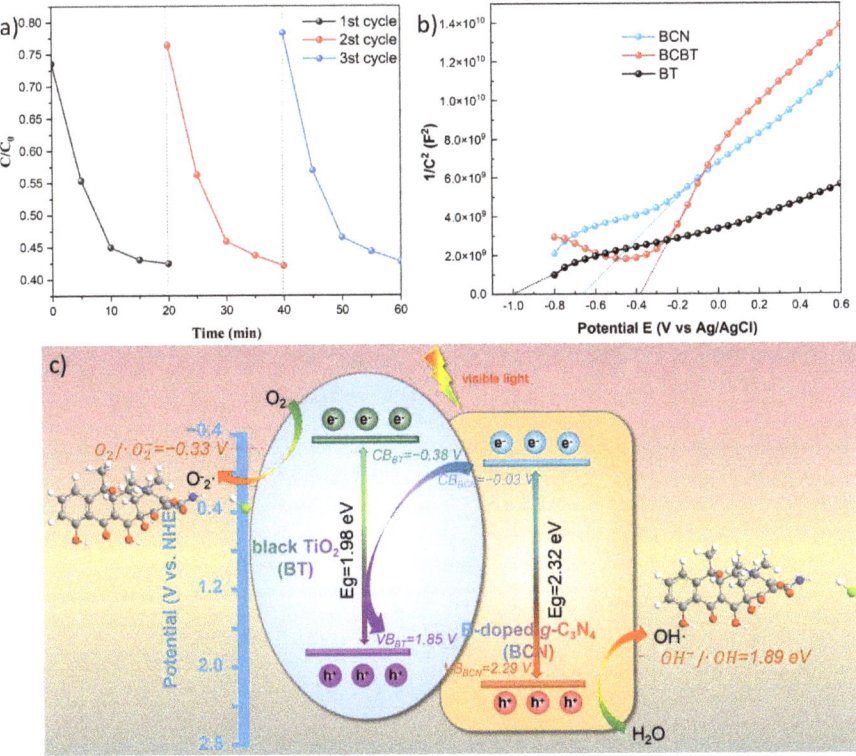

Figure 6. Recycling experiments (**a**), Mott–Schottky plots (**b**) of BCBT, and the presented mechanism (**c**) for photocatalytic TC elimination by BCBT photocatalyst.

4. Conclusions

In conclusion, the successful synthesis of the Z-scheme black TiO_2/B-doped g-C_3N_4 heterojunction photocatalyst and evaluation of the photocatalytic processes were accomplished. Black TiO_2/B-doped g-C_3N_4 had a higher photocatalytic activity than black TiO_2 and B-doped g-C_3N_4. The TC removal ratio for BCBT reached up to 65% within 30 min, which was much higher than that for pure BT and BCN. The abundance of OVs and B-doped elements in BCBT was largely responsible for its outstanding photocatalytic activity. These led to effective photogenerated carrier separation and enough visible light absorption, which improved BCBT's photocatalytic efficiency.

Supplementary Materials: The following supporting information can be downloaded at: https://www.mdpi.com/article/10.3390/nano13030518/s1, Figure S1: EIS plot of BT, BCN and BCBT, respectively.

Author Contributions: Methodology, Y.W., Y.J. and Y.Y.; software, K.X.; formal analysis, K.X.; investigation, K.X.; resources, L.F.; data curation, Y.W. and Y.J.; writing—original draft preparation, Y.W.; supervision, L.F. and H.J.; funding acquisition, Y.W. All authors have read and agreed to the published version of the manuscript.

Funding: The financial support from the Heilongjiang Provincial Natural Science Foundation of China (LH2020E127)", "Fundamental Research Funds in Heilongjiang Provincial Universities (135309347)", "Key research and development guidance projects in Heilongjiang Province (GZ20210034)", "Opening Foundation of Heilongjiang Provincial Key Laboratory of Polymeric Composition materials (CLKFKT2021B3)", "Undergraduate Training Programs for Innovation and Entrepreneurship of Qiqihar University (YJSCX2021039)" and Key Laboratory of Functional Inorganic Material Chemistry (Heilongjiang University), Ministry of Education.

Institutional Review Board Statement: Not applicable.

Informed Consent Statement: Not applicable.

Data Availability Statement: The raw/processed data required to reproduce these findings cannot be shared at this time due to technical or time limitations.

Conflicts of Interest: The authors declare no conflict of interest.

References

1. Liu, X.; Iocozzia, J.; Wang, Y.; Cui, X.; Chen, Y.; Zhao, S.; Li, Z.; Lin, Z. Noble Metal–Metal Oxide Nanohybrids with Tailored Nanostructures for Efficient Solar Energy Conversion, Photocatalysis and Environmental Remediation. *Energy Environ. Sci.* **2017**, *10*, 402–434. [CrossRef]
2. Meng, N.; Ren, J.; Liu, Y.; Huang, Y.; Petit, T.; Zhang, B. Engineering Oxygen-Containing and Amino Groups into Two-Dimensional Atomically-Thin Porous Polymeric Carbon Nitrogen for Enhanced Photocatalytic Hydrogen Production. *Energy Environ. Sci.* **2018**, *11*, 566–571. [CrossRef]
3. Rahman, M.Z.; Kwong, C.W.; Davey, K.; Qiao, S.Z. 2D Phosphorene as a Water Splitting Photocatalyst: Fundamentals to Applications. *Energy Environ. Sci.* **2016**, *9*, 709–728. [CrossRef]
4. Wu, C.; Xing, Z.; Yang, S.; Li, Z.; Zhou, W. Nanoreactors for Photocatalysis. *Coordin. Chem. Rev.* **2023**, *477*, 214939. [CrossRef]
5. Fang, B.; Xing, Z.; Sun, D.; Li, Z.; Zhou, W. Hollow Semiconductor Photocatalysts for Solar Energy Conversion. *Adv. Powder Mater.* **2022**, *1*, 100021. [CrossRef]
6. Fujishima, A.; Honda, K. Electrochemical Photolysis of Water at a Semiconductor Electrode. *Nature* **1972**, *238*, 37–38. [CrossRef]
7. Yu, C.; Zhou, W.; Liu, H.; Liu, Y.; Dionysiou, D.D. Design and Fabrication of Microsphere Photocatalysts for Environmental Purification and Energy Conversion. *Chem. Eng. J.* **2016**, *287*, 117–129. [CrossRef]
8. Pi, Y.; Li, X.; Xia, Q.; Wu, J.; Li, Y.; Xiao, J.; Li, Z. Adsorptive and Photocatalytic Removal of Persistent Organic Pollutants (POPs) in Water by Metal-Organic Frameworks (MOFs). *Chem. Eng. J.* **2018**, *337*, 351–371. [CrossRef]
9. Fang, B.; Xing, Z.; Kong, W.; Li, Z.; Zhou, W. Electron Spin Polarization-Mediated Charge Separation in Pd/CoP@CoNiP Superstructures toward Optimized Photocatalytic Performance. *Nano Energy* **2022**, *101*, 107616. [CrossRef]
10. Zhou, W.; Sun, F.; Pan, K.; Tian, G.; Jiang, B.; Ren, Z.; Tian, C.; Fu, H. Well-Ordered Large-Pore Mesoporous Anatase TiO_2 with Remarkably High Thermal Stability and Improved Crystallinity: Preparation, Characterization, and Photocatalytic Performance. *Adv. Funct. Mater.* **2011**, *21*, 1922–1930. [CrossRef]
11. Hosseini, S.M.; Ghiaci, M.; Kulinich, S.A.; Wunderlich, W.; Ghaziaskar, H.S.; Koupaei, A.J. Ethyl Benzene Oxidation under Aerobic Conditions Using Cobalt Oxide Imbedded in Nitrogen-Doped Carbon Fiber Felt Wrapped by Spiral TiO_2-SiO_2. *Appl. Catal. A-Gen.* **2022**, *630*, 118456. [CrossRef]
12. Hosseini, S.M.; Ghiaci, M.; Kulinich, S.A.; Wunderlich, W.; Monjezi, B.H.; Ghorbani, Y.; Ghaziaskar, H.S.; Javaheri Koupaei, A. Au-Pd Nanoparticles Enfolded in Coil-like TiO_2 Immobilized on Carbon Fibers Felt as Recyclable Nanocatalyst for Benzene Oxidation under Mild Conditions. *Appl. Surf. Sci.* **2020**, *506*, 144644. [CrossRef]
13. Chen, X.; Liu, L.; Yu, P.Y.; Mao, S.S. Increasing Solar Absorption for Photocatalysis with Black Hydrogenated Titanium Dioxide Nanocrystals. *Science* **2011**, *331*, 746–750. [CrossRef]
14. Chen, X.; Liu, L.; Huang, F. Black Titanium Dioxide (TiO_2) Nanomaterials. *Chem. Soc. Rev.* **2015**, *44*, 1861–1885. [CrossRef]
15. Li, Z.; Li, H.; Wang, S.; Yang, F.; Zhou, W. Mesoporous Black TiO_2/MoS_2/Cu_2S Hierarchical Tandem Heterojunctions toward Optimized Photothermal-Photocatalytic Fuel Production. *Chem. Eng. J.* **2022**, *427*, 131830. [CrossRef]
16. Li, Z.; Wang, S.; Wu, J.; Zhou, W. Recent Progress in Defective TiO_2 Photocatalysts for Energy and Environmental Applications. *Renew. Sustain. Energy Rev.* **2022**, *156*, 111980. [CrossRef]
17. Wang, X.; Maeda, K.; Thomas, A.; Takanabe, K.; Xin, G.; Carlsson, J.M.; Domen, K.; Antonietti, M. A Metal-Free Polymeric Photocatalyst for Hydrogen Production from Water under Visible Light. *Nat. Mater.* **2009**, *8*, 76–80. [CrossRef]

18. Ong, W.-J.; Tan, L.-L.; Ng, Y.H.; Yong, S.-T.; Chai, S.-P. Graphitic Carbon Nitride (g-C$_3$N$_4$)-Based Photocatalysts for Artificial Photosynthesis and Environmental Remediation: Are We a Step Closer To Achieving Sustainability? *Chem. Rev.* **2016**, *116*, 7159–7329. [CrossRef]
19. Fu, J.; Yu, J.; Jiang, C.; Cheng, B. G-C$_3$N$_4$-Based Heterostructured Photocatalysts. *Adv. Energy Mater.* **2018**, *8*, 1701503. [CrossRef]
20. Mamba, G.; Mishra, A.K. Graphitic Carbon Nitride (g-C$_3$N$_4$) Nanocomposites: A New and Exciting Generation of Visible Light Driven Photocatalysts for Environmental Pollution Remediation. *Appl. Catal. B-Environ.* **2016**, *198*, 347–377. [CrossRef]
21. Hosseini, S.M.; Ghiaci, M.; Kulinich, S.A.; Wunderlich, W.; Farrokhpour, H.; Saraji, M.; Shahvar, A. Au-Pd@g-C$_3$N$_4$ as an Efficient Photocatalyst for Visible-Light Oxidation of Benzene to Phenol: Experimental and Mechanistic Study. *J. Phys. Chem. C* **2018**, *122*, 27477–27485. [CrossRef]
22. Xiang, Q.; Yu, J.; Jaroniec, M. Preparation and Enhanced Visible-Light Photocatalytic H$_2$-Production Activity of Graphene/C$_3$N$_4$ Composites. *J. Phys. Chem. C* **2011**, *115*, 7355–7363. [CrossRef]
23. Zhang, Y.; Mori, T.; Ye, J.; Antonietti, M. Phosphorus-Doped Carbon Nitride Solid: Enhanced Electrical Conductivity and Photocurrent Generation. *J. Am. Chem. Soc.* **2010**, *132*, 6294. [CrossRef]
24. Liu, G.; Niu, P.; Sun, C.; Smith, S.C.; Chen, Z.; Lu, G.Q.; Cheng, H.-M. Unique Electronic Structure Induced High Photoreactivity of Sulfur-Doped Graphitic C$_3$N$_4$. *J. Am. Chem. Soc.* **2010**, *132*, 11642–11648. [CrossRef] [PubMed]
25. Zhang, M.; Yang, L.; Wang, Y.; Li, L.; Chen, S. High Yield Synthesis of Homogeneous Boron Doping C$_3$N$_4$ Nanocrystals with Enhanced Photocatalytic Property. *Appl. Surf. Sci.* **2019**, *489*, 631–638. [CrossRef]
26. Zhao, S.; Liu, J.; Li, C.; Ji, W.; Yang, M.; Huang, H.; Liu, Y.; Kang, Z. Tunable Ternary (N, P, B)-Doped Porous Nanocarbons and Their Catalytic Properties for Oxygen Reduction Reaction. *ACS Appl. Mater. Interfaces* **2014**, *6*, 22297–22304. [CrossRef] [PubMed]
27. Kang, Q.; Cao, J.; Zhang, Y.; Liu, L.; Xu, H.; Ye, J. Reduced TiO$_2$ Nanotube Arrays for Photoelectrochemical Water Splitting. *J. Mater. Chem. A* **2013**, *1*, 5766. [CrossRef]
28. Kawaguchi, M.; Kawashima, T.; Nakajima, T. Syntheses and Structures of New Graphite-like Materials of Composition BCN(H) and BC$_3$N(H). *Chem. Mater.* **1996**, *8*, 1197–1201. [CrossRef]
29. Song, L.; Ci, L.; Lu, H.; Sorokin, P.B.; Jin, C.; Ni, J.; Kvashnin, A.G.; Kvashnin, D.G.; Lou, J.; Yakobson, B.I.; et al. Large Scale Growth and Characterization of Atomic Hexagonal Boron Nitride Layers. *Nano Lett.* **2010**, *10*, 3209–3215. [CrossRef]
30. Zhao, D.; Wang, Y.; Dong, C.-L.; Huang, Y.-C.; Chen, J.; Xue, F.; Shen, S.; Guo, L. Boron-Doped Nitrogen-Deficient Carbon Nitride-Based Z-Scheme Heterostructures for Photocatalytic Overall Water Splitting. *Nat. Energy* **2021**, *6*, 388–397. [CrossRef]
31. Yu, H.; Chen, F.; Li, X.; Huang, H.; Zhang, Q.; Su, S.; Wang, K.; Mao, E.; Mei, B.; Mul, G.; et al. Synergy of Ferroelectric Polarization and Oxygen Vacancy to Promote CO$_2$ Photoreduction. *Nat. Commun* **2021**, *12*, 4594. [CrossRef] [PubMed]
32. Xu, Y.; Li, H.; Sun, B.; Qiao, P.; Ren, L.; Tian, G.; Jiang, B.; Pan, K.; Zhou, W. Surface Oxygen Vacancy Defect-Promoted Electron-Hole Separation for Porous Defective ZnO Hexagonal Plates and Enhanced Solar-Driven Photocatalytic Performance. *Chem. Eng. J.* **2020**, *379*, 122295. [CrossRef]
33. Yang, D.; Xu, Y.; Pan, K.; Yu, C.; Wu, J.; Li, M.; Yang, F.; Qu, Y.; Zhou, W. Engineering Surface Oxygen Vacancy of Mesoporous CeO$_2$ Nanosheets Assembled Microspheres for Boosting Solar-Driven Photocatalytic Performance. *Chin. Chem. Lett.* **2022**, *33*, 378–384. [CrossRef]
34. Zhou, W.; Li, W.; Wang, J.-Q.; Qu, Y.; Yang, Y.; Xie, Y.; Zhang, K.; Wang, L.; Fu, H.; Zhao, D. Ordered Mesoporous Black TiO$_2$ as Highly Efficient Hydrogen Evolution Photocatalyst. *J. Am. Chem. Soc.* **2014**, *136*, 9280–9283. [CrossRef]
35. Sinhamahapatra, A.; Jeon, J.-P.; Yu, J.-S. A New Approach to Prepare Highly Active and Stable Black Titania for Visible Light-Assisted Hydrogen Production. *Energy Environ. Sci.* **2015**, *8*, 3539–3544. [CrossRef]
36. Lu, Z.; Zeng, L.; Song, W.; Qin, Z.; Zeng, D.; Xie, C. In Situ Synthesis of C-TiO$_2$/g-C$_3$N$_4$ Heterojunction Nanocomposite as Highly Visible Light Active Photocatalyst Originated from Effective Interfacial Charge Transfer. *Appl. Catal. B-Environ.* **2017**, *202*, 489–499. [CrossRef]
37. Kumar Ray, S.; Dhakal, D.; Gyawali, G.; Joshi, B.; Raj Koirala, A.; Wohn Lee, S. Transformation of Tetracycline in Water during Degradation by Visible Light Driven Ag Nanoparticles Decorated α-NiMoO$_4$ Nanorods: Mechanism and Pathways. *Chem. Eng. J.* **2019**, *373*, 259–274. [CrossRef]
38. Song, W.; Zhao, H.; Ye, J.; Kang, M.; Miao, S.; Li, Z. Pseudocapacitive Na$^+$ Insertion in Ti–O–C Channels of TiO$_2$–C Nanofibers with High Rate and Ultrastable Performance. *ACS Appl. Mater. Interfaces* **2019**, *11*, 17416–17424. [CrossRef]
39. Bao, S.; Liang, H.; Li, C.; Bai, J. The Synthesis and Enhanced Photocatalytic Activity of Heterostructure BiOCl/TiO$_2$ Nanofibers Composite for Tetracycline Degradation in Visible Light. *J. Disper. Sci. Technol.* **2021**, *42*, 2000–2013. [CrossRef]
40. Huang, X.; Guo, F.; Li, M.; Ren, H.; Shi, Y.; Chen, L. Hydrothermal Synthesis of ZnSnO$_3$ Nanoparticles Decorated on G-C$_3$N$_4$ Nanosheets for Accelerated Photocatalytic Degradation of Tetracycline under the Visible-Light Irradiation. *Sep. Purif. Technol.* **2020**, *230*, 115854. [CrossRef]
41. Bao, S.; Liu, H.; Liang, H.; Li, C.; Bai, J. Electrospinned Silk-Ribbon-like Carbon-Doped TiO$_2$ Ultrathin Nanosheets for Enhanced Visible-Light Photocatalytic Activity. *Colloid. Surface A* **2021**, *616*, 126289. [CrossRef]
42. Geng, R.; Yin, J.; Zhou, J.; Jiao, T.; Feng, Y.; Zhang, L.; Chen, Y.; Bai, Z.; Peng, Q. In Situ Construction of Ag/TiO$_2$/g-C$_3$N$_4$ Heterojunction Nanocomposite Based on Hierarchical Co-Assembly with Sustainable Hydrogen Evolution. *Nanomaterials* **2020**, *10*, 1. [CrossRef] [PubMed]

43. Wang, Y.; Liu, M.; Wu, C.; Gao, J.; Li, M.; Xing, Z.; Li, Z.; Zhou, W. Hollow Nanoboxes $Cu_{2-x}S@ZnIn_2S_4$ Core-Shell S-Scheme Heterojunction with Broad-Spectrum Response and Enhanced Photothermal-Photocatalytic Performance. *Small* **2022**, *18*, 2202544. [CrossRef] [PubMed]
44. Sun, B.; Zhou, W.; Li, H.; Ren, L.; Qiao, P.; Li, W.; Fu, H. Synthesis of Particulate Hierarchical Tandem Heterojunctions toward Optimized Photocatalytic Hydrogen Production. *Adv. Mater.* **2018**, *30*, 1804282. [CrossRef]
45. Chen, P.; Wang, F.; Chen, Z.-F.; Zhang, Q.; Su, Y.; Shen, L.; Yao, K.; Liu, Y.; Cai, Z.; Lv, W.; et al. Study on the Photocatalytic Mechanism and Detoxicity of Gemfibrozil by a Sunlight-Driven TiO_2/Carbon Dots Photocatalyst: The Significant Roles of Reactive Oxygen Species. *Appl. Catal. B-Environ.* **2017**, *204*, 250–259. [CrossRef]
46. Huang, H.; He, Y.; Li, X.; Li, M.; Zeng, C.; Dong, F.; Du, X.; Zhang, T.; Zhang, Y. $Bi_2O_2(OH)(NO_3)$ as a Desirable $[Bi_2O_2]^{2+}$ Layered Photocatalyst: Strong Intrinsic Polarity, Rational Band Structure and {001} Active Facets Co-Beneficial for Robust Photooxidation Capability. *J. Mater. Chem. A* **2015**, *3*, 24547–24556. [CrossRef]

Disclaimer/Publisher's Note: The statements, opinions and data contained in all publications are solely those of the individual author(s) and contributor(s) and not of MDPI and/or the editor(s). MDPI and/or the editor(s) disclaim responsibility for any injury to people or property resulting from any ideas, methods, instructions or products referred to in the content.

Review

Recent Advances in Black TiO₂ Nanomaterials for Solar Energy Conversion

Lijun Liao [†], Mingtao Wang [†], Zhenzi Li, Xuepeng Wang * and Wei Zhou *

Shandong Provincial Key Laboratory of Molecular Engineering, School of Chemistry and Chemical Engineering, Qilu University of Technology (Shandong Academy of Sciences), Jinan 250353, China
* Correspondence: wxpchem@hotmail.com (X.W.); wzhou@qlu.edu.cn (W.Z.)
† These authors contributed equally to this work.

Abstract: Titanium dioxide (TiO_2) nanomaterials have been widely used in photocatalytic energy conversion and environmental remediation due to their advantages of low cost, chemical stability, and relatively high photo-activity. However, applications of TiO_2 have been restricted in the ultraviolet range because of the wide band gap. Broadening the light absorption of TiO_2 nanomaterials is an efficient way to improve the photocatalytic activity. Thus, black TiO_2 with extended light response range in the visible light and even near infrared light has been extensively exploited as efficient photocatalysts in the last decade. This review represents an attempt to conclude the recent developments in black TiO_2 nanomaterials synthesized by modified treatment, which presented different structure, morphological features, reduced band gap, and enhanced solar energy harvesting efficiency. Special emphasis has been given to the newly developed synthetic methods, porous black TiO_2, and the approaches for further improving the photocatalytic activity of black TiO_2. Various black TiO_2, doped black TiO_2, metal-loaded black TiO_2 and black TiO_2 heterojunction photocatalysts, and their photocatalytic applications and mechanisms in the field of energy and environment are summarized in this review, to provide useful insights and new ideas in the related field.

Keywords: photocatalysis; black TiO_2; doping; heterojunction; solar energy conversion

Citation: Liao, L.; Wang, M.; Li, Z.; Wang, X.; Zhou, W. Recent Advances in Black TiO₂ Nanomaterials for Solar Energy Conversion. *Nanomaterials* **2023**, *13*, 468. https://doi.org/10.3390/nano13030468

Academic Editor: Chiara Maccato

Received: 29 December 2022
Revised: 16 January 2023
Accepted: 21 January 2023
Published: 24 January 2023

Copyright: © 2023 by the authors. Licensee MDPI, Basel, Switzerland. This article is an open access article distributed under the terms and conditions of the Creative Commons Attribution (CC BY) license (https://creativecommons.org/licenses/by/4.0/).

1. Introduction

With the rapid development of industry and human society, fossil fuels, including coal, natural gas, and petroleum, have been excessively consumed in the last decades, thereby creating energy crises and environmental pollutions. Searching for alternative energy resources and improving the living environment have become urgent issues all over the world. Photocatalytic technology, which can employ the inexhaustible solar energy to H_2 generation from water splitting [1–3], CO_2 reduction to small sustainable fuels [4–6], and environmental pollutant degradation [7–9], has been developed and attracted much attention due to a series of excellent physical and chemical characteristics, such as low energy consumption, simple operation, no secondary pollution, low cost, and sustainability [10]. Since 1972, TiO_2 has been used as a photocatalyst and developed rapidly in the field of energy conversion and environmental remediation [1]. Three main kinds of TiO_2, including anatase, rutile, and brookite, can be distinguished according to their different crystal structures [11–13]. Anatase and rutile are the most frequently investigated TiO_2 photocatalysts because of their superior photocatalytic activity under UV irradiation than brookite. The photocatalytic performance and properties of TiO_2 are severely influenced by its preparation, morphology, and dimensions. Serga et al. reported an extraction-pyrolytic method for the synthesis of nanocrystalline TiO_2 powders using valeric acid as an extractant [14]. This method can be applied for the fabrication of anatase, rutile, or mixed anatase-rutile TiO_2 powders [14]. Poly (titanium dioxide) is found to have a significant influence on the component compatibility and relaxation behavior of interpenetrating polymer networks [15]. TiO$_2$ photocatalysts treated at 800 °C in hydrogen atmosphere for 1 h showed higher visible photocatalytic activity for C-H/C-H coupling of dipyrromethanes with azines than commercial TiO$_2$ (P25) [16]. TiO$_2$ nanosheets were proved to exhibit superior photocatalytic activity for CO$_2$ reduction than the nanoparticle, thanks to its much higher surface area and surface activity [17]. In addition, the effects of the particle size of TiO$_2$ on photocatalytic pollutant removal were thoroughly investigated by Kim et al. [18]. The photocatalytic degradation efficiency for methylene blue can be effectively improved by controlling the particle size and TiO$_2$ concentration in the reaction mixture [18].

However, due to the wide band gap of TiO$_2$ (anatase: 3.2 eV, rutile: 3.0 eV), it can only absorb the ultraviolet part of sunlight (less than 5%), resulting in low light utilization efficiency and low photocatalytic activity [19]. In addition, the high photo-generated electron-hole recombination rate of TiO$_2$ materials leads to low quantum efficiency [20]. Therefore, improving the quantum efficiency and photocatalytic activity have always been a concern in the field of photocatalysis.

Previously, the doping modification of TiO$_2$, including metal ions (such as Co, Ni, Pt, etc.) or nonmetallic ion (N, H, S, etc.), was introduced to directly modify the TiO$_2$ surface electronic properties and broaden the absorption of light, thus improving the efficiency of the charge separation on the TiO$_2$ surface [21,22]. Later, TiO$_2$ nanomaterials were also combined with other semiconductors to form heterojunctions, thereby enhancing the photo-induced charge separation and migration efficiency, and greatly reducing the corresponding recombination rate [23,24]. In 2011, Chen et al. reported that the hydrogenation strategy can reduce the band gap of TiO$_2$, change its color from white to black, expand the light response range to the visible/near infrared region, and improve the visible light catalytic performance [25]. This partially reduced TiO$_2$ is coined as "black TiO$_2$" in the study [25]. Since then, studies on black TiO$_2$ in various fields of photocatalysis, including energy conversion and pollutant removal, have been growing over the last decade. The above-mentioned modification methods (such as doping, heterojunction, etc.) were subsequently applied to black TiO$_2$ to narrow the band gap, thereby further promoting the visible light absorption and charge separation efficiency.

In this review, the structure properties, synthesis routes, and applications of black TiO$_2$-based nanomaterials in the environmental and energy fields, such as photocatalytic water splitting and the photodegradation of organic pollutants, are summarized. As shown in Figure 1, the above aspects will be concluded and discussed from the perspectives of black TiO$_2$, doped black TiO$_2$, metal-loaded black TiO$_2$, and black TiO$_2$ heterojunction. Finally, the status quo of black TiO$_2$ materials is reviewed, and the future development prospects and challenges are proposed.

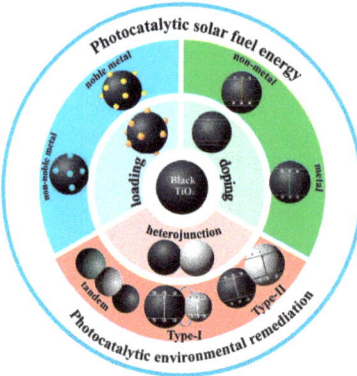

Figure 1. Outline diagram of the types of black TiO$_2$ nanomaterials.

2. Morphology and Structural Properties of Black TiO$_2$

The beauty of nanomaterials is that the (photo)catalytic activity is highly influenced by their morphology and structural properties, including the crystal structure, presence of vacancies, and partially phase transformation. Scanning electron microscopy (SEM) and transmission electron microscopy (TEM) were frequently utilized to investigate the morphology of black TiO$_2$. Black TiO$_2$ with different morphologies, such as nanospheres [26–34], nanotubes [35–38], nanoarrays [39–43], nanowires [44–47], nanoplates [48,49], nanosheets [50,51], nanobelts [52,53], nanocages [54], nanoflowers [55], nanofibers [56], hollow shells [57–59], films [60,61], and nanolaces [62], were synthesized via varied synthetic approaches. Commercial TiO$_2$ materials are often directly used to produce black TiO$_2$ nanospheres. Biswas et al. obtained black TiO$_{2-x}$ nanoparticles at high temperatures via NaBH$_4$ reduction and studied their light absorption ability after reduction [28]. The band gap of black TiO$_2$ nanospheres was 2.54 eV, which was much lower than the pristine commercial anatase (3.27 eV) [28]. Katal et al. prepared black TiO$_2$ nanoparticles at high temperatures under a vacuum atmosphere and investigated the color change and shrinkage of reduced TiO$_2$ pellets over temperature [63]. The reduction process was performed by sintering commercial P25 pellets under a vacuum condition at different temperatures (500, 600, 700, 800 °C) for 3 h [63]. The color of the white P25 pellets changed into pale yellow after calcination at 500 °C in the air condition [63]. Its color became darker after sintering at 500 °C under the vacuum condition [63]. Black TiO$_2$ pellets were obtained after calcination at temperatures higher than 500 °C, and the size of the pellets became smaller after treatment at 700 °C [63]. The phase transformation from anatase to rutile was observed in TiO$_2$ after high temperature calcination [63]. The corresponding visible change and red-shift in UV-vis absorption spectra are presented in Figure 2A [63]. The band gap energy gradually decreased with temperature from 3.1 eV for P25, to 2.24 eV for BT-800 [63].

The synthesis of black TiO$_2$ with unique morphology, such as nanoflowers, tubes, and wires, usually necessitates specific synthetic procedure for TiO$_2$ nanomaterials, including hydrothermal, solvothermal treatment, anodization, etc. Lim et al. prepared partially reduced hollow TiO$_2$ nanowires (R-HTNWs) using the hydrothermal method and the subsequent treatment with NaBH$_4$ under the nitrogen atmosphere [47]. The local distribution of Ti^{3+} species (oxygen vacancies) in reduced hollow TiO$_2$ nanowires was confirmed to be primarily present in the surface region compared to the core using electron energy loss spectroscopy (EELS) [47]. In addition, trace impurities including B, Na, N from NaBH$_4$, and nitrogen were located mostly at the surface and the distorted rutile structure region of R-HTNWs [47]. The SEM, TEM image, and EELS Ti L$_{2,3}$ data are illustrated in Figure 2B [47]. Ti^{3+} present on the surface of TiO$_2$ could be stabilized by the surface impurities [47]. Black TiO$_2$ materials generally possessed certain amounts of oxygen vacancies, which can be confirmed by X-ray photoelectron spectroscopy (XPS). The concentration of oxygen vacancies was normally controlled by the different thermal treatment time or temperature [41,45,51]. However, there is lack of precise, quantitative characterization techniques for oxygen vacancies present on the black TiO$_2$ surface. The band gap parameters of black TiO$_2$ were usually measured and calculated by XPS and UV-vis spectroscopy measurements. The decrease in Eg of black TiO$_2$ was assumed to be related to the surface disorder, including the presence of Ti^{3+} and oxygen vacancies [47].

The mesoporous structure of TiO$_2$ can increase the surface area and phase stability. Zhou et al. synthesized mesoporous black TiO$_2$ hollow spheres (MBTHSs) via the combination of a template-free solvothermal method and amine molecules encircling strategy, and the subsequent atmospheric hydrogenation process [64]. The wall thickness and diameter of MBTHSs could be tuned by adjusting the solvothermal reaction time and the Ti precursor concentration, respectively [64]. Ti^{3+} species were proved to be mainly present in the bulk but not on the surface of MBTHSs via XPS measurements [64]. The light absorption of MBTHSs was effectively extended to the visible light range compared with the pristine TiO_2 [64]. The synthetic procedure, SEM image, and UV-vis absorption properties are present in Figure 2C [64]. The anatase phase remained unchanged after hydrogenation [64]. The band gap of mesoporous TiO_2 was largely reduced to 2.59 eV after hydrogenation [64]. The black TiO_2 consisted of mesoporous structure with cylindrical channels providing the relatively high surface area of ~124 m^2 g^{-1} [64]. They also fabricated the heterojunctions of γ-Fe_2O_3 nanosheets/mesoporous black TiO_2 hollow sphere to enhance the charge separation and photocatalytic tetracycline degradation efficiency [59]. Porous black TiO_2 photocatalysts tended to appear in three dimensional structures, such as foams, pillars, and hollow structures. Zhang et al. synthesized the 3D macro-mesoporous black TiO_2 foams via freeze-drying, cast molding technology, and high-temperature surface hydrogenation [65]. The large, closed pores were generated using polyacrylamide as the organic template, while plenty of open pores were formed in the frameworks and on the surface of the black TiO_2 thanks to the water evaporation in the freeze-drying process [65]. This black TiO_2 material exhibited a self-floating amphiphilic property and an enhanced solar energy harvesting efficiency [65]. Zhou et al. prepared porous black TiO_2 pillars through an oil bath reaction and high-temperature hydrogenation reduction [66]. The porous structure and mesopores of black TiO_2 pillars were clearly observed by the Scanning electron microscope and transmission electron microscope [66]. The enhanced photocatalytic performance was attributed to more active surface sites offered by the porous pillar structure and the self-doped Ti^{3+} [66]. The hollow structured black TiO_2 with plenty of pore channels and an exposed surface also showed an enhanced photocatalytic efficiency [57]. The pores of the porous black TiO_2, generally located in its whole frameworks with open pores connected with surface, providing abundant active sites and surface defects, thus promoting the photocatalytic performance. Ethylenediamine was often utilized to maintain the porous structure of black TiO_2 and to prevent its phase transformation from anatase to rutile.

In addition to oxygen vacancies, disordered structures and surface amorphization in black TiO_2 may have significant impacts on its photoresponsive properties. Kang et al. prepared black TiO_2 with amorphous domains through a glycol-assisted solvothermal method and subsequent calcination [67]. Oxygen vacancies were introduced in the amorphous domains of the black TiO_2 nanosheets [67]. Figure 2D shows the color and optical absorption property changes [67]. The color of the brown TiO_2 turned into black after 2 h of calcination at 350 °C under Ar [67]. The light absorption of black TiO_2 was significantly extended to the near-infrared region [67]. Oxygen vacancies were confirmed to be present in the subsurface of black TiO_2 by first-principle calculations [67]. Table 1 summarizes the properties of some black TiO_2 nanomaterials with varied morphology. The color of most reduced TiO_2 is black. The morphology of black TiO_2 is not determined by the reduction process.

Table 1. Properties of black TiO_2 nanomaterials.

Crystal Phase	Reducing Agent	Color	Morphology	Oxygen Vacancies	Ref.
Anatase with minor rutile	$NaBH_4$	Black	Nanospheres	✔	[26]
Anatase and rutile	Hydrogen radicals	Black	Nanospheres	✔	[27]
Anatase	$NaBH_4$	Black	Nanoparticles	✔	[28]
Anatase and rutile	Hydrogen	Black	Microspheres	✔	[29]
Anatase and rutile	PulsedLaser Ablation in Liquid	Black	Core-shell microspheres	✔	[30]

Table 1. Cont.

Crystal Phase	Reducing Agent	Color	Morphology	Oxygen Vacancies	Ref.
Anatase	Urea	Black	Nanoparticles	✔	[31]
Anatase	NaBH$_4$	Black	Nanospheres	✔	[32]
Anatase	NaBH$_4$	Black	Nanospheres	✔	[33]
Anatase	NaBH$_4$	Black	Nanospheres	✔	[34]
Anatase	Electrochemical reduction	Black	Nanotubes	✔	[35]
Anatase	CaH$_2$	Black	Nanotubes	✘	[36]
Anatase	Hydrogen	Black	Nanotubes	✔	[37]
Anatase	Ag particles	Black	Nanotubes	✔	[38]
Anatase with minor rutile	Electrochemical reduction	Black	Nanotubes	✔	[39]
Anatase and rutile	Electrochemical reduction	Blue and black	Nanotube arrays	N/S	[40]
Anatase and rutile	Electrochemical reduction	Dark yellow	Nanotube arrays	✔	[41]
Anatase and rutile	aluminothermic reduction	Black	Nanotubes	✔	[42]
Anatase	Electrochemical reduction	Black	Nanotube arrays	✔	[43]
Anatase	Hydrogen	Black	Nanowires	✔	[44]
Anatase	NaBH$_4$	Black	Nanowires	✔	[45]
Anatase	Glycerol	Black	Nanoparticles and nanowires	✔	[46]
Rutile	NaBH$_4$	N/S	Hollow nanowires	✔	[47]
Anatase	H$_2$S and SO$_2$	Black	Nanoplatelets	✔	[48]
Anatase	NaBH$_4$	Black	Nanoplates	✔	[49]
Anatase	Hydrogen	Black	Nanosheets	✔	[50]
Anatase	Aluminothermic reduction	Black	Nanosheet array films	✔	[51]
Anatase	Hydrogen	Black	Nanobelts	✔	[52]
Anatase	NaBH$_4$	Black	Nanobelts	✔	[53]
Anatase	Hydrogen	Black	Nanocages	✔	[54]
Anatase	Hydrogen	Black	3D nanoflower	✔	[55]
Anatase and rutile	NaBH$_4$	Black	Nanofiber	✔	[56]
Anatase	NaBH$_4$	Black	Hollow shell	✔	[57]
Anatase	Annealing in vacuum	Black	Core-shell	✔	[58]
Anatase	Hydrogen	Black	Mesoporous hollow sphere	✔	[59]
Anatase and rutile	Electrochemical	Black	Films	✔	[60]
Anatase	H$_2$ plasma treatment	Black	Mesoporous films	✔	[61]
Anatase	Hydrogen	Black	Hierarchical nanolace films	✔	[62]

Notes: N/S: not studied; ✔: yes; ✘: no.

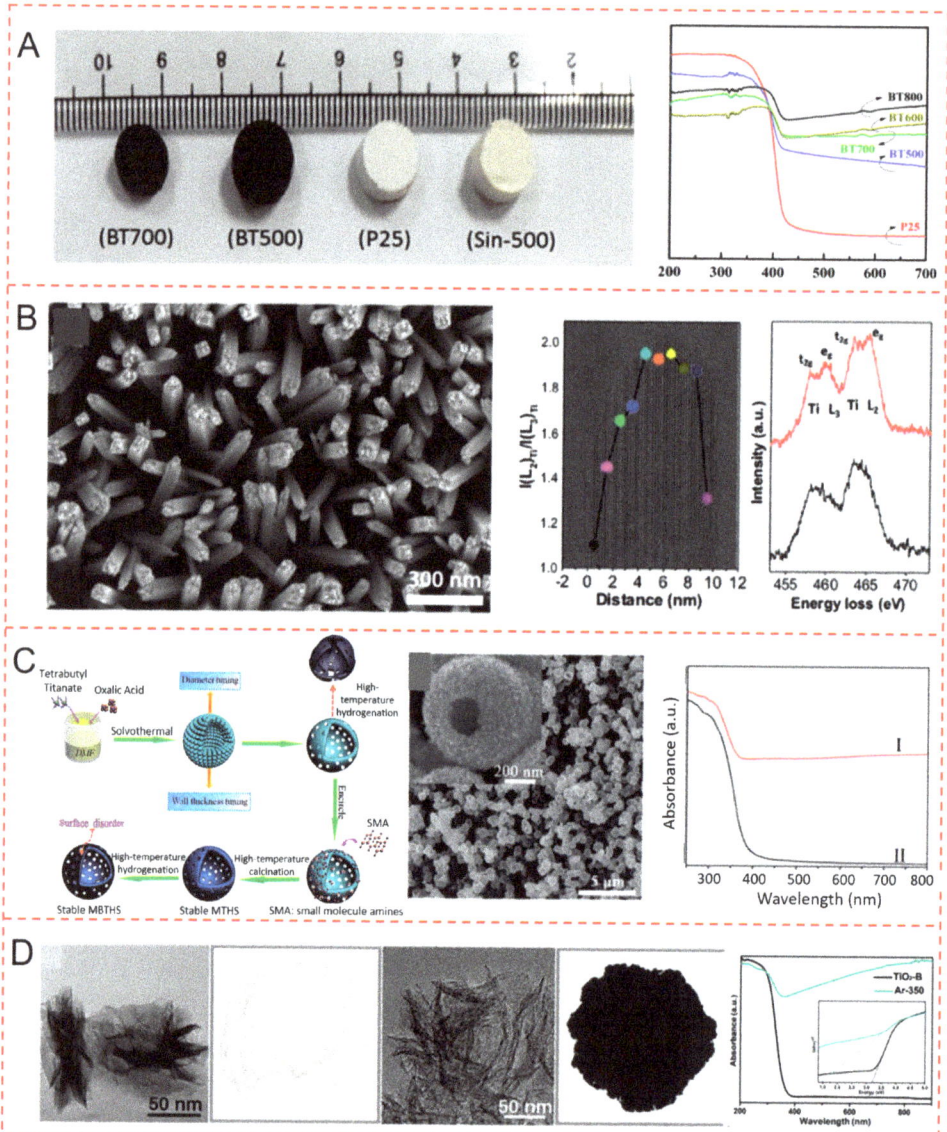

Figure 2. (**A**) Picture and UV-vis spectra of reduced P25 pellets at various temperatures. Reprinted with permission for ref. [63]. Copyright 2018, American Chemical Society. (**B**) SEM and TEM image and overlaid $I(L_2)_{Ti}/I(L_3)_{Ti}$ values determined from the EELS Ti $L_{2,3}$ data at each position of R-HTNWs. Reprinted with permission for ref. [47]. Copyright 2019, Wiley-VCH. (**C**) Schematic illustration of the synthetic procedure, SEM image, for MBTHSs and UV-vis absorption properties (I: mesoporous black TiO_2 hollow spheres; II: mesoporous TiO_2 hollow spheres). Reprinted with permission for ref. [64]. Copyright 2016, The Royal Society of Chemistry. (**D**) TEM images, and optical photographs of (black) TiO_2, and UV-vis spectra. Reprinted with permission for ref. [67] Copyright 2021, Wiley-VCH.

3. Synthesis of Black TiO$_2$

Currently, various methods were used to synthesize black TiO$_2$, which can be divided into two main approaches: high temperature hydrogenation reduction and solid phase reduction [68]. The high temperature hydrogenation method often uses hydrogen or hydrogen-contained gas mixtures to treat samples at high temperatures [69,70]. The materials used in the solid phase reduction method are generally NaBH$_4$ [71], CaH$_2$ [72], Mg powder [72], or other reducibility materials [72]. The reduction method can be expressed in reaction Equation (1):

$$TiO_2 + A \rightarrow TiO_{2-x} + AO_x \tag{1}$$

In addition, researchers also use hot wire annealing [73], laser irradiation [74,75], anode reduction [76], and other methods to synthesize black TiO$_2$ [77,78].

3.1. High Temperature Hydrogenation

Hydrogen reduction involves the reduction of pure H$_2$ gas, H$_2$/Ar, or H$_2$/N$_2$ mixture at high or low pressures [69], which is a simple, effective, and straightforward method.

Zhou et al. successfully prepared the ordered mesoporous black TiO$_2$ material by hydrogenation at high temperature (500 °C) under atmospheric pressure (Figure 3), which had a larger specific surface area and pore size compared with the pristine titanium dioxide [79]. As shown in Figure 3, after hydrogenation at high temperature, the regular hexagonal channel of the obtained black TiO$_2$ was completely maintained [79]. It can be seen from the XRD in Figure 3 that there was no phase change in black TiO$_2$ compared with the original materials, thus proving the high thermal stability of the sample prepared by this method [79]. Notably, it can also be clearly seen that its crystallinity decreased, proved by the XRD intensity, thereby indicating that the surface disorder of TiO$_2$ has been created after the hydrogenation process [79]. The color of the white TiO$_2$ turned into black after 3 h of hydrogenation [79].

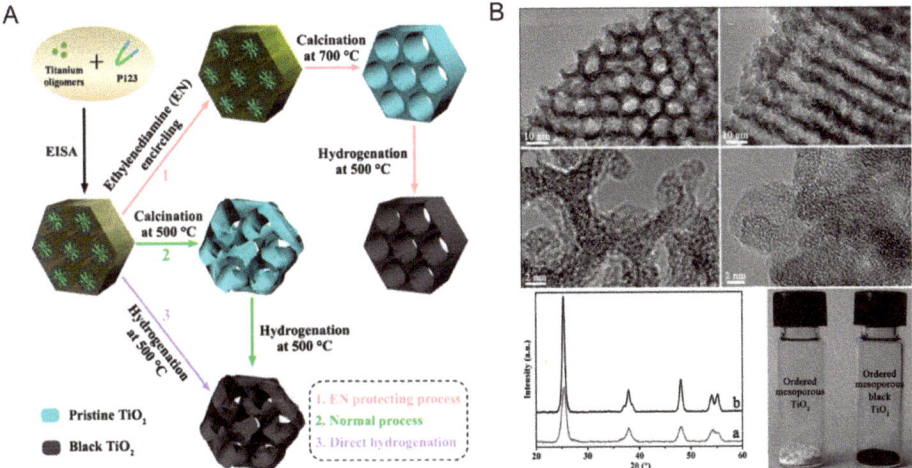

Figure 3. (**A**) Schematic synthesis process for the ordered mesoporous black TiO$_2$ materials. (**B**) Representative TEM images along [100] and [110] planes, HRTEM images of the ordered mesoporous black TiO$_2$ materials, and X-ray diffraction patterns and the photos of the ordered mesoporous black TiO$_2$ materials (a) and ordered mesoporous TiO$_2$ materials (b). Reprinted with permission for ref. [79]. Copyright 2014, American Chemical Society.

Black TiO$_2$ with different morphologies can be obtained via hydrogenation. Yang et al. prepared one-dimensional black TiO$_2$ nanotubes by the hydrogenation method, with an inner diameter of 7 nm and a wall thickness of 6 nm, as presented in Figure 4A [80].

Spherical and lamellar structures have also received much attention due to their large specific surface areas. As shown in Figure 4B, after simple hydrogenation reduction, Li et al. successfully prepared black TiO_2 nanospheres and observed the mesoporous structure in the TEM image [81]. Although the crystal surface structure of anatase became slightly disordered after hydrogenation, its special lattice fringes (d = 0.35 nm) did not change [81]. Black TiO_2 nanotubes with the mesoporous nanosheet structure were successfully prepared by the hydrogen reduction method by Zhang et al. [82]. Ethylenediamine coating method was used before hydrogenation [82]. The original morphology of TiO_2 was completely retained [82]. Wu et al. also synthesized two dimensional ultrathin mesoporous black TiO_2 nanosheet materials using the similar ethylenediamine encircling strategy (Figure 5) with 4 h hydrogenation reaction at 500 °C [83].

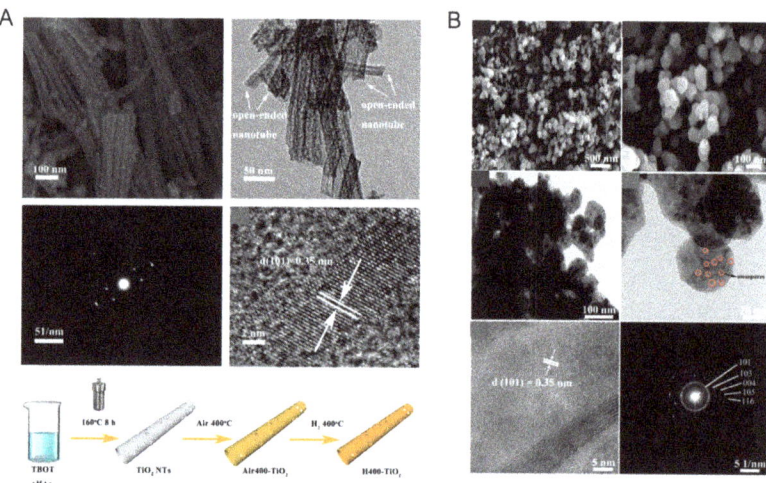

Figure 4. (**A**) SEM, TEM, SAED, HRTEM images and schematic illustration of the formation process of H400-TiO_2. Reprinted with permission for ref. [80]. Copyright 2021, Elsevier. (**B**) Representative SEM, TEM, HRTEM images and the corresponding elected area electron diffraction (SAED) pattern of the mesoporous TiO_2 nanospheres after surface hydrogenation. Reprinted with permission for ref. [81]. Copyright 2021, Elsevier.

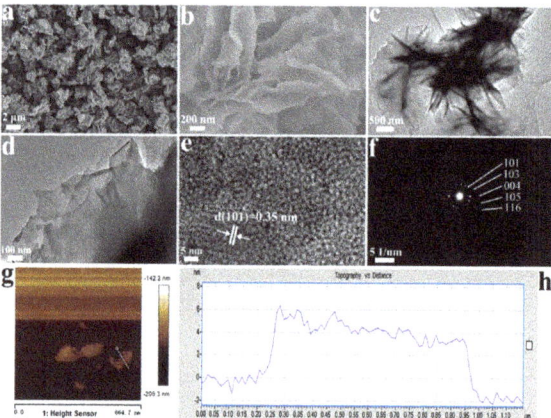

Figure 5. (**a,b**) SEM images, (**c,d**) TEM images, (**e**) HRTEM, (**f**) SAED pattern, (**g**) AFM topography image, and (**h**) the corresponding height information of the 2D ultrathin nanosheets. Reprinted with permission for ref. [83]. Copyright 2020, The Royal Society of Chemistry.

In addition to the method of hydrogen reduction at high temperature and atmospheric pressure, researchers also use the high pressure method. Wu et al. prepared black anatase TiO_2 in a two-step process [84]. The sample was degassed at 200 °C firstly, then was heated to 400 °C, and hydrogenated under high pressure (5-bar) for 24 h. The black sample was finally obtained after cooling to room temperature. The black TiO_2 was successfully prepared by Hamad et al. at a high pressure (8 bar) and relatively mild temperature [85]. The hydrogenation time was longer than in other similar research (1–5 days) [85]. The synthesized samples were uniform and stable in size, and showed higher photocatalytic activity compared with the pristine white TiO_2 [85]. Mixed gases with H_2 were also used as reducing agents in the synthesis of black TiO_2. Cai et al. successfully prepared black TiO_2 with the surface disorder structure using H_2/N_2 mixed gas with 10% content of hydrogen [86].

3.2. Solid Phase Reduction

Compared with the high temperature hydrogenation, the solid phase reduction method has certain advantages. The high temperature hydrogenation process normally starts from the outside to the inside with a relatively moderate reaction rate, while the solid phase reduction method can provide a more complete and intense reaction and may produce a series of doping at the same time. The defect is a double-edged sword. Too many defects may be detrimental to the photocatalytic performance, so the proportion and dosage of reductants in solid phase reactions should be reasonably controlled.

Xiao et al. prepared the black TiO_2 by the solid-state chemical reduction strategy by mixing the sample with sodium borohydride in a certain proportion [87]. Then, the mixture was heated in a tubular furnace with N_2 atmosphere [87]. Finally, the resulting sample was washed with deionized water to remove the unreacted sodium borohydride [87]. As shown in Figure 6, the color of the sample was getting darker with the temperature [87]. The absorption of the visible light was much enhanced after the reduction of TiO_2 [87].

Zhu et al. showed that CaH_2 can also be used as a constant reducing agent to prepare black TiO_2 [88]. The reduction process was conducted at varied temperatures [88]. It was found that the obtained black TiO_2 after reduction treatment at 400 °C had the best absorption of sunlight (over 80%), which was 11 times that of the pristine TiO_2 [88]. This simple method provides an alternative for improving the absorption of visible light on the TiO_2 surface.

Sinhamahapatra et al. reported the reduction of TiO_2 particles to black TiO_2 by magnesium thermal reduction method, which was inspired by the Kroll process, for the first time [89]. The synthetic procedure of this method was approximately identical to the method of sodium borohydride reduction [89]. TiO_2 and magnesium powder were thoroughly mixed first, and then heated in a tube furnace at 650 °C with 5% H_2/Ar for 5 h [89]. The obtained samples were placed in HCl solution for 24 h, and then washed with water to remove the acid, and finally dried at 80 °C [89].

3.3. Hot-Wire Annealing Method

In addition to the high temperature hydrogenation reduction and solid-phase reduction, researchers have also explored some other methods to synthesize black TiO_2, which has made the method of preparing black TiO_2 diversified. Wang et al. proposed a simple and direct hot-wire annealing (HWA) method [73]. The titanium dioxide nanorods were treated with highly active atomic hydrogen simply generated by hot wire [73]. The reduction mechanism was similar to that of the high temperature hydrogenation [73]. The resulted black TiO_2 nanorods had better stability and higher photocurrent density compared with the traditional hydrogenation method [73]. In addition, it had no damage to the photoelectric chemical devices [73].

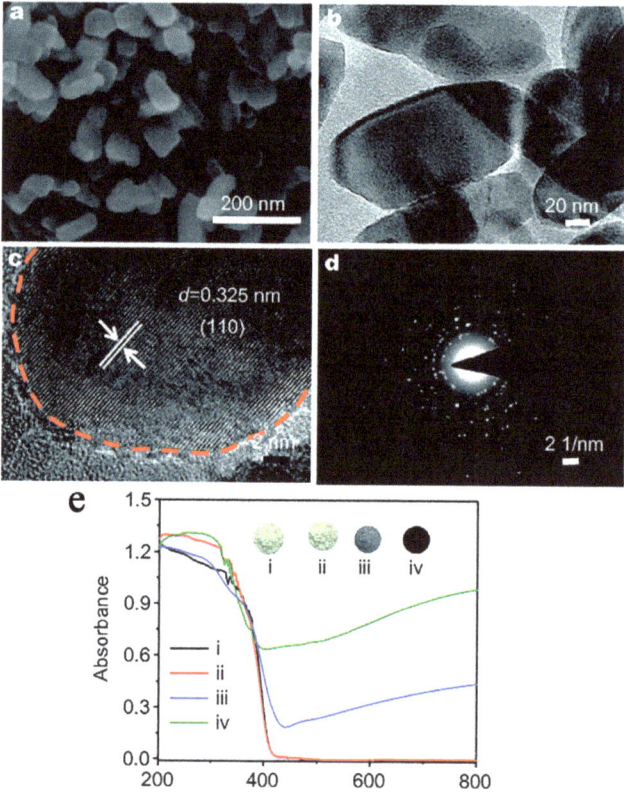

Figure 6. SEM (**a**), TEM (**b**), HRTEM images (**c**), the corresponding selected-area electron diffraction pattern (**d**) of the hydrogenated rutile TiO_2 (300 °C) and the UV/vis absorption spectra (**e**) of the pristine rutile TiO_2 (i) and the hydrogenated rutile TiO_2 under 250 °C (ii), 300 °C (iii), and 350 °C (iv). Reprinted with permission for ref. [87]. Copyright 2018, Springer.

3.4. Anode Oxidation Method

The introduction of crystal defects to titanium dioxide can effectively extend the light absorption range to the visible light region without side effects. Anode oxidation is a simple and efficient method to synthesize defective black TiO_2. Dong et al. successfully prepared black TiO_2 using a two-step anode oxidation method [76]. The first step was to anodize Ti foil in the ethylene glycol solution with a certain proportion of NH_4F and distilled water, and the corresponding voltage was set at 60 V [76]. After 10 h of oxidation, an oxide layer was obtained [76]. Subsequently, the Ti foil was purified to remove organic impurities, and treated at high temperature (450 °C) for 1 h to form black TiO_2 [76].

3.5. Plasma Treatment

Zhu et al. prepared black TiO_2 nanoparticles via the one-step solution plasma method under mild conditions [27]. The structural disorder layer was assumed to be formed in TiO_2 after the solution plasma process [27]. The light absorption of TiO_2 in the visible and near infrared range was significantly enhanced after the plasma treatment, thus increasing its activity in the water evaporation under solar illumination [27]. Teng et al. prepared black TiO_2 using P25 as the precursor system, hydrogen plasma, and a hot filament chemical vapor deposition (HFCVD) device with H_2 as the reducing gas [77]. The visible and near-infrared light absorption of TiO_2 were much enhanced after the surface reduction [77]. Oxygen vacancies and Ti-H bonds were formed on the black TiO$_2$ surface, thereby improving the photocatalytic activity [77].

3.6. Gel Combustion

Ullattil et al. prepared black anatase TiO$_{2-x}$ photocatalysts through a one-pot gel combustion process using titanium butoxide, diethylene glycol, and water as precursors [90]. Plenty of Ti^{3+} and oxygen vacancies existed in the synthesized black anatase TiO$_2$ nanocrystals confirmed by XPS measurements [90]. The light absorption of TiO$_2$ was extended from UV to the near-infrared range [90]. Campbell et al. also synthesized black TiO$_2$ via the sol-gel combustion method using titanium tetraisopropoxide as the precursor [91]. The light absorption ability was significantly enhanced compared to commercial TiO$_2$ [91]. The obtained black TiO$_2$ with the high surface area demonstrated much improved photocatalytic degradation efficiency of the organic dye under the visible light irradiation [91].

4. Strategies for Promoting Photocatalytic Activity of Black TiO$_2$

Researchers have been trying to use metal and non-metal doping methods to prepare the modified TiO$_2$ with better light absorption ability and photocatalytic activity. The introduction of metal ions and non-metallic elements into the TiO$_2$ lattice can expand its absorption range to the visible light, thus enhancing the photocatalytic performance [92]. In recent years, doped black TiO$_2$ has also been widely explored to narrow its band gap, thereby improving its optical properties in the visible light region, and enhancing its photocatalytic activity in various reactions.

4.1. Metallic Doped Black TiO$_2$

It was found that by doping different metals in TiO$_2$, Ti^{4+} in TiO$_2$ lattice was replaced [93]. New impurity levels would be introduced in the band gap of TiO$_2$ [93]. The band gap would be narrowed by the doping process, thus improving the separation efficiency of the photoelectron-hole of TiO$_2$, increasing the quantum yield, and expanding the light absorption to the visible light region [92]. Photocatalytic degradation, hydrogen production capacity, and light energy conversion can be significantly improved [92]. Previously, various metal elements, including Cu, Co, Mn, Fe, Mo, etc., had been used to produce the doped TiO$_2$ via different approaches [93]. Lately, some of the metal elements, such as Al, Ni, Na, etc., were also utilized to dope black TiO$_2$ for achieving the narrower band gap and better photocatalytic performance [47,94,95].

Yi et al. prepared the amorphous Al-Ti-O nanostructure in black TiO$_2$ via a scalable and low-cost strategy [94]. The commercial TiO$_2$ and Al powders were mixed and then grinded in an agate mortar at room temperature for 0-50 min [94]. The color of the light gray TiO$_2$ turned into gray after 2 min of milling [94]. Its color became much darker after the longer milling time [94]. Black Al-Ti-O oxide samples were obtained after milling for more than 5 min [94]. The color changes, UV-Vis-NIR diffuse reflectance spectra, and TEM image of the samples were shown in Figure 7 [94]. The crystalline Al and anatase TiO$_2$ were transformed into amorphous Al-black TiO$_2$ after the ball milling [94]. Al-black TiO$_2$ after 20 min milling exhibited the best light absorption in the visible light and near infrared region [94].

Zhang et al. prepared Ni^{2+}-doped porous black TiO$_2$ photocatalysts through the combination of the sol-gel method and in situ solid-state chemical reduction process [95]. The reduction approach was performed by heating the mixture of Ni-doped TiO$_2$ and NaBH$_4$ at 350 °C under Ar atmosphere for 1 h [95]. The color of the white as-made TiO$_2$ became yellowish after Ni doping [95]. Black Ni-doped TiO$_2$ was obtained after the reduction with NaBH$_4$ [95]. Figure 8 shows the optical properties and the band gap of different materials [95]. The light absorption of TiO$_2$ was extended to the visible light range after Ni doping, and further enhanced after the chemical reduction, which was attributed to the generation of oxygen vacancies, Ti^{3+}, and Ni^{2+} [95]. The band gap of the black Ni-doped TiO$_2$ was only 1.96 eV [95].

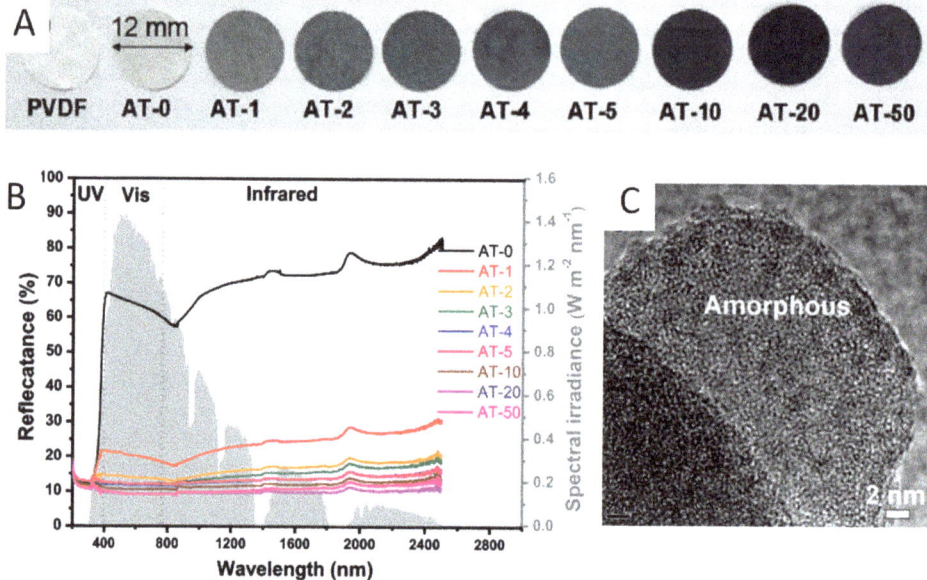

Figure 7. (**A**) the set of the AT-x (Al-Ti-O-milling time) membranes; (**B**) UV-Vis-NIR diffuse reflectance spectra of AT-x; (**C**) TEM image of Al-black TiO_2. Reprinted with permission for ref. [94]. Copyright 2017, Elsevier.

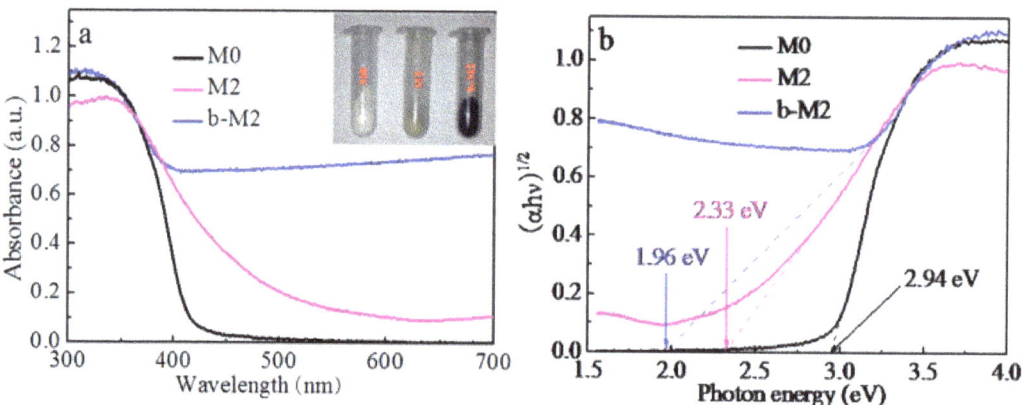

Figure 8. (**a**) UV-vis absorption spectra of TiO_2 (M0), Ni-doped TiO_2 (M2), and black Ni-TiO_2 (b-M2); (**b**) band gap for M0, M2, and b-M2 samples, respectively. Reprinted with permission for ref. [95]. Copyright 2015, The Royal Society of Chemistry.

Zhang et al. reported Ti^{3+} self-doped black TiO_2 nanotubes with mesoporous nanosheet structure via a two-step approach consisting of the solvothermal reaction and hydrogenation process [82]. The appearance of the white TiO_2 turned into black after hydrogenation at 600 °C for 2 h [82]. The optical absorption was significantly extended to the range of 400–800 nm after hydrogenation [82]. The band gap of pristine TiO_2 decreased from 3.2 eV to 2.87 eV after the surface hydrogenation [82].

4.2. Non-Metallic Doped Black TiO$_2$

The doping mechanism of nonmetallic elements can be explained as follows: the doped elements act as overlapping impurity levels in the valence band inside the photocatalyst crystal, thereby reducing the band gap of semiconductors and promoting the migration of photogenerated electrons to the active site. The doping of non-metal elements in the crystal lattice of TiO$_2$ can slow down the electron-hole pair recombination rate, which is an effective modification way to improve the photocatalytic activity of TiO$_2$.

The nitrogen atom, which has five outer shell electrons, has a similar radius to oxygen. The introduction of N into TiO$_2$ enhances its visible light photocatalytic activity, which is proved to be the most ideal non-metallic doping element in a large number of studies [96,97]. Since the 2p orbital of N has a similar energy level to that of the oxygen atom and is easy to hybridize, the researchers found that the doping of N can improve the defects of TiO$_2$ and broaden the response range of the absorption spectra [98]. The N-Ti-O bond generated by doping the crystal can change the energy level structure of TiO$_2$ and improve the quantum efficiency [99]. In addition, N can also replace O in the lattice with the formation of the Ti-N bond, which increases the absorption of the visible light by TiO$_2$ and improves the photocatalytic efficiency of TiO$_2$ [96].

Zhou et al. successfully synthesized the nitrogen-doped black titanium dioxide nanocatalyst by calcining white TiO$_2$ with or without urea at varied temperatures under the different atmosphere [31]. The N-doped TiO$_2$ using urea as N precursor has a better visible light absorption, narrower band gap and the most effective excitation charge separation, and higher photocatalytic activity [31]. Liu et al. prepared N-doped black TiO$_2$ spheres via a two-step process consisting of the solvothermal reaction and calcination in the nitrogen atmosphere [100]. The black N-TiO$_2$ photocatalysts were obtained after heat treatment at 500 °C in N$_2$ atmosphere for 3 h [100]. Ammonium chloride was used as the nitrogen source during the synthetic process [100]. The obtained black N-TiO$_2$ with a moderate mole ratio of ammonium chloride to TiO$_2$ (2:1) had the narrowest band gap and the highest photocatalytic pollutant degradation efficiency [100].

Gao et al. prepared black TiO$_2$ nanotube arrays with dual defects consisting of bulk N doping and surface oxygen vacancies [101]. Urea was utilized as the N precursor during the anodic oxidation process [101]. Black N-doped TiO$_2$ nanotube arrays were obtained after calcination at 600 °C in the Ar atmosphere using aluminum powder [101]. The doping of N generated a new energy level and shortened the carrier migration distance [101]. The synergistic effect of the two defects established an internal electric field, promoted the transfer of charge, and achieved the balance between kinetics and thermodynamics, thereby enhancing the photocatalytic hydrogen production efficiency [101].

Cao et al. successfully synthesized N and Ti^{3+} co-doped mesoporous black TiO$_2$ hollow spheres (N-TiO$_{2-x}$) by a step-by-step method [102]. The prepared three different kinds of TiO$_2$ had similar XRD peaks, indicating no impurities was formed during the synthetic process, as presented in Figure 9a [102]. The broader peaks of the black TiO$_2$ may be attributed to the lattice distortion caused by N doping [102]. Raman spectra (Figure 9b) showed that the main phase of the titanium dioxide hollow sphere was anatase [102]. The co-doping of Ti^{3+} and N in N-TiO$_{2-x}$ resulted in a certain amount of attenuation [102]. The absorption in the visible light was much enhanced after the co-doping of N and Ti^{3+} in TiO$_2$ [102]. The color of the samples gradually became darker after the nitrogen and Ti^{3+} doping [102]. In addition, the band gap of the black N-doped TiO$_2$ was much smaller than the pristine TiO$_2$ (Figure 9c,d) [102]. The charge separation and photocatalytic activity for photocatalytic pollutant removal and hydrogen generation were much improved after the co-doping of N and Ti^{3+} in the lattice of TiO$_2$ (Figure 9e) [102].

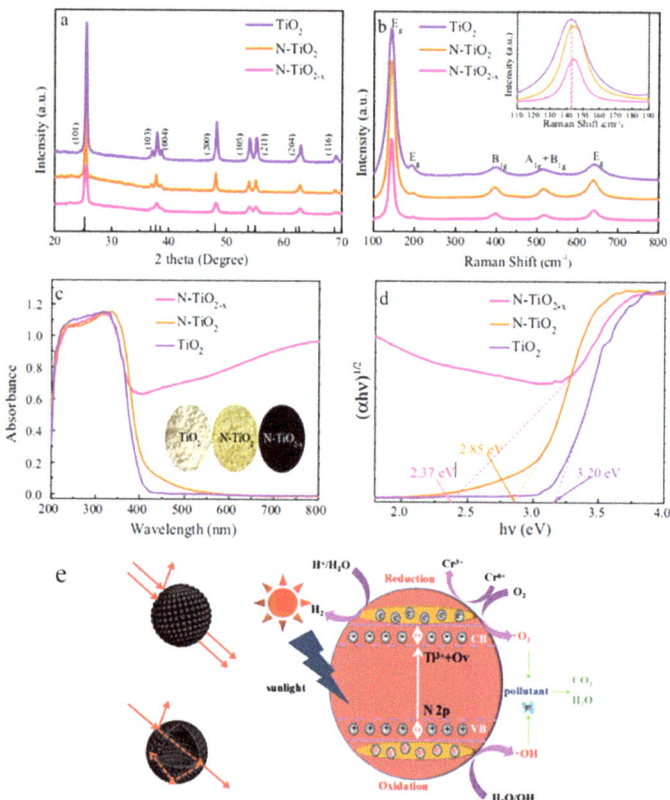

Figure 9. XRD patterns (**a**), Raman spectra (**b**), UV-vis adsorption spectra (**c**), and determination of the indirect interband transition energies (**d**) of TiO_2, $N-TiO_2$, and $N-TiO_{2-x}$, respectively. Schematic diagrams of light pathways in nanoparticles, hollow spheres, and schematic illustration for the solar-driven photocatalytic mechanism of $N-TiO_{2-x}$ (**e**). Reprinted with permission for ref. [102]. Copyright 2017, Elsevier.

4.3. Metal-Loaded Black TiO_2

Metal nanoparticles, such as Ag, Cu, Pt, etc., can generate the surface plasmon resonance (SPR) effect, thus improving the UV-vis absorption ability of photocatalysts [103]. The introduction of metal nanoparticles or clusters to the black TiO_2 photocatalyst surface could further expand its light absorption range and enhance the photo-induced charge separation and transfer efficiency, thus improving the photocatalytic performance [38,104–109]. Silver nanoparticles or clusters have been extensively explored to construct Ag/black TiO_2 photocatalysts due to the relatively lower cost of Ag than other noble metals and the SPR effect. Jiang et al. prepared the Ag-decorated 3D urchinlike $N-TiO_{2-x}$ via a facile photo-deposition method combined with a reduction process, as presented in Figure 10 [104]. The $AgNO_3$ solution was used as the Ag precursor and deposited onto the $N-TiO_2$ surface under UV illumination at the wavelength of 365 nm for 30 min [104]. Notably, the unique 3D urchinlike structure was retained after the Ag deposition and $NaBH_4$ reduction process [104]. The light absorption of photocatalysts in the visible light range was further enhanced after the Ag deposition, with a much smaller band gap (2.61 eV) [104]. The $Ag/N-TiO_{2-x}$ photocatalysts presented the most excellent photocatalytic H_2 production rate (186.2 μmol h^{-1} g^{-1}) [104].

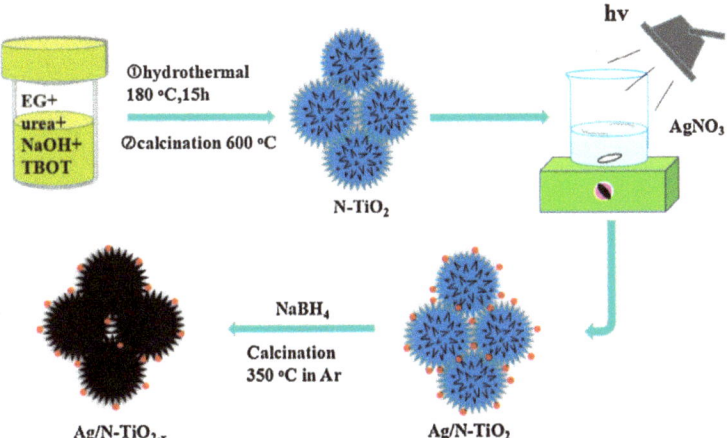

Figure 10. Schematic illustration for preparation of 3D urchinlike Ag/N-TiO$_{2-x}$. Reprinted with permission for ref. [104]. Copyright 2018, Elsevier.

Li et al. constructed Ag nanoparticle-decorated black TiO$_2$ foams through the wet impregnation and high temperature surface hydrogenation process [105]. Ag nanoparticles were formed in the open pores of black TiO$_2$ foams after the hydrogen atmosphere reduction, thus decreasing its surface area [105]. The synthetic process and UV-vis absorption spectra of Ag-black TiO$_2$ foams were presented in Figure 11 [105]. The Ag-black TiO$_2$ foams with varied amounts of silver showed apparent absorption at around 500 nm thanks to the SPR effect of Ag nanoparticles [105]. The Ag-black TiO$_2$ foams containing 3 wt.% Ag nanoparticles exhibit the highest photocatalytic efficiency for atrazine removal [105]. Excess amounts of Ag nanoparticles in black TiO$_2$ foams would decrease its photocatalytic performance due to the aggregation of Ag nanoparticles [105]. Ag nanoparticles were also decorated onto black TiO$_2$ nanorods [106,109] and nanotubes surface [38], to further improve its photocatalytic performance using the SPR effect. In addition, NiS and Pt nanoparticles were co-decorated onto the surface of black TiO$_2$ nanotubes via the solvothermal and photo-deposition approach, respectively [109]. The SPR effect of Pt nanoparticles effectively improved the light absorption ability, thus enhancing the photocatalytic water splitting [109]. Wang et al. successfully deposited Pt single atoms onto the black TiO$_{2-x}$/Cu$_x$O surface assisted by the presence of surface oxygen vacancies [110]. The deposition of Pt single atoms further improved the light absorption of black TiO$_{2-x}$/Cu$_x$O in the entire visible region [110]. Cu, which was a much cheaper candidate than noble metal, was also used for the surface decoration of black TiO$_2$ surface as a SPR effect metal, thereby improving its photothermal effect [108].

4.4. Construction of Black TiO$_2$ Based Heterojunction Photocatalysts

Although the visible light absorption ability of TiO$_2$ has been much enhanced after the hydrogenation or reduction process, the charge separation and transfer efficiency of black TiO$_2$ is still far from satisfactory for photocatalytic applications. In addition to the surface modification with metal nanoparticles, the construction of black TiO$_2$-based junctions is an efficient way to improve the photo-generated charge separation and migration efficiency. The types of black TiO$_2$-based heterojunctions can be divided into three main categories, including type II heterojunctions [59,111–119], Z-scheme heterojunctions [120], and tandem heterojunctions [29,121,122]. Tan et al. fabricated the Ti^{3+}-TiO$_2$/g-C$_3$N$_4$ nanosheets heterojunctions through a facile calcinations-sonication assisted approach [118]. The photo absorption of Ti^{3+}-TiO$_2$ in the visible light range had been evidently enhanced after the coupling with meso-g-C$_3$N$_4$ [118]. The synthetic procedure of Ti^{3+}-TiO$_2$/g-C$_3$N$_4$ nanosheets, the UV-vis absorption spectra, and the band gaps of different samples were shown in

Figure 12 [118]. The separation and migration of the photo-induced charge carrier had been effectively improved due to the construction of heterojunctions [118]. Ti^{3+}-TiO_2/g-C_3N_4 nanosheets exhibited the highest photocatalytic H_2 evolution rate and phenol degradation efficiency [118]. A type II heterojunction with enhanced charge separation and transfer efficiency had been proposed [118].

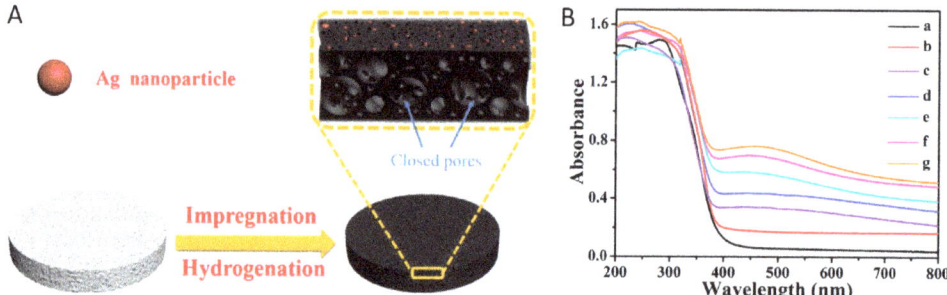

Figure 11. (**A**) Schematic view for the synthetic process of Ag/black TiO_2 foams; (**B**) The UV-vis absorption spectra of black TiO_2 foams (a) and Ag-black TiO_2 foams with different Ag contents of 0.5 (b), 1 (c), 2 (d), 3 (e), 4 (f), and 5 wt.% (g). Reprinted with permission for ref. [105]. Copyright 2018, Elsevier.

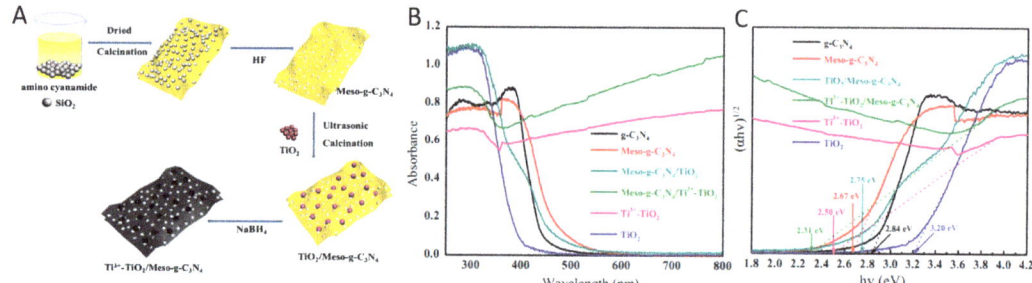

Figure 12. (**A**) The synthetic process of Ti^{3+}-TiO_2/g-C_3N_4 nanosheets heterojunctions; (**B**) UV-vis diffuse reflectance spectra; (**C**) the corresponding calculated band gaps. Reprinted with permission for ref. [118]. Copyright 2018, Elsevier.

Ren et al. prepared magnetic γ-Fe_2O_3/black TiO_2 heterojunctions via the metal-ion intervened hydrothermal method and high temperature hydrogenation process [59]. In addition, α-Fe_2O_3 nanosheets were transformed to the surface defected γ-Fe_2O_3 after the hydrogenation process [59]. The light utilization of black TiO_2 in visible light even near the infrared region has been improved after the combination with γ-Fe_2O_3 [59]. Figure 13 presented the UV-vis absorption spectra and the proposed band structure and charge transfer mechanism of γ-Fe_2O_3/black TiO_2 heterojunctions [59]. The fabrication of the type II heterojunctions efficiently enhanced the photo-generated charge separation and transfer process [59]. The photocatalytic degradation of tetracycline on the heterojunction photocatalysts surface had been much improved, compared with the pristine TiO_2 [59]. In addition, Bi_2MoO_6 [111], CdS [115], CeO_2 [117], $SrTiO_3$ [119], etc. had also been used for the construction of type II heterojunctions with black TiO_2 with much improved photocatalytic efficiency.

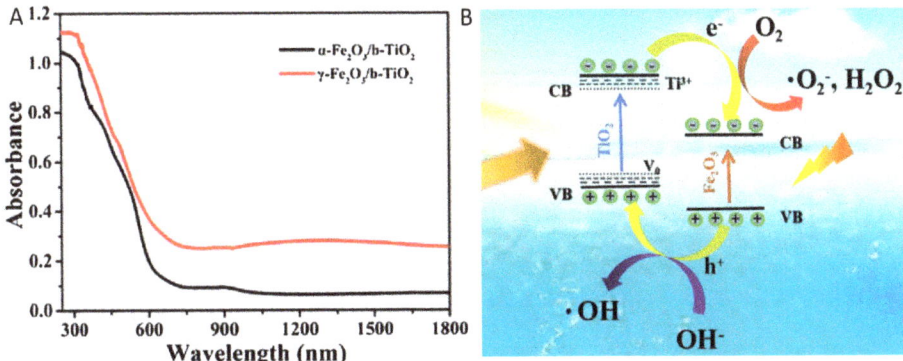

Figure 13. (**A**) UV-vis spectra; (**B**) Proposed band alignment and charge transfer mechanism. Reprinted with permission for ref. [59]. Copyright 2019, Elsevier.

Sun et al. synthesized the CdS quantum dots/defective ZnO_{1-x}-TiO_{2-x} Z scheme heterojunction via the combination of hydrothermal synthesis, chemical reduction, and electroless planting process [120]. The visible light absorption was enhanced by the formation of ZnO-TiO_2 heterojunction, and further improved after combining with CdS [120]. The formation process of the heterojunction, UV-vis spectra, and proposed charge transfer mechanism are shown in Figure 14 [120]. The charge separation efficiency and photocatalytic organic pollutant removal rate were much improved upon the formation of CdS QDs/defective ZnO_{1-x}-TiO_{2-x} Z scheme heterojunction [120].

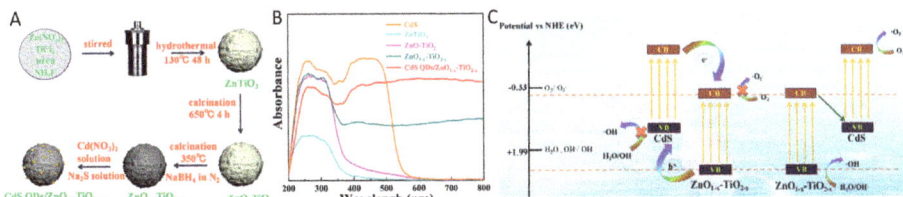

Figure 14. (**A**) Formation process of CdS QDs/ZnO_{1-x}-TiO_{2-x}; (**B**) UV-vis spectra; (**C**) Proposed band alignment and charge transfer mechanism. Reprinted with permission for ref. [120]. Copyright 2020, Elsevier.

To further promote the visible light utilization and photo-induced charge separation and transfer efficiency, black TiO_2-based tandem heterojunction photocatalysts have been proposed by researchers for photocatalytic hydrogen production. Sun et al. prepared a hierarchical hollow black TiO_2/MoS_2/CdS tandem heterojunction photocatalyst through the combination of the solvothermal method and high-temperature hydrogenation treatment [122]. The black TiO_2/MoS_2 heterojunction effectively enhanced the photon absorption in visible light and the near infrared region [122]. The tandem system further promoted visible light utilization compared to other combinations [122]. The schematic view of the construction of black TiO_2/MoS_2/CdS, UV-vis absorption spectra, and photocatalytic hydrogen evolution rate are illustrated in Figure 15. The charge separation and migration efficiency were much promoted by the formation of black TiO_2/MoS_2/CdS tandem heterojunction [122]. The tandem heterojunction exhibited the highest photocatalytic hydrogen production rate under AM 1.5 illumination [122]. In addition, a sandwich-like mesoporous black TiO_2/MoS_2/black TiO_2 nanosheet photocatalyst was proposed for visible light photocatalytic hydrogen generation with a much-promoted photo-generated charge transfer efficiency [121]. The mesoporous TiO_2 and Cu_2S were also combined with MoS_2 to synthesize the hierarchical tandem heterojunctions [29]. The near-infrared energy utilization was enhanced by the tandem system, thereby promoting the photothermal effect [29]. The visible light photocatalytic H_2 generation rate was significantly improved, achieving 3376.7 μmol h^{-1} g^{-1}, which was approximately 16 times that of black TiO$_2$ [29].

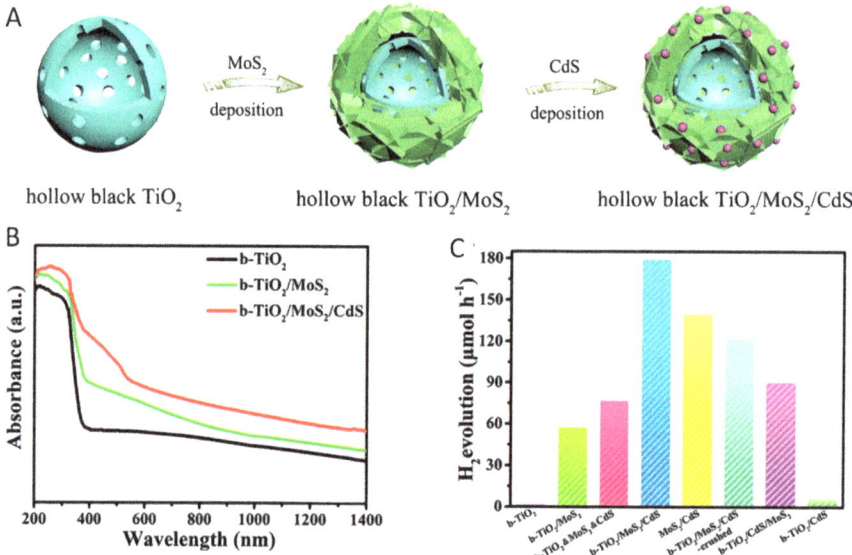

Figure 15. (**A**) Schematic illustration for the formation of TiO$_2$/MoS$_2$/CdS, (**B**) UV-vis spectra, (**C**) photocatalytic H$_2$ evolution measurements. Reprinted with permission for ref. [122]. Copyright 2018, Wiley-VCH.

5. Applications of Black TiO$_2$

The absorption of light can be extended from the ultraviolet light to visible light, and even to near-infrared light by changing the white TiO$_2$ into black TiO$_2$ via various methods. This strategy can be utilized for enhancing the photocatalytic activity in the visible light range. Changing the color of TiO$_2$ from white to black is one of most efficient ways for improving its photocatalytic efficiency in various fields, such as photocatalytic water splitting, photocatalytic pollutant degradation, etc.

5.1. Photocatalytic Water Splitting

Black TiO$_2$ has a modified band structure, thereby improving the charge separation and migration efficiency. Its enhanced photocatalytic performance has been extensively investigated in water splitting. Black mesoporous TiO$_2$ synthesized by Zhou et al. has excellent hydrogen production performance [79]. As shown in Figure 16A, black TiO$_2$ had a higher rate of hydrogen production than original TiO$_2$ under the condition of AM 1.5G and had excellent hydrogen production ability in visible light [79]. In addition, almost no attenuation was detected during photocatalytic measurements after 10 cycles [79]. As presented in Figure 16B, its apparent quantum efficiency at each single wavelength was much higher than that of the original sample [79]. The mesoporous black TiO$_2$ photocatalysts showed remarkable photocatalytic stability in 10 cycling hydrogen evolution measurements within 30 h (each cycling test was conducted in the presence of fresh 1 mL methanol) [79]. The black TiO$_2$ nanotubes prepared by Yang et al. showed an excellent photocatalytic hydrogen production performance (9.8 mmol h^{-1}g^{-1}) through high temperature hydrogenation [80]. The enhanced photocatalytic activity could be attributed to two aspects: (1) the special one-dimensional hollow tube structure improved the charge separation efficiency; (2) the high temperature hydrogenation strategy improved its ability for sunlight utilization [80]. The photocatalytic activity of the black TiO$_2$ nanotubes for H$_2$ production remained stable for 5 cycles in 15 h using H$_2$PtCl$_6$ as the co-catalyst and methanol as the sacrificing agent [80].

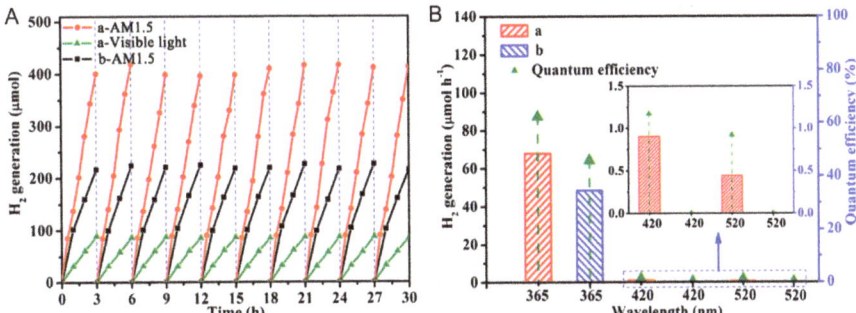

Figure 16. Photocatalytic hydrogen evolution of the ordered mesoporous black TiO$_2$ (a) and pristine ordered mesoporous TiO$_2$ materials (b). (**A**) Cycling tests of photocatalytic hydrogen generation under AM 1.5 and visible light irradiation. (**B**) The photocatalytic hydrogen evolution rates under single-wavelength light and the corresponding QE. The inset enlarges the QE of single-wavelength light at 420 and 520 nm. Reprinted with permission for ref. [79]. Copyright 2014, American Chemical Society.

Two-dimensional lamellar structures with plenty of active sites are often used in TiO$_2$ photocatalysis because of their large specific surface area. The black TiO$_2$ nanosheets prepared by Zhang et al. shortened the band gap to 2.85 eV, thereby broadening the light response to the visible light region [123]. The hydrogen production rate was up to 165 μmol h^{-1} 0.05 g^{-1}, which was twice as much as that of the original sample [123]. The chemical stability, light corrosion resistance, and photocatalytic activity for H$_2$ generation were confirmed by 5 cycling tests in 25 h, using H$_2$PtCl$_6$ and methanol as the co-catalyst and sacrificing agent, respectively [123]. Similarly, Wu et al. synthesized another black TiO$_2$ nanosheets, which had a high hydrogen production rate of 3.73 mmol h^{-1} g^{-1} [83]. This photocatalyst exhibited an unchanged photocatalytic H$_2$ evolution rate in 5 cycling measurements with 15 h [83]. Li et al. designed and synthesized black TiO$_2$ nanospheres by the self-assembly solvothermal method combined with the hydrogenation strategy [81]. The charge separation efficiency had been effectively improved after the hydrogenation process, confirmed by experimental results and DFT calculations [81]. The photocatalytic performance for H$_2$ formation was also repeated 5 times in 15 h and remained stable in the 5 cycles [81].

The black rutile TiO$_2$ prepared by Xiao et al. by the solid-phase reduction method showed much-enhanced hydrogen production performance, stability, and high apparent quantum efficiency, which was about 1.5 times that of the original sample (Figure 17) [87]. In addition, the defects in TiO$_2$ could be regulated by varying the hydrogenation temperature, and the optimal hydrogenation temperature was proved to be 300 °C [87]. The stability of photocatalytic activity was verified by 5 cycling hydrogen formation measurements in 12 h under AM 1.5 illumination [87]. No obvious decrease in the activity was observed in the cycling tests [87]. The maximum hydrogen production rate of black TiO$_2$ synthesized by Sinhamahapatra et al. using the controllable magnesium thermal reduction method was 43 mmol h^{-1} g^{-1} in the full solar wavelength range with excellent stability, which was better than the black TiO$_2$ material previously reported [89]. The black TiO$_2$ nanoparticles presented great stability in the photocatalytic hydrogen evolution confirmed by 10 cycling measurements, which were conducted for 10 consecutive days using the same solution [89]. The aging of the black TiO$_2$ materials was generally not mentioned in the reported publications. The photocatalytic stability of the hydrogen generation was mostly measured in 5 cycling tests within 15 h using the same solution. Some photocatalysts were tested in 10 repeated photocatalytic hydrogen formation measurements for more than 20 h. Black TiO$_2$ photocatalysts usually showed good stability and light corrosion resistance in the photocatalytic H$_2$ evolution reaction, providing the possibility of long-term usage of black TiO$_2$ photocatalysts for H$_2$ production.

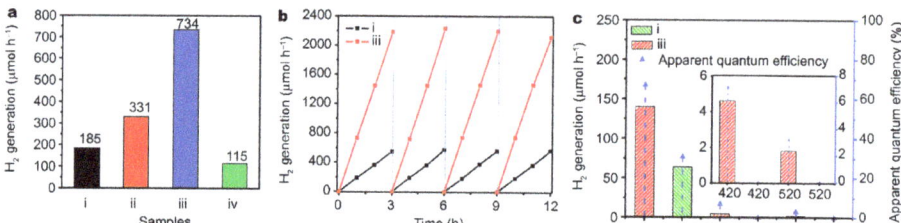

Figure 17. (a) Photocatalytic hydrogen evolution rates, (b) cycling tests of photocatalytic hydrogen generation under AM 1.5 irradiation, (c) the single wavelength photocatalytic hydrogen evolution rates of pristine TiO$_2$ (i), and hydrogenated TiO$_2$ at 250 °C (ii), 300 °C (iii), and 350 °C (iv), respectively. Reprinted with permission for ref. [87]. Copyright 2018, Springer.

5.2. Photocatalytic Degradation of Pollutants

In addition to the hydrogen production, pollutant degradation is also one of the main applications of photocatalysis. Titanium dioxide was often used in the photocatalytic degradation of organic dye and pesticides [124–126]. Black TiO$_2$ with an enhanced light absorption ability would have much improved the photocatalytic pollutant removal efficiency. The black TiO$_2$ obtained by CaH$_2$ reduction not only presented an enhanced hydrogen generation rate, which was 1.7 times that of the original sample, but also achieved a huge improvement in the degradation of pollutants with the complete removal of methyl orange within 8 min [72]. Hamad et al. synthesized black TiO$_2$ using a new method of controlled hydrolysis [85]. The oxygen vacancy concentration was significantly increased with a much-reduced band gap, thereby showing an excellent organic pollutant degradation rate under visible light irradiation [85].

The oxygen vacancy plays an important role in photocatalysis. Black TiO$_2$ prepared by Teng et al. via vapor deposition had a high photocatalytic oxidation activity for organic pollutants in the water, due to the formation of Ti-H bonds and a large number of oxygen vacancies [77]. All pollutants (rhodamine B) could be completely degraded within 50 min detected by the UV-vis spectrophotometer [77]. The defective TiO$_{2-x}$ prepared by the anodic oxidation method was characterized by the electron paramagnetic resonance spectroscopy, confirming the existence of oxygen vacancies and the extension of the absorption from the ultraviolet to visible light region [76]. This black TiO$_2$ material showed excellent photocatalytic degradation activity for rhodamine B under 400–500 nm light irradiation [76].

The black TiO$_2$-based heterojunction could significantly improve its photocatalytic efficiency in pollutant remediation thanks to the enhanced charge separation and transfer efficiency. Jiang et al. prepared black TiO$_2$/Cu$_2$O/Cu composites via in-situ photodeposition and the solid reduction method [127]. The light energy harvesting in the visible and infrared range was much enhanced after the formation of the composites [127]. The photocatalytic efficiency of the composites for Rhodamine B degradation was improved compared with the commercial P25, due to the enhanced charge separation efficiency [127]. Qiang et al. synthesized the RuTe$_2$/black TiO$_2$ photocatalyst through gel calcination and the microwave-assisted process [128]. The light absorption range of the as-made composites was enlarged compared to the pristine TiO$_2$. The photocatalytic efficiency of the diclofenac degradation was 1.2 times higher than the pure black TiO$_2$ [128]. The stability of RuTe$_2$/black TiO$_2$ for the photocatalytic diclofenac degradation was confirmed via 5 repeated experiments [128].

Tetracycline is a toxic antibiotic which is difficult to remove. Li et al. synthesized black TiO_2 modified with Ag/La presented an improved visible light photocatalytic performance for the tetracycline degradation [26]. The photocatalytic stability and reusability of black TiO_2-based photocatalysts were studied via 5 cycling tests without apparent deactivation [26]. Wu et al. reported that the synthesized black anatase TiO_2 exhibited impressive photocatalytic degradation of tetracycline [84]. Its degradation efficiency of tetracycline was 66.2% under the visible light illumination, which was higher than that of the white titanium dioxide and doped titanium dioxide [84]. In addition, $·O_2^-$ and h^+ were found to play important roles in the degradation process, which was different from the original TiO_2, providing new insights for environmental protection [84]. The stability of the photocatalytic tetracycline degradation was measured in four repeated experiments within 960 min without apparent deactivation after four cycles [84]. Table 2 summarizes the applications and photocatalytic stability of the black TiO_2 nanomaterials. The long-term photocatalytic stability of pollutant removal is often overlooked and unverified in most reported research. Therefore, researchers should pay more attention to aging and photocatalytic stability in pollutant degradation in the future.

Table 2. Applications and properties of black TiO_2 nanomaterials.

Black TiO_2	Reducing Agent	Oxygen Vacancy	Application	Numbers of Cycling Tests	Ref.
Anatase with minor rutile	$NaBH_4$	✔	Tetracycline degradation	5	[26]
Anatase and rutile	H_2	✔	Rhodamine B degradation	N/S	[77]
Anatase	H_2	✔	H_2 evolution	10	[79]
Anatase	H_2	✔	H_2 evolution	5	[80]
Anatase	H_2	✔	H_2 evolution	5	[81]
Anatase	H_2	✔	H_2 evolution	5	[83]
Anatase	H_2	✔	Tetracycline degradation	4	[84]
Anatase	H_2	N/S	Orange G degradation	N/S	[85]
Rutile	$NaBH_4$	✔	H_2 evolution	4	[87]
Anatase with minor rutile	H_2	✔	H_2 evolution	10	[89]
Anatase	Anodization	✔	Rhodamine B degradation	N/S	[76]
Anatase	H_2	N/S	H_2 evolution	5	[123]
Anatase and rutile	$NaBH_4$	✔	Rhodamine B degradation	4	[127]
Anatase	H_2	✔	Diclofenac degradation	5	[128]

Notes: N/S: not studied; ✔: yes.

6. Summary and Outlooks

The utilization and conversion of sunlight in a more efficient way has gained much interest due to the energy crisis and global warming effect. The construction of black TiO_2-based materials is proved to be an effective approach for promoting visible light utilization. Black TiO_2 with various morphologies, such as nanospheres, nanotubes, nanowires, etc., have been rationally designed. The properties of black TiO_2, such as the surface area, are easily affected by its morphology. Although the photon absorption of TiO_2 in the visible light region can be effectively increased after the surface reduction process, its photocatalytic performance still needs to be improved for practical applications. The photocatalytic activity of black TiO_2 photocatalysts can be further enhanced by three main methods: element doping, decoration with metal nanoparticles, and fabrication of heterojunctions. The introduction of metal or nonmetal elements into the black TiO_2 lattice can create new impurities, thus narrowing its band gap. The SPR effect caused by metal decoration on the black TiO_2 surface can efficiently improve its visible light utilization. The fabrication of black TiO_2 heterojunctions, including type II, Z scheme, and tandem heterojunctions, can significantly enhance the photo-induced charge separation and transfer efficiency, thereby promoting the photocatalytic performance. The photocatalytic activity of black TiO_2-based materials are mainly evaluated in the photocatalytic hydrogen production and pollutant removal. Although these black TiO_2-based nanomaterials exhibit excellent photocatalytic activity in the visible light region, technologies for enhancing light harvesting in near-infrared should be developed. In addition, the enhancement of the photo-generated charge separation and transfer should be further reinforced to meet the standard for practical application. Applications of black TiO_2-based materials in industrial and outdoor fields, such as self-cleaning surfaces, should also be investigated in future.

Author Contributions: Conceptualization, W.Z.; writing—original draft preparation, L.L. and M.W.; writing—review and editing, Z.L., X.W. and W.Z.; Supervision, Funding acquisition, W.Z. All authors have read and agreed to the published version of the manuscript.

Funding: We gratefully acknowledge the support of the National Natural Science Foundation of China (52172206), the Natural Science Foundation of Shandong Province (ZR2022QD062), the Special Fund for Taishan Scholars Project, and the Development Plan of Youth Innovation Team in Col-leges and Universities of Shandong Province.

Data Availability Statement: Not applicable.

Conflicts of Interest: The authors declare no conflict of interest.

References

1. Fujishima, A.; Honda, K. Electrochemical photolysis of water at a semiconductor electrode. *Nature* **1972**, *238*, 37–38. [CrossRef]
2. Xu, Y.; Kraft, M.; Xu, R. Metal-free carbonaceous electrocatalysts and photocatalysts for water splitting. *Chem. Soc. Rev.* **2016**, *45*, 3039–3052. [CrossRef] [PubMed]
3. Yao, B.; Zhang, J.; Fan, X.; He, J.; Li, Y. Surface Engineering of Nanomaterials for Photo-Electrochemical Water Splitting. *Small* **2019**, *15*, 1803746. [CrossRef]
4. Xu, S.; Carter, E.A. Theoretical Insights into Heterogeneous (Photo)electrochemical CO_2 Reduction. *Chem. Rev.* **2019**, *119*, 6631–6669. [CrossRef]
5. Zhang, B.; Sun, L. Artificial photosynthesis: Opportunities and challenges of molecular catalysts. *Chem. Soc. Rev.* **2019**, *48*, 2216–2264. [CrossRef] [PubMed]
6. Diercks, C.S.; Liu, Y.; Cordova, K.E.; Yaghi, O.M. The role of reticular chemistry in the design of CO_2 reduction catalysts. *Nat. Mater.* **2018**, *17*, 301–307. [CrossRef] [PubMed]
7. Hoffmann, M.R.; Martin, S.T.; Choi, W.; Bahnemann, D.W. Environmental Applications of Semiconductor Photocatalysis. *Chem. Rev.* **1995**, *95*, 69–96. [CrossRef]
8. Wang, H.; Zhang, L.; Chen, Z.; Hu, J.; Li, S.; Wang, Z.; Liu, J.; Wang, X. Semiconductor heterojunction photocatalysts: Design, construction, and photocatalytic performances. *Chem. Soc. Rev.* **2014**, *43*, 5234–5244. [CrossRef]
9. Calvete, M.J.F.; Piccirillo, G.; Vinagreiro, C.S.; Pereira, M.M. Hybrid materials for heterogeneous photocatalytic degradation of antibiotics. *Coord. Chem. Rev.* **2019**, *395*, 63–85. [CrossRef]
10. Schneider, J.; Matsuoka, M.; Takeuchi, M.; Zhang, J.; Horiuchi, Y.; Anpo, M.; Bahnemann, D.W. Understanding TiO_2 Photocatalysis: Mechanisms and Materials. *Chem. Rev.* **2014**, *114*, 9919–9986. [CrossRef]
11. Honda, M.; Ochiai, T.; Listiani, P.; Yamaguchi, Y.; Ichikawa, Y. Low-Temperature Synthesis of Cu-Doped Anatase TiO_2 Nanostructures via Liquid Phase Deposition Method for Enhanced Photocatalysis. *Materials* **2023**, *16*, 639. [CrossRef]
12. Boytsova, O.; Zhukova, I.; Tatarenko, A.; Shatalova, T.; Beiltiukov, A.; Eliseev, A.; Sadovnikov, A. The Anatase-to-Rutile Phase Transition in Highly Oriented Nanoparticles Array of Titania with Photocatalytic Response Changes. *Nanomaterials* **2022**, *12*, 4418. [CrossRef]
13. Di Paola, A.; Bellardita, M.; Palmisano, L. Brookite, the Least Known TiO_2 Photocatalyst. *Catalysts* **2013**, *3*, 36–73. [CrossRef]
14. Serga, V.; Burve, R.; Krumina, A.; Romanova, M.; Kotomin, E.A.; Popov, A.I. Extraction–Pyrolytic Method for TiO_2 Polymorphs Production. *Crystals* **2021**, *11*, 431. [CrossRef]
15. Tsebriienko, T.; Popov, A.I. Effect of Poly (Titanium Oxide) on the Viscoelastic and Thermophysical Properties of Interpenetrating Polymer Networks. *Crystals* **2021**, *11*, 794. [CrossRef]
16. Trestsova, M.A.; Utepova, I.A.; Chupakhin, O.N.; Semenov, M.V.; Pevtsov, D.N.; Nikolenko, L.M.; Tovstun, S.A.; Gadomska, A.V.; Shchepochkin, A.V.; Kim, G.A.; et al. Oxidative C-H/C-H Coupling of Dipyrromethanes with Azines by TiO_2-Based Photocatalytic System. Synthesis of New BODIPY Dyes and Their Photophysical and Electrochemical Properties. *Molecules* **2021**, *26*, 5549. [CrossRef]

17. Karawek, A.; Kittipoom, K.; Tansuthepverawongse, L.; Kitjanukit, N.; Neamsung, W.; Lertthanaphol, N.; Chanthara, P.; Ratchahat, S.; Phadungbut, P.; Kim-Lohsoontorn, P.; et al. The Photocatalytic Conversion of Carbon Dioxide to Fuels Using Titanium Dioxide Nanosheets/Graphene Oxide Heterostructure as Photocatalyst. *Nanomaterials* **2023**, *13*, 320. [CrossRef]
18. Kim, S.-Y.; Lee, T.-G.; Hwangbo, S.-A.; Jeong, J.-R. Effect of the TiO_2 Colloidal Size Distribution on the Degradation of Methylene Blue. *Nanomaterials* **2023**, *13*, 302. [CrossRef] [PubMed]
19. Chen, X.; Mao, S.S. Titanium Dioxide Nanomaterials: Synthesis, Properties, Modifications, and Applications. *Chem. Rev.* **2007**, *107*, 2891–2959. [CrossRef]
20. Dahl, M.; Liu, Y.; Yin, Y. Composite Titanium Dioxide Nanomaterials. *Chem. Rev.* **2014**, *114*, 9853–9889. [CrossRef]
21. Kumaravel, V.; Mathew, S.; Bartlett, J.; Pillai, S. Photocatalytic hydrogen production using metal doped TiO_2: A review of recent advances. *Appl. Catal. B Environ.* **2019**, *244*, 1021–1064. [CrossRef]
22. Asahi, R.; Morikawa, T.; Irie, H.; Ohwaki, T. Nitrogen-Doped Titanium Dioxide as Visible-Light-Sensitive Photocatalyst: Designs, Developments, and Prospects. *Chem. Rev.* **2014**, *114*, 9824–9852. [CrossRef]
23. Wei, L.; Yu, C.; Zhang, Q.; Liu, H.; Wang, Y. TiO_2-based heterojunction photocatalysts for photocatalytic reduction of CO_2 into solar fuels. *J. Mater. Chem. A* **2018**, *6*, 22411–22436. [CrossRef]
24. Liu, J.; Luo, Z.; Mao, X.; Dong, Y.; Peng, L.; Sun-Waterhouse, D.; Kennedy, J.V.; Waterhouse, G.I.N. Recent Advances in Self-Supported Semiconductor Heterojunction Nanoarrays as Efficient Photoanodes for Photoelectrochemical Water Splitting. *Small* **2022**, *18*, 2204552. [CrossRef]
25. Chen, X.; Liu, L.; Yu, P.Y.; Mao, S.S. Increasing Solar Absorption for Photocatalysis with Black Hydrogenated Titanium Dioxide Nanocrystals. *Science* **2011**, *331*, 746–750. [CrossRef]
26. Li, C.; Sun, T.; Zhang, D.; Zhang, X.; Qian, Y.; Zhang, Y.; Lin, X.; Liu, J.; Zhu, L.; Wang, X.; et al. Fabrication of ternary Ag/La-black TiO_{2-x} photocatalyst with enhanced visible-light photocatalytic activity for tetracycline degradation. *J. Alloy. Compd.* **2022**, *891*, 161960. [CrossRef]
27. Zhu, S.; Yu, Z.; Zhang, L.; Watanabe, S. Solution Plasma-Synthesized Black TiO_2 Nanoparticles for Solar–Thermal Water Evaporation. *ACS Appl. Nano Mater.* **2021**, *4*, 3940–3948. [CrossRef]
28. Biswas, S.; Lee, H.; Prasad, M.; Sharma, A.; Yu, J.; Sengupta, S.; Deo Pathak, D.; Sinhamahapatra, A. Black TiO_{2-x} Nanoparticles Decorated with Ni Nanoparticles and Trace Amounts of Pt Nanoparticles for Photocatalytic Hydrogen Generation. *ACS Appl. Nano Mater.* **2021**, *4*, 4441–4451. [CrossRef]
29. Li, Z.; Li, H.; Wang, S.; Yang, F.; Zhou, W. Mesoporous black $TiO_2/MoS_2/Cu_2S$ hierarchical tandem heterojunctions toward optimized photothermal-photocatalytic fuel production. *Chem. Eng. J.* **2022**, *427*, 131830. [CrossRef]
30. Balati, A.; Tek, S.; Nash, K.; Shipley, H. Nanoarchitecture of TiO_2 microspheres with expanded lattice interlayers and its heterojunction to the laser modified black TiO_2 using pulsed laser ablation in liquid with improved photocatalytic performance under visible light irradiation. *J. Colloid Interface Sci.* **2019**, *541*, 234–248. [CrossRef]
31. Zhou, L.; Cai, M.; Zhang, X.; Cui, N.; Chen, G.; Zou, G. In-situ nitrogen-doped black TiO_2 with enhanced visible-light-driven photocatalytic inactivation of Microcystis aeruginosa cells: Synthesization, performance, and mechanism. *Appl. Catal. B Environ.* **2020**, *272*, 119019. [CrossRef]
32. Zada, I.; Zhang, W.; Sun, P.; Imtiaz, M.; Iqbal, N.; Ghani, U.; Naz, R.; Zhang, Y.; Li, Y.; Gu, J.; et al. Superior photothermal black TiO_2 with random size distribution as flexible film for efficient solar steam generation. *Appl. Mater. Today* **2020**, *20*, 100669. [CrossRef]
33. Li, F.; Wang, S.; Yin, H.; Chen, Y.; Zhou, Y.; Huang, J.; Ai, S. Photoelectrochemical Biosensor for DNA Formylation Detection in Genomic DNA of Maize Seedlings Based on Black TiO_2-Enhanced Photoactivity of MoS_2/WS_2 Heterojunction. *ACS Sens.* **2020**, *5*, 1092–1101. [CrossRef]
34. Chen, Y.; Yin, H.; Li, F.; Zhou, J.; Wang, L.; Wang, J.; Ai, S. Polydopamine-sensitized WS_2/black-TiO_2 heterojunction for histone acetyltransferase detection with enhanced visible-light-driven photoelectrochemical activity. *Chem. Eng. J.* **2020**, *393*, 124707. [CrossRef]
35. Cheng, Y.; Gao, J.; Shi, Q.; Li, Z.; Huang, W. In situ electrochemical reduced Au loaded black TiO_2 nanotubes for visible light photocatalysis. *J. Alloy. Compd.* **2022**, *901*, 163562. [CrossRef]
36. Touni, A.; Liu, X.; Kang, X.; Carvalho, P.A.; Diplas, S.; Both, K.G.; Sotiropoulos, S.; Chatzitakis, A. Galvanic Deposition of Pt Nanoparticles on Black TiO2 Nanotubes for Hydrogen Evolving Cathodes. *ChemSusChem* **2021**, *14*, 4993–5003. [CrossRef] [PubMed]
37. Lim, J.; Yang, Y.; Hoffmann, M.R. Activation of Peroxymonosulfate by Oxygen Vacancies-Enriched Cobalt-Doped Black TiO_2 Nanotubes for the Removal of Organic Pollutants. *Environ. Sci. Technol.* **2019**, *53*, 6972–6980. [CrossRef]
38. Qiao, P.; Sun, B.; Li, H.; Pan, K.; Tian, G.; Wang, L.; Zhou, W. Surface Plasmon Resonance-Enhanced Visible-NIR-Driven Photocatalytic and Photothermal Catalytic Performance by Ag/Mesoporous Black TiO_2 Nanotube Heterojunctions. *Chem. Asian J.* **2019**, *14*, 177–186. [CrossRef]
39. Gao, J.; Xue, J.; Jia, S.; Shen, Q.; Zhang, X.; Jia, H.; Liu, X.; Li, Q.; Wu, L. Self-Doping Surface Oxygen Vacancy-Induced Lattice Strains for Enhancing Visible Light-Driven Photocatalytic H_2 Evolution over Black TiO_2. *ACS Appl. Mater. Interfaces* **2021**, *13*, 18758–18771. [CrossRef]
40. Shim, Y.; Lim, J.; Hong, S. Black-TiO_2 based photoelectrochemical oxidation of flue-gas desulfurization wastewater for effective reuse in flow-electrode CDI. *Desalination* **2022**, *538*, 115899. [CrossRef]

41. Chen, J.; Fu, Y.; Sun, F.; Hu, Z.; Wang, X.; Zhang, T.; Zhang, F.; Wu, X.; Chen, H.; Cheng, G.; et al. Oxygen vacancies and phase tuning of self-supported black TiO_{2-x} nanotube arrays for enhanced sodium storage. *Chem. Eng. J.* **2020**, *400*, 125784. [CrossRef]
42. Gao, J.; Shen, Q.; Guan, R.; Xue, J.; Liu, X.; Jia, H.; Li, Q.; Wu, Y. Oxygen vacancy self-doped black TiO_2 nanotube arrays by aluminothermic reduction for photocatalytic CO_2 reduction under visible light illumination. *J. CO2 Util.* **2020**, *35*, 205–215. [CrossRef]
43. Li, Z.; Bian, H.; Xiao, X.; Shen, J.; Zhao, C.; Lu, J.; Li, Y. Defective Black TiO_2 Nanotube Arrays for Enhanced Photocatalytic and Photoelectrochemical Applications. *ACS Appl. Nano Mater.* **2019**, *2*, 7372–7378. [CrossRef]
44. Zhang, D.; Cong, T.; Xia, L.; Pan, L. Growth of black TiO_2 nanowire/carbon fiber composites with dendritic structure for efficient visible-light-driven photocatalytic degradation of methylene blue. *J. Mater. Sci.* **2019**, *54*, 7576–7588. [CrossRef]
45. Yang, L.; Peng, Y.; Yang, Y.; Liu, J.; Li, Z.; Ma, Y.; Zhang, Z.; Wei, Y.; Li, S.; Huang, Z.; et al. Green and Sensitive Flexible Semiconductor SERS Substrates: Hydrogenated Black TiO_2 Nanowires. *ACS Appl. Nano Mater.* **2018**, *1*, 4516–4527. [CrossRef]
46. Nawaz, R.; Sahrin, N.T.; Haider, S.; Ullah, H.; Junaid, M.; Akhtar, M.S.; Khan, S. Photocatalytic performance of black titanium dioxide for phenolic compounds removal from oil refinery wastewater: Nanoparticles vs. nanowires. *Appl. Nanosci.* **2022**, *12*, 3499–3515. [CrossRef]
47. Lim, J.; Kim, S.; Armengol, R.A.; Kasian, O.; Choi, P.; Stephenson, L.T.; Gault, B.; Scheu, C. Atomic-Scale Mapping of Impurities in Partially Reduced Hollow TiO_2 Nanowires. *Angew. Chem. Int. Ed.* **2020**, *59*, 5651–5655. [CrossRef]
48. Xue, X.; Chen, H.; Xiong, Y.; Chen, R.; Jiang, M.; Fu, G.; Xi, Z.; Zhang, X.; Ma, J.; Fang, W.; et al. Near-Infrared-Responsive Photo-Driven Nitrogen Fixation Enabled by Oxygen Vacancies and Sulfur Doping in Black $TiO_{2-x}S_y$ Nanoplatelets. *ACS Appl. Mater. Interfaces* **2021**, *13*, 4975–4983. [CrossRef]
49. Sun, L.; Xie, J.; Zhang, L.; Jiang, R.; Wu, J.; Fan, L.; Shao, R.; Chen, Z.; Jin, Z. 2D black TiO_{2-x} nanoplate-decorated Ti_3C_2 MXene hybrids for ultrafast and elevated stable lithium storage. *FlatChem* **2020**, *20*, 100152. [CrossRef]
50. Zhu, G.; Ma, L.; Lin, H.; Zhao, P.; Wang, L.; Hu, Y.; Chen, R.; Chen, T.; Wang, Y.; Tie, Z.; et al. High-performance Li-ion capacitor based on black-TiO_{2-x}/graphene aerogel anode and biomass-derived microporous carbon cathode. *Nano Res.* **2019**, *12*, 1713–1719. [CrossRef]
51. Zhang, W.; Xue, J.; Shen, Q.; Jia, S.; Gao, J.; Liu, X.; Jia, H. Black single-crystal TiO_2 nanosheet array films with oxygen vacancy on {001} facets for boosting photocatalytic CO_2 reduction. *J. Alloys Compd.* **2021**, *870*, 159400. [CrossRef]
52. Zou, L.; Zhu, Y.; Hu, Z.; Cao, X.; Cen, W. Remarkably improved photocatalytic hydrogen evolution performance of crystalline TiO_2 nanobelts hydrogenated at atmospheric pressure with the assistance of hydrogen spillover. *Catal. Sci. Technol.* **2022**, *12*, 5575–5585. [CrossRef]
53. Zhang, T.; Xing, Z.; Xiu, Z.; Li, Z.; Yang, S.; Zhu, Q.; Zhou, W. Surface defect and rational design of TiO_{2-x} nanobelts/ g-C_3N_4 nanosheets/ CdS quantum dots hierarchical structure for enhanced visible-light-driven photocatalysis. *Int. J. Hydrog. Energ.* **2019**, *44*, 1586–1596.
54. Zhao, Q.; Bi, R.; Cui, J.; Yang, X.; Zhang, L. TiO_{2-x} Nanocages Anchored in N-Doped Carbon Fiber Films as a Flexible Anode for High-Energy Sodium-Ion Batteries. *ACS Appl. Energy Mater.* **2018**, *1*, 4459–4466. [CrossRef]
55. Yu, Z.; Xun, S.; Jing, M.; Chen, H.; Song, W.; Chao, Y.; Rahmani, M.; Ding, Y.; Hua, M.; Liu, J.; et al. Construction of 3D TiO_2 nanoflower for deep catalytic oxidative desulfurization in diesel: Role of oxygen vacancy and Ti^{3+}. *J. Hazard. Mater.* **2022**, *440*, 129859. [CrossRef]
56. Choi, J.U.; Kim, Y.G.; Jo, W.K. Multiple photocatalytic applications of non-precious Cu-loaded g-C_3N_4/hydrogenated black TiO_2 nanofiber heterostructure. *Appl. Surf. Sci.* **2019**, *473*, 761–769. [CrossRef]
57. Yan, Z.; Huang, W.; Jiang, X.; Gao, J.; Hu, Y.; Zhang, H.; Shi, Q. Hollow structured black TiO_2 with thickness-controllable microporous shells for enhanced visible-light-driven photocatalysis. *Microporous Mesoporous Mater.* **2021**, *323*, 111228. [CrossRef]
58. Shi, X.; Liu, Z.; Li, X.; You, W.; Shao, Z.; Che, R. Enhanced dielectric polarization from disorder-engineered Fe_3O_4@black TiO_{2-x} heterostructure for broadband microwave absorption. *Chem. Eng. J.* **2021**, *419*, 130020. [CrossRef]
59. Ren, L.; Zhou, W.; Sun, B.; Li, H.; Qiao, P.; Xu, Y.; Wu, J.; Lin, K.; Fu, H. Defects-engineering of magnetic γ-Fe_2O_3 ultra-thin nanosheets/mesoporous black TiO_2 hollow sphere heterojunctions for efficient charge separation and the solar-driven photocatalytic mechanism of tetracycline degradation. *Appl. Catal. B Environ.* **2019**, *240*, 319–328. [CrossRef]
60. Varnagiris, S.; Medvids, A.; Lelis, M.; Milcius, D.; Antuzevics, A. Black carbon-doped TiO_2 films: Synthesis, characterization and photocatalysis. *J. Photochem. Photobiol. A Chem.* **2019**, *382*, 111941. [CrossRef]
61. Islam, S.Z.; Reed, A.; Nagpure, S.; Wanninayake, N.; Browning, J.F.; Strzalka, J.; Kim, D.Y.; Rankin, S.E. Hydrogen incorporation by plasma treatment gives mesoporous black TiO_2 thin films with visible photoelectrochemical water oxidation activity. *Microporous Mesoporous Mater.* **2018**, *261*, 35–43. [CrossRef]
62. Chahrour, K.M.; Yam, F.K.; Eid, A.M.; Nazeer, A.A. Enhanced photoelectrochemical properties of hierarchical black TiO_{2-x} nanolaces for Cr (VI) photocatalytic reduction. *Int. J. Hydrog. Energ.* **2020**, *45*, 22674–22690. [CrossRef]
63. Katal, R.; Salehi, M.; Hossein, M.; Farahani, D.A.; Masudy-Panah, S.; Ong, S.L.; Hu, J. Preparation of a New Type of Black TiO_2 under a Vacuum Atmosphere for Sunlight Photocatalysis. *ACS Appl. Mater. Interfaces* **2018**, *10*, 35316–35326. [CrossRef] [PubMed]
64. Hu, W.; Zhou, W.; Zhang, K.; Zhang, X.; Wang, L.; Jiang, B.; Tian, G.; Zhao, D.; Fu, H. Facile strategy for controllable synthesis of stable mesoporous black TiO_2 hollow spheres with efficient solar-driven photocatalytic hydrogen evolution. *J. Mater. Chem. A* **2016**, *4*, 7495–7502. [CrossRef]

65. Zhang, K.; Zhou, W.; Zhang, X.; Sun, B.; Wang, L.; Pan, K.; Jiang, B.; Tian, G.; Fu, H. Self-floating amphiphilic black TiO_2 foams with 3D macro-mesoporous architectures as efficient solar-driven photocatalysts. *Appl. Catal. B Environ.* **2017**, *206*, 336–343. [CrossRef]
66. Zhou, G.; Meng, H.; Cao, Y.; Kou, X.; Duan, S.; Fan, L.; Xiao, M.; Zhou, F.; Li, Z.; Xing, Z. Surface plasmon resonance-enhanced solar-driven photocatalytic performance from Ag nanoparticles-decorated Ti^{3+} self-doped porous black TiO_2 pillars. *J. Ind. Eng. Chem.* **2018**, *64*, 188–193. [CrossRef]
67. Kang, J.; Zhang, Y.; Chai, Z.; Qiu, X.; Cao, X.; Zhang, P.; Teobaldi, G.; Liu, L.; Guo, L. Amorphous Domains in Black Titanium Dioxide. *Adv. Matter.* **2021**, *33*, 2100407. [CrossRef]
68. Ullattil, S.G.; Narendrannth, S.B.; Pillai, S.C.; Periyat, P. Black TiO_2 Nanomaterials: A Review of Recent Advances. *Chem. Eng. J.* **2018**, *343*, 708–736. [CrossRef]
69. Zheng, P.; Zhang, L.; Zhang, X.; Ma, Y.; Qian, J.; Jiang, Y.; Li, H. Hydrogenation of TiO_2 nanosheets and nanoparticles: Typical reduction stages and orientation-related anisotropic disorder. *J. Mater. Chem. A* **2021**, *9*, 22603–22614. [CrossRef]
70. Ioannidou, E.; Ioannidi, A.; Frontistis, Z.; Antonopoulou, M.; Tselios, C.; Tsikritzis, D.; Konstantinou, I.; Kennou, S.; Kondarides, D.I.; Mantzavinos, D. Correlating the properties of hydrogenated titania to reaction kinetics and mechanism for the photocatalytic degradation of bisphenol A under solar irradiation. *Appl. Catal. B Environ.* **2016**, *188*, 65–76. [CrossRef]
71. Koohgard, M.; Hosseini-Sarvari, M. Black TiO_2 nanoparticles with efficient photocatalytic activity under visible light at low temperature: Regioselective C-N bond cleavage toward the synthesis of thioureas, sulfonamides, and propargylamines. *Catal. Sci. Technol.* **2020**, *10*, 6825–6839. [CrossRef]
72. Andronic, L.; Enesca, A. Black TiO_2 Synthesis by Chemical Reduction Methods for Photocatalysis Applications. *Front. Chem.* **2020**, *8*, 565489. [CrossRef]
73. Wang, X.; Mayrhofer, L.; Hoefer, M.; Estrade, S.; Lopez-Conesa, L.; Zhou, H.; Lin, Y.; Peiró, F.; Fan, Z.; Shen, H.; et al. Facile and Efficient Atomic Hydrogenation Enabled Black TiO_2 with Enhanced Photo-Electrochemical Activity via a Favorably Low-Energy-Barrier Pathway. *Adv. Energy Mater.* **2019**, *9*, 1900725. [CrossRef]
74. Zimbone, M.; Cacciato, G.; Sanz, R.; Carles, R.; Gulino, A.; Privitera, V.; Grimaldi, M.G. Black TiO_x photocatalyst obtained by laser irradiation in water. *Catal. Commun.* **2016**, *84*, 11–15. [CrossRef]
75. Yao, D.; Hu, Z.; Zheng, L.; Chen, S.; Lü, W.; Xu, H. Laser-engineered black rutile TiO_2 photoanode for CdS/CdSe-sensitized quantum dot solar cells with a significant power conversion efficiency of 9.1%. *Appl. Surf. Sci.* **2023**, *608*, 155230. [CrossRef]
76. Dong, J.; Han, J.; Liu, Y.; Nakajima, A.; Matsushita, S.; Wei, S.; Gao, W. Defective Black TiO_2 Synthesized via Anodization for Visible-Light Photocatalysis. *ACS Appl. Mater. Interfaces* **2014**, *6*, 1385–1388. [CrossRef]
77. Teng, F.; Li, M.; Gao, C.; Zhang, G.; Zhang, P.; Wang, Y.; Chen, L.; Xie, E. Preparation of black TiO_2 by hydrogen plasma assisted chemical vapor deposition and its photocatalytic activity. *Appl. Catal. B Environ.* **2014**, *148–149*, 339–343. [CrossRef]
78. Huang, H.; Zhang, H.; Ma, Z.; Liu, Y.; Zhang, X.; Han, Y.; Kang, Z. Si quantum dot-assisted synthesis of mesoporous black TiO_2 nanocrystals with high photocatalytic activity. *J. Mater. Chem. A* **2013**, *1*, 4162–4166. [CrossRef]
79. Zhou, W.; Li, W.; Wang, J.; Qu, Y.; Yang, Y.; Xie, Y.; Zhang, K.; Wang, L.; Fu, H.; Zhao, D. Ordered Mesoporous Black TiO_2 as Highly Efficient Hydrogen Evolution Photocatalyst. *J. Am. Chem. Soc.* **2014**, *136*, 9280–9283. [CrossRef]
80. Yang, W.; Li, M.; Pan, K.; Guo, L.; Wu, J.; Li, Z.; Yang, F.; Lin, K.; Zhou, W. Surface engineering of mesoporous anatase titanium dioxide nanotubes for rapid spatial charge separation on horizontal-vertical dimensions and efficient solar-driven photocatalytic hydrogen evolution. *J. Colloid Interface Sci.* **2021**, *586*, 75–83. [CrossRef]
81. Li, Z.; Wang, S.; Xie, Y.; Yang, W.; Tao, B.; Lu, J.; Wu, J.; Qu, Y.; Zhou, W. Surface defects induced charge imbalance for boosting charge separation and solar-driven photocatalytic hydrogen evolution. *J. Colloid Interface Sci.* **2021**, *596*, 12–21. [CrossRef] [PubMed]
82. Zhang, X.; Hu, W.; Zhang, K.; Wang, J.; Sun, B.; Li, H.; Qiao, P.; Wang, L.; Zhou, W. Ti^{3+} Self-Doped Black TiO_2 Nanotubes with Mesoporous Nanosheet Architecture as Efficient Solar-Driven Hydrogen Evolution Photocatalysts. *ACS Sustain. Chem. Eng.* **2017**, *5*, 6894–6901. [CrossRef]
83. Wu, J.; Qiao, P.; Li, H.; Xu, Y.; Yang, W.; Yang, F.; Lin, K.; Pan, K.; Zhou, W. Engineering surface defects on two-dimensional ultrathin mesoporous anatase TiO_2 nanosheets for efficient charge separation and exceptional solar-driven photocatalytic hydrogen evolution. *J. Mater. Chem. C* **2020**, *8*, 3476–3482. [CrossRef]
84. Wu, S.; Li, X.; Tian, Y.; Lin, Y.; Hu, Y.H. Excellent photocatalytic degradation of tetracycline over black anatase-TiO_2 under visible light. *Chem. Eng. J.* **2021**, *406*, 126747. [CrossRef]
85. Hamad, H.; Bailón-García, E.; Maldonado-Hódar, F.J.; Pérez-Cadenas, A.F.; Carrasco-Marín, F.; Morales-Torres, S. Synthesis of Ti_xO_y nanocrystals in mild synthesis conditions for the degradation of pollutants under solar light. *Appl. Catal. B Environ.* **2019**, *241*, 385–392. [CrossRef]
86. Cai, J.; Wu, M.; Wang, Y.; Zhang, H.; Meng, M.; Tian, Y.; Li, X.; Zhang, J.; Zheng, L.; Gong, J. Synergetic Enhancement of Light Harvesting and Charge Separation over Surface-Disorder-Engineered TiO_2 Photonic Crystals. *Chem* **2017**, *2*, 877–892. [CrossRef]
87. Xiao, F.; Zhou, W.; Sun, B.; Li, H.; Qiao, P.; Ren, L.; Zhao, X.; Fu, H. Engineering oxygen vacancy on rutile TiO_2 for efficient electron-hole separation and high solar-driven photocatalytic hydrogen evolution. *Sci. China Mater.* **2018**, *61*, 822–830. [CrossRef]
88. Zhu, G.; Yin, H.; Yang, C.; Cui, H.; Wang, Z.; Xu, J.; Lin, T.; Huang, F. Black Titania for Superior Photocatalytic Hydrogen Production and Photoelectrochemical Water Splitting. *ChemCatChem* **2015**, *7*, 2614–2619. [CrossRef]

89. Sinhamahapatra, A.; Jeon, J.P.; Yu, J.S. A new approach to prepare highly active and stable black titania for visible light-assisted hydrogen production. *Energy Environ. Sci.* **2015**, *8*, 3539–3544. [CrossRef]
90. Ullattil, S.G.; Periyat, P. A 'one pot' gel combustion strategy towards Ti^{3+} self-doped 'black' anatase TiO_{2-x} solar photocatalyst. *J. Mater. Chem. A* **2016**, *4*, 5854–5858. [CrossRef]
91. Campbell, L.; Nguyen, S.H.; Webb, H.K.; Eldridge, D.S. Photocatalytic disinfection of S. aureus using black TiO_{2-x} under visible light. *Catal. Sci. Technol.* **2023**, *13*, 62–71. [CrossRef]
92. Basavarajappa, P.S.; Patil, S.B.; Ganganagappa, N.; Reddy, K.R.; Raghu, A.V.; Reddy, C.V. Recent progress in metal-doped TiO_2, non-metal doped/codoped TiO_2 and TiO_2 nanostructured hybrids for enhanced photocatalysis. *Int. J. Hydrog. Energy* **2020**, *45*, 7764–7778. [CrossRef]
93. Nah, Y.C.; Paramasivam, I.; Schmuki, P. Doped TiO_2 and TiO_2 Nanotubes: Synthesis and Applications. *ChemPhysChem* **2010**, *11*, 2698–2713. [CrossRef]
94. Yi, L.; Ci, S.; Luo, S.; Shao, P.; Hou, Y.; Wen, Z. Scalable and low-cost synthesis of black amorphous Al-Ti-O nanostructure for high-efficient photothermal desalination. *Nano Energy* **2017**, *41*, 600–608. [CrossRef]
95. Zhang, H.; Xing, Z.; Zhang, Y.; Li, Z.; Wu, X.; Liu, C.; Zhu, Q.; Zhou, W. Ni^{2+} and Ti^{3+} co-doped porous black anatase TiO_2 with unprecedented-high visible-light-driven photocatalytic degradation performance. *RSC Adv.* **2015**, *5*, 107150–107157. [CrossRef]
96. Ansari, S.A.; Khan, M.M.; Ansari, M.O.; Cho, M.H. Nitrogen-doped titanium dioxide (N-doped TiO_2) for visible light photocatalysis. *New J. Chem.* **2016**, *40*, 3000–3009. [CrossRef]
97. Petala, A.; Tsikritzis, D.; Kollia, M.; Ladas, S.; Kennou, S.; Kondarides, D.I. Synthesis and characterization of N-doped TiO_2 photocatalysts with tunable response to solar radiation. *Appl. Surf. Sci.* **2014**, *305*, 281–291. [CrossRef]
98. Asahi, R.; Morikawa, T.; Ohwaki, T.; Aoki, K.; Taga, Y. Visible-Light Photocatalysis in Nitrogen-Doped Titanium Oxides. *Science* **2001**, *293*, 269–271. [CrossRef]
99. Hu, Z.; Xu, T.; Liu, P.; Oeser, M. Microstructures and optical performances of nitrogen-vanadium co-doped TiO_2 with enhanced purification efficiency to vehicle exhaust. *Environ. Res.* **2021**, *193*, 110560. [CrossRef]
100. Liu, H.; Fan, H.; Wu, R.; Tian, L.; Yang, X.; Sun, Y. Nitrogen-doped black TiO_2 spheres with enhanced visible light photocatalytic performance. *SN Appl. Sci.* **2019**, *1*, 487. [CrossRef]
101. Gao, J.; Xue, J.; Shen, Q.; Liu, T.; Zhang, X.; Liu, X.; Jia, H.; Li, Q.; Wu, Y. A promoted photocatalysis system trade-off between thermodynamic and kinetic via hierarchical distribution dual-defects for efficient H_2 evolution. *Chem. Eng. J.* **2022**, *431*, 133281. [CrossRef]
102. Cao, Y.; Xing, Z.; Hu, M.; Li, Z.; Wu, X.; Zhao, T.; Xiu, Z.; Yang, S.; Zhou, W. Mesoporous black N-TiO_{2-x} hollow spheres as efficient visible-light-driven photocatalysts. *J. Catal.* **2017**, *356*, 246–254. [CrossRef]
103. Wang, S.; Gao, Y.; Miao, S.; Liu, T.; Li, R.; Mu, L.; Li, R.; Fan, F.; Li, C. Positioning the Water Oxidation Reaction Sites in Plasmonic Photocatalysts. *J. Am. Chem. Soc.* **2017**, *139*, 11771–11778. [CrossRef]
104. Jiang, J.; Xing, Z.; Li, M.; Li, Z.; Yin, J.; Kuang, J.; Zou, J.; Zhu, Q.; Zhou, W. Plasmon Ag decorated 3D urchinlike N-TiO_{2-x} for enhanced visible-light-driven photocatalytic performance. *J. Colloid Interface Sci.* **2018**, *521*, 102–110. [CrossRef] [PubMed]
105. Li, H.; Shen, L.; Zhang, K.; Sun, B.; Ren, L.; Qiao, P.; Pan, K.; Wang, L.; Zhou, W. Surface plasmon resonance-enhanced solar-driven photocatalytic performance from Ag nanoparticle-decorated self-floating porous black TiO_2 foams. *Appl. Catal. B: Environ.* **2018**, *220*, 111–117. [CrossRef]
106. Kuang, J.; Xing, Z.; Yin, J.; Li, Z.; Zhu, Q.; Zhou, W. Surface plasma Ag-decorated single-crystalline TiO_{2-x}(B) nanorod/defect-rich g-C_3N_4 nanosheet ternary superstructure 3D heterojunctions as enhanced visible-light-driven photocatalyst. *J. Colloid Interface Sci.* **2019**, *542*, 63–72. [CrossRef]
107. Li, M.; Xing, Z.; Jiang, J.; Li, Z.; Yin, J.; Kuang, J.; Tan, S.; Zhu, Q.; Zhou, W. Surface plasmon resonance-enhanced visible-light-driven photocatalysis by Ag nanoparticles decorated S-TiO_{2-x} nanorods. *J. Taiwan Inst. Chem. Eng.* **2018**, *82*, 198–204. [CrossRef]
108. Xiao, Y.; Wang, K.; Yang, Z.; Xing, Z.; Li, Z.; Pan, K.; Zhou, W. Plasma Cu-decorated TiO_{2-x}/CoP particle-level hierarchical heterojunctions with enhanced hotocatalytic-photothermal performance. *J. Hazard. Mater.* **2021**, *414*, 125487. [CrossRef]
109. Wang, S.; Sun, H.; Qiao, P.; Li, Z.; Xie, Y.; Zhou, W. NiS/Pt nanoparticles co-decorated black mesoporous TiO_2 hollow nanotube assemblies as efficient hydrogen evolution photocatalysts. *Appl. Mater. Today* **2021**, *22*, 100977. [CrossRef]
110. Wang, C.; Li, J.; Paineau, E.; Remita, H.; Ghazzal, M.N. Pt Atomically Dispersed in Black $TiO2-x$/CuxO with Chiral-Like Nanostructure for Visible-Light H2 Generation. *Sol. RRL* **2023**, 2200929, early view. [CrossRef]
111. Yin, J.; Xing, Z.; Kuang, J.; Li, Z.; Zhu, Q.; Zhou, W. Dual oxygen vacancy defects-mediated efficient electron-hole separation via surface engineering of Ag/Bi_2MoO_6 nanosheets/TiO_2 nanobelts ternary heterostructures. *J. Ind. Eng. Chem.* **2019**, *78*, 155–163. [CrossRef]
112. Liu, X.; Xing, Z.; Zhang, Y.; Li, Z.; Wu, X.; Tan, S.; Yu, X.; Zhu, Q.; Zhou, W. Fabrication of 3D flower-like black N-TiO_{2-x}@MoS_2 for unprecedented-high visible-light-driven photocatalytic performance. *Appl. Catal. B Environ.* **2017**, *201*, 119–127. [CrossRef]
113. Shen, L.; Xing, Z.; Zou, J.; Li, Z.; Wu, X.; Zhang, Y.; Zhu, Q.; Yang, S.; Zhou, W. Black TiO_2 nanobelts/g-C_3N_4 nanosheets Laminated Heterojunctions with Efficient Visible-Light-Driven Photocatalytic Performance. *Sci. Rep.* **2017**, *7*, 41978. [CrossRef]
114. Hu, M.; Xing, Z.; Cao, Y.; Li, Z.; Yan, X.; Xiu, Z.; Zhao, T.; Yang, S.; Zhou, W. Ti^{3+} self-doped mesoporous black TiO_2/SiO_2/g-C_3N_4 sheets heterojunctions as remarkable visible-lightdriven photocatalysts. *Appl. Catal. B Environ.* **2018**, *226*, 499–508. [CrossRef]

115. Zhao, T.; Xing, Z.; Xiu, Z.; Li, Z.; Shen, L.; Cao, Y.; Hu, M.; Yang, S.; Zhou, W. CdS quantum dots/Ti^{3+}-TiO_2 nanobelts heterojunctions as efficient visible-light-driven photocatalysts. *Mater. Res. Bull.* **2018**, *103*, 114–121. [CrossRef]
116. Sun, B.; Zhou, W.; Li, H.; Ren, L.; Qiao, P.; Xiao, F.; Wang, L.; Jiang, B.; Fu, H. Magnetic Fe_2O_3/mesoporous black TiO_2 hollow sphere heterojunctions with wide-spectrum response and magnetic separation. *Appl. Catal. B Environ.* **2018**, *221*, 235–242. [CrossRef]
117. Xiu, Z.; Xing, Z.; Li, Z.; Wu, X.; Yan, X.; Hu, M.; Cao, Y.; Yang, S.; Zhou, W. Ti^{3+}-TiO_2/Ce^{3+}-CeO_2 Nanosheet heterojunctions as efficient visible-light-driven photocatalysts. *Mater. Res. Bull.* **2018**, *100*, 191–197. [CrossRef]
118. Tan, S.; Xing, Z.; Zhang, J.; Li, Z.; Wu, X.; Cui, J.; Kuang, J.; Zhu, Q.; Zhou, W. Ti^{3+}-TiO_2/g-C_3N_4 mesostructured nanosheets heterojunctions as efficient visible-light-driven photocatalysts. *J. Catal.* **2018**, *357*, 90–99. [CrossRef]
119. Kuang, J.; Xing, Z.; Yin, J.; Li, Z.; Zhu, Q.; Zhou, W. Assembly of surface-defect single-crystalline strontium titanate nanocubes acting as molecular bricks onto surface-defect single-crystalline titanium dioxide (B) nanorods for efficient visible-light-driven photocatalytic performance. *J. Colloid Interface Sci.* **2019**, *537*, 441–449. [CrossRef]
120. Sun, D.; Chi, D.; Yang, Z.; Xing, Z.; Chen, P.; Li, Z.; Pan, K.; Zhou, W. CdS quantum dots modified surface oxygen vacancy defect ZnO_{1-x}-TiO_{2-x} solid solution sphere as Z-Scheme heterojunctions for efficient visible light-driven photothermal-photocatalytic performance. *J. Alloys Compd.* **2020**, *826*, 154218. [CrossRef]
121. Liu, X.; Xing, Z.; Zhang, H.; Wang, W.; Zhang, Y.; Li, Z.; Wu, X.; Yu, X.; Zhou, W. Fabrication of 3D Mesoporous Black TiO_2/MoS_2/TiO_2 Nanosheets for Visible-Light-Driven Photocatalysis. *ChemSusChem* **2016**, *9*, 1118–1124. [CrossRef]
122. Sun, B.; Zhou, W.; Li, H.; Ren, L.; Qiao, P.; Li, W.; Fu, H. Synthesis of Particulate Hierarchical Tandem Heterojunctions toward Optimized Photocatalytic Hydrogen Production. *Adv. Mater.* **2018**, *30*, 1804282. [CrossRef]
123. Zhang, K.; Zhou, W.; Zhang, X.; Qu, Y.; Wang, L.; Hu, W.; Pan, K.; Li, M.; Xie, Y.; Jiang, B.; et al. Large-scale synthesis of stable mesoporous black TiO_2 nanosheets for efficient solar-driven photocatalytic hydrogen evolution via an earth-abundant low-cost biotemplate. *RSC Adv.* **2016**, *6*, 50506–50512. [CrossRef]
124. Lee, B.T.; Han, J.K.; Gain, A.K.; Lee, K.H.; Saito, F. TEM microstructure characterization of nano TiO_2 coated on nano ZrO_2 powders and their photocatalytic activity. *Mater. Lett.* **2006**, *60*, 2101–2104. [CrossRef]
125. Wafi, M.A.E.; Ahmed, M.A.; Abdel-Samad, H.S.; Medien, H.A.A. Exceptional removal of methylene blue and p-aminophenol dye over novel TiO_2/RGO nanocomposites by tandem adsorption-photocatalytic processes. *Mater. Sci. Energy Technol.* **2022**, *5*, 217–231. [CrossRef]
126. Zeshan, M.; Bhatti, I.A.; Mohsin, M.; Iqbal, M.; Amjed, N.; Nisar, J.; AlMasoud, N.; Alomar, T.S. Remediation of pesticides using TiO_2 based photocatalytic strategies: A review. *Chemosphere* **2022**, *300*, 134525. [CrossRef]
127. Jiang, X.; Fuji, M. In-Situ Preparation of Black TiO_2/Cu_2O/Cu Composites as an Efficient Photocatalyst for Degradation Pollutants and Hydrogen Production. *Catal. Lett.* **2022**, *152*, 3272–3283. [CrossRef]
128. Qiang, C.; Li, N.; Zuo, S.; Guo, Z.; Zhan, W.; Li, Z.; Ma, J. Microwave-assisted synthesis of $RuTe_2$/black TiO_2 photocatalyst for enhanced diclofenac degradation: Performance, mechanistic investigation and intermediates analysis. *Sep. Purif. Technol.* **2022**, *283*, 120214. [CrossRef]

Disclaimer/Publisher's Note: The statements, opinions and data contained in all publications are solely those of the individual author(s) and contributor(s) and not of MDPI and/or the editor(s). MDPI and/or the editor(s) disclaim responsibility for any injury to people or property resulting from any ideas, methods, instructions or products referred to in the content.

Article

Bioactive Coatings Based on Nanostructured TiO₂ Modified with Noble Metal Nanoparticles and Lysozyme for Ti Dental Implants

Emilian Chifor [1,2], Ion Bordeianu [1], Crina Anastasescu [3,*], Jose Maria Calderon-Moreno [3], Veronica Bratan [3], Diana-Ioana Eftemie [3], Mihai Anastasescu [3,*], Silviu Preda [3,*], Gabriel Plavan [4], Diana Pelinescu [5], Robertina Ionescu [5], Ileana Stoica [5], Maria Zaharescu [3] and Ioan Balint [3]

[1] Faculty of Medicine of the Ovidius University, Aleea Universitatii nr.1, 900470 Constanţa, Romania
[2] "Strungareata" SRL, Strada Garii nr. 24, 800217 Galati, Romania
[3] "Ilie Murgulescu" Institute of Physical Chemistry of the Romanian Academy, 202 Spl. Independentei, 060021 Bucharest, Romania
[4] Faculty of Biology, "Alexandru Ioan Cuza" University, 700505 Iasi, Romania
[5] Faculty of Biology, Intrarea Portocalilor 1-3, Sector 5, 060101 Bucharest, Romania
* Correspondence: canastasescu@yahoo.com (C.A.); manastasescu_ro@yahoo.com (M.A.); predas01@yahoo.co.uk (S.P.)

Abstract: This work presents the synthesis of nanostructured TiO₂ modified with noble metal nanoparticles (Au, Ag) and lysozyme and coated on titanium foil. Moreover, the specific structural and functional properties of the resulting inorganic and hybrid materials were explored. The purpose of this study was to identify the key parameters for developing engineered coatings on titanium foil appropriate for efficient dental implants with intrinsic antibacterial activity. TiO₂ nanoparticles obtained using the sol–gel method were deposited on Ti foil and modified with Au/Ag nanoparticles. Morphological and structural investigations (scanning electron and atomic force microscopies, X-ray diffraction, photoluminescence, and UV–Vis spectroscopies) were carried out for the characterization of the resulting inorganic coatings. In order to modify their antibacterial activity, which is essential for safe dental implants, the following aspects were investigated: (a) singlet oxygen ($^1O_2$) generation by inorganic coatings exposed to visible light irradiation; (b) the antibacterial behavior emphasized by titania-based coatings deposited on titanium foil (TiO₂/Ti foil; Au–TiO₂/Ti foil, Ag–TiO₂/Ti foil); (c) the lysozyme bioactivity on the microbial substrate (*Micrococcus lysodeicticus*) after its adsorption on inorganic surfaces (Lys/TiO₂/Ti foil; Lys/Au–TiO₂/Ti foil, Lys/Ag–TiO₂/Ti foil); (d) the enzymatic activity of the above-mentioned hybrids materials for the hydrolysis reaction of a synthetic organic substrate usually used for monitoring the lysozyme biocatalytic activity, namely, 4-Methylumbelliferyl β-D-N,N′,N″-triacetylchitotrioside [4-MU-β-(GlcNAc)₃]. This was evaluated by identifying the presence of a fluorescent reaction product, 7-hydroxy-4-metyl coumarin (4-methylumbelliferone).

Keywords: TiO₂ coatings for titanium implant; noble metal nanoparticles; photosensitive materials; antibacterial activity; hybrid materials; biocatalytic activity of lysozyme

Citation: Chifor, E.; Bordeianu, I.; Anastasescu, C.; Calderon-Moreno, J.M.; Bratan, V.; Eftemie, D.-I.; Anastasescu, M.; Preda, S.; Plavan, G.; Pelinescu, D.; et al. Bioactive Coatings Based on Nanostructured TiO₂ Modified with Noble Metal Nanoparticles and Lysozyme for Ti Dental Implants. *Nanomaterials* 2022, 12, 3186. https://doi.org/10.3390/nano12183186

Academic Editor: Wei Zhou

Received: 21 August 2022
Accepted: 12 September 2022
Published: 14 September 2022

Publisher's Note: MDPI stays neutral with regard to jurisdictional claims in published maps and institutional affiliations.

Copyright: © 2022 by the authors. Licensee MDPI, Basel, Switzerland. This article is an open access article distributed under the terms and conditions of the Creative Commons Attribution (CC BY) license (https://creativecommons.org/licenses/by/4.0/).

1. Introduction

Titanium-based materials are commonly used for biomedical applications, especially in the dentistry and orthopedic fields, as they are characterized by a wide spectrum of morphological, compositional, and functional parameters [1–3]. The intensive development of innovative and valuable dental implants relies on certain prerequisite features and standards of the envisaged materials, such as appropriate mechanical resistance, biocompatibility [1], and the ability to promote osseointegration [2]. Moreover, numerous research studies and studies of medical technologies focus on materials that have their own intrinsic antibacterial capacities, which are able to develop bioactive interfaces with the contacting tissues. These are focused on the prevention of bacterial adhesion and biofilm formation [3] on the implant surface, leading to peri-implant disease and implant failure. Therefore, many studies are concerned with modifying the implant roughness by electrochemical [4], acid etching [5,6] and various blasting procedures [7,8]. The resulting rough surfaces are beneficial for faster healing, cell adhesion, and the osteogenesis process [9]. Similar advances can be achieved by the chemical modification of the titanium surface, leading to a wide range of bioactive coatings, such calcium phosphate ceramics, especially hydroxyapatite [10], titania [11] silica [12,13], zirconia [14], zinc oxide [15], and their composites/mixtures [16].

A great number of self-disinfecting inorganic coatings for titanium implants are based on metal and oxide nanoparticles, which are activated by contact with the appropriate media or by light irradiation [17]. An example of such a coating is a Ag–TiO_2 layer, which was shown to be active under visible light against the vesicular stomatitis virus [18]. These approaches complete conventional medical procedures in terms of decontaminating the implanted materials and protecting the injured tissues from bacterial colonization, which is common after implantation procedures [19]. According to the literature data, the main mechanisms exhibited by the inorganic nanomaterials to induce pathogen extinction are related to the release of metallic ions, electrostatic interaction with the bacteria membrane, and generating reactive oxygen species (singlet oxygen, superoxide anion, hydroxyl radical, hydrogen peroxide) (ROS) [20]. Photogenerated ROS were investigated as a potential solution in the recent pandemic context due to their wide antimicrobial effect, even against bacterial biofilms and viruses [17].

This work investigates the pathways to developing and characterizing effective titanium coatings appropriate for bioactive dental implants that are able to display self-decontamination prior to implantation through singlet oxygen ($^1O_2$) generation under visible irradiation, antibacterial effect revealed by the interaction with *Micrococcus lysodeicticus* both for inorganic nanostructured layers and the resulting organic/inorganic hybrid systems (Lys/TiO_2/Ti foil; Lys/Au–TiO_2/Ti foil, Lys/Ag–TiO_2/Ti foil) after lysozyme adsorption on the inorganic surface, and biocatalytic effectiveness of the loaded lysozyme for the hydrolysis of a synthetic substrate (4-Methylumbelliferyl β-D-N,N′,N″-triacetylchitotrioside) used previously for monitoring the enzymatic activity of lysozyme after immobilization on solid carriers [21,22].

Moreover, it is important to investigate the engineered TiO_2-based coatings and their lysozyme loading capacity to better understand the usual in vivo-developed TiO_2 layer on titanium implants, including their interaction with lysozyme usually found in saliva [23,24]. Although lysozyme is an antibacterial enzyme that is common in nature, present in human body [25], and frequently used as a model protein for fundamental and applicative research studies [26,27] due to its appropriate dimensions and relative stability after immobilization on different solid supports, its action mechanism remains poorly understood.

The aim of the present work was to propose an innovative approach for the dual modification of a sol–gel TiO_2 layer covering titanium, both with Au/Ag NPs and lysozyme, in order to achieve active nanostructures for safe dental implants with self-disinfecting, antibacterial, and biocatalytic features.

2. Materials and Methods
Material Synthesis

Development of bare and metal-modified TiO_2 coatings (Au–TiO_2/Ti foil, Ag–TiO_2/Ti foil, TiO_2/Ti foil)

Titania precursor sol was prepared according to our previous work [28], using titanium isopropoxide (97% Aldrich), isopropyl alcohol (99.7% Lachner), and 2,4-Pentadione (99% Alfa Aesar), cast by spin coating on Ti foil (0.1 mm thickness, Goodfellow Metals, metal plate samples of 0.7×1.5 cm^2) previously subjected to etching with nitric acid (65% Lachner) for 1 h. In this sense, five successive depositions were made using 10 μL of the synthesized sol subjected to 500 rpm for 60 s (VTC 100 PA, Vacuum Spin Coater,

MTI Corporation, UK). After five deposition cycles, thermal treatment in air at 250 °C for 3 h was performed. Aqueous solution (7.5 µL, 3 mM) of gold chloride trihydrate (MP Biomedicals, LLC) or aqueous solution (7.5 µL, 3 mM) of silver nitrate (99.9% Wako) was further added by drop casting on the above-mentioned coatings, which were dried at 80 °C and subsequently treated at 400 °C for 3 h.

Lysozyme adsorption on inorganic coatings

The coated metal plates (TiO_2/Ti, Au–TiO_2/Ti, Ag–TiO_2/Ti) were introduced into 4 mL potassium phosphate buffer (PBS, pH 6.5) containing lysozyme (0.4 mg/mL) and gently shaken for 1 h at 25 °C. After removing the supernatant, the plates were washed twice (with PBS and ultrapure water) and dried in a vacuum. Lysozyme (from chicken egg white) and potassium phosphate buffer were contained using a Lysozyme Activity Kit (LY0100) (Sigma Aldrich).

Scanning Electron Microscopy (SEM)

SEM images were obtained in a high0resolution microscope (FEI Quanta 3D FEG model) equipped with the Octane Elect X-ray EDS system, in a high vacuum, using an acceleration voltage of 30 kV for both SEM (secondary electrons detection mode) and EDS measurements. NPs size distribution of Au and Ag was estimated with an SPIP (Scanning Probe Image Processor, v. 4.6.0).

Atomic Force Microscopy (AFM)

Atomic force microscopy (AFM) measurements were performed with an XE–100 from Park Systems, equipped with XY/Z decoupled scanners, by selecting the "non-contact" mode. This working mode was preferred due to the minimization of tip-sample interaction. All AFM measurements were performed with NCHR tips produced by Nanosensors, with a typical radius of curvature of ~8 nm, a length of ~125 µm, a width of 30 µm, an elasticity constant of ~42 N/m, and a resonance frequency of ~330 kHz. The AFM images were processed with the XEI program (v 1.8.0), produced by the same company (Park Systems).

X-ray diffraction (XRD)

The measurements were performed using the Rigaku Ultima IV equipment, with Cu K_α radiation and a fixed power source (40 kV and 30 mA). The diffractometer was set in the grazing incidence X-ray diffraction (GIXD) condition with the fixed incidence angle set at $\alpha = 0.5°$. The films were scanned at a rate of $1°$/min over a range of $2\theta = 20–90°$.

UV–Vis Spectroscopy (UV–Vis)

Diffuse reflectance UV–Vis spectra were recorded with a Perkin Elmer Lambda 35 spectrophotometer with a spectral range of 200–1000 nm. The registered reflectance data were transformed into absorption spectra using the Kubelka–Munk function.

Photoluminescence Spectroscopy (PL)

Photoluminescence data were registered with a Carry Eclipse fluorescence spectrometer (Agilent Technologies) equipped with thin film accessories. The working parameters were as follows: a scan rate of 120 nm min^{-1}, slits were set at 20 nm both in excitation and emission, and measurements were performed at room temperature for λ_{exc} = 270 nm.

ROS (singlet oxygen $^1O_2$) identification

Measurements for singlet oxygen identification were performed in quartz cuvettes containing methanolic solution (5 µM) of SOSG (Singlet Oxygen Sensor Green–Thermo Fisher Scientific/Invitrogen). The interest sample was also placed in the cuvette with the coated side in front of a solar simulator (PECEL equipped with a cut of filter for $\lambda > 420$ nm) in order to expose the oxygen singlet ($^1O_2$) under light. Due to its reaction with SOSG (anthracene component), endoperoxide formation occurred. Its presence was further evidenced by a photoluminescence signal peaked around 530 nm for λ_{exc} = 488 nm at every 10 min after light exposure.

Antibacterial activity of inorganic coatings (TiO_2/Ti, Ag–TiO_2/Ti, Au–TiO_2/Ti) was tested against *Micrococcus lysodeicticus*.

The samples of interest (0.3 × 0.3 cm^2) were introduced into 1 mL LB medium and 30 µL *Micrococcus (M.) lysodeikticus* ATCC 4698 cell suspension 0.01% *w/v* in potassium phosphate buffer (Lysozyme Activity Kit –LY0100 Sigma Aldrich) and incubated for 24 h at 37 °C. In order to determine the cell growth, the optical density (OD) at 600 nm was recorded using the multireader Bio Tek Sybergy HTX, Agilent US. The determination of OD values represents a rapid method for microbial cell quantification. All tests were performed in triplicate.

Lysozyme (Lys/TiO$_2$/Ti, Lys/Ag–TiO$_2$/Ti, Lys/Au–TiO$_2$/Ti) activity assays on microbial substrate (*Micrococcus lysodeicticus*)

The samples of interest (0.3 × 0.3 cm^2) were introduced into 1 mL *M. lysodeikticus* cell suspension (ATCC 4698 from Lysozyme Activity Kit, LY0100 Sigma Aldrich, 0.01% w/v in potassium phosphate buffer) and incubated at 25 °C. The decrease in absorbance at 450 nm was monitored at 5, 10 min, 1 h, and 24 h with the multireader Bio Tek Sybergy HTX, Agilent US. The recorded results were compared with the *M. lysodeikticus* cell suspension (negative control) and lysozyme solution (300 U/mL, positive control). All tests were performed in triplicate.

Lysozyme (Lys/TiO$_2$/Ti, Lys/Ag–TiO$_2$/Ti, Lys/Au–TiO$_2$/Ti) activity assays on synthetic substrate [4-MU-β-(GlcNAc)$_3$]

Biocatalytic activity of hybrid systems (Lys/TiO$_2$/Ti, Lys/Ag–TiO$_2$/Ti, Lys/Au–TiO$_2$/Ti) was tested for 4-methylumbelliferyl β-D-N,N′,N″-triacetylchitotrioside [4-MU-β-(GlcNAc)$_3$] hydrolysis reaction. The formation of a fluorescent reaction product, namely, 7-hydroxy-4-metyl coumarin was monitored by spectroscopy fluorescence measurements, with a PL emission signal being registered at 450 nm for λ_{exc} = 355 nm.

In order to perform the enzymatic assay, hybrid organic/inorganic samples (0.7 × 1.5 cm^2) were kept for 3 h at 30 °C in 3 mL buffered solution of 4-MU-β-(GlcNAc)$_3$ (0.01 mg/mL, pH 7). The release in the solution of the fluorescent reaction product was proved by photoluminescence measurements for λ_{exc} = 355 nm, with an emission peak centered at 447 nm being present. The measurements were conducted with a Carry Eclipse fluorescence spectrometer (Agilent Technologies). The enzymatic activity tests were performed in triplicate.

3. Results

3.1. SEM

SEM investigations allow the Au- and Ag-modified TiO$_2$ films covering the Ti foil to be explored.

The microstructural study by SEM revealed the formation of a titania film with Au-faceted particles, from 30 to 400 nm in size (Figure 1a,b). The cross-section image (Figure 1b) of the Au–TiO$_2$ film showed the presence of Au particles embedded inside the film, with a uniform distribution. SEM micrographs of the Ag–TiO$_2$ film (Figure 1d,e) showed the presence of Ag nanoparticles with relatively uniform sizes, i.e., ~50–100 nm, at the film surface (Figure 1d), while the cross-section (Figure 1e) at the film edge showed aggregates of a few nanoparticles embedded inside the film. The NPs size distribution of the Au (Figure 1c) and Ag (Figure 1f) showed that the Au NPs exhibited a larger size (with the majority centered on 100 nm) in comparison with Ag NPs, with an average size of ~60 nm.

In addition, from Figure 1b,e, the substrate for the TiO$_2$ layer (bottom right corner) can be observed, namely, the titanium foil exposing a rough and defective surface. This is due to the treatment with nitric acid, which was meant to increase the adhesion of the sol-containing titanium precursor. According to the literature [29], the acid etching procedure is also used to decontaminate the implant surface. Au and Ag nanoparticles cover the titania surface and can also be identified in the layer thickness, displaying a composite structure. The two-stage thermal treatment applied to the investigated coatings was responsible for the nanoparticle mixture, i.e., the 250 °C air treatment of the Ti foil coated with the titanium containing sol produced an amorphous structure. This was converted into a crystalline phase (anatase) at the second stage of thermal treatment (at 450 °C) also involving the metal precursors.

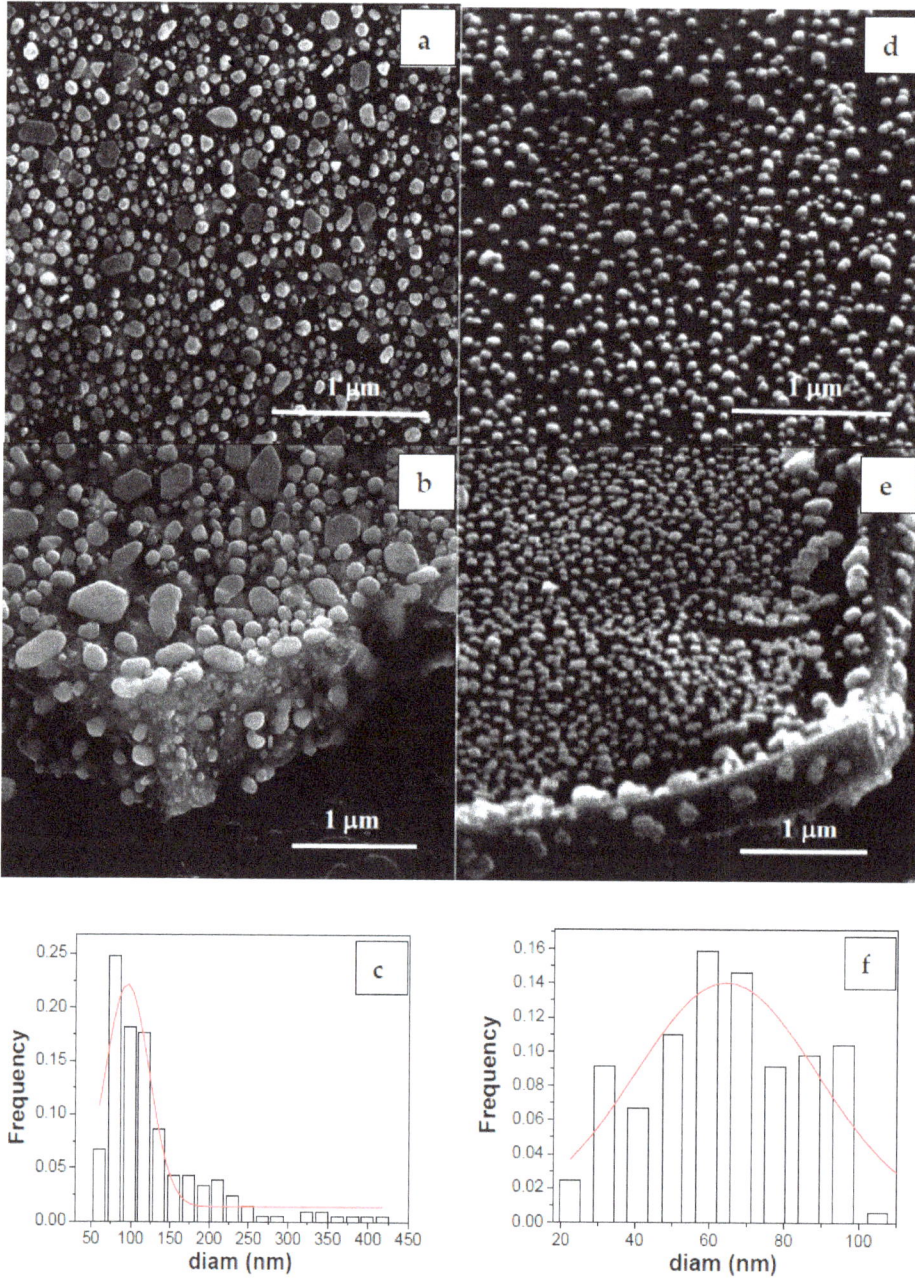

Figure 1. SEM micrographs of Au–TiO$_2$ film (**a,b**) and Ag–TiO$_2$ film (**d,e**), showing surface Au top view (**a**) and Ag tilted view (**d**) together with cross-section edge views (**b,e**) of the films. The particle size distribution for Au is presented in (**c**) while the Ag NPs histogram distribution is shown in (**f**).

Chemical elemental analysis by EDS (Figure 2) confirmed the presence of Au and Ag particles decorating the titania films. Taking into account the atomic weights of Ag, Au, O, and Ti, the EDS-calculated Ag/TiO$_2$ volume ratio was found to be ~30 +/− 8%, while for the Au/TiO$_2$, the evaluated volume ratio was ~12 +/− 5%. In fact, the functional "fingerprint" of the metallic nanoparticles dispersed on the titania layer surface should be further explored in terms of the antibacterial activity of the synthesized coatings, a hindering effect of bacterial growth being presumable. The volume ratio of the metal particles and the TiO$_2$ matrix was evaluated from EDS measurements.

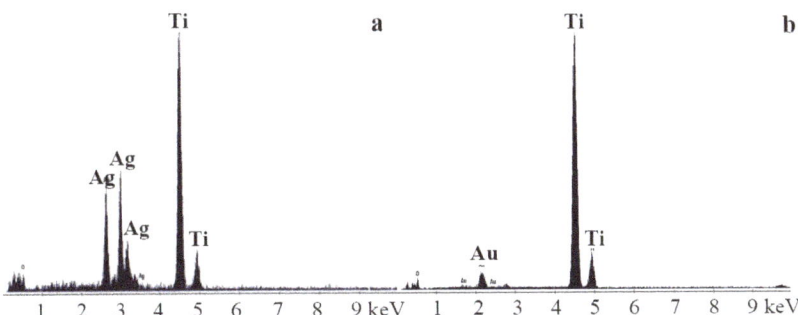

Figure 2. EDS spectra of the Ag–TiO$_2$ film (**a**) and Au–TiO$_2$ film (**b**).

3.2. AFM

Figure 3 comparatively presents the morphology of all investigated samples by AFM at the scales of 2 × 2 µm^2 (left column) and 1 × 1 µm^2 (right column). Thus, Figure 3a,b show the topography of the bare Ti foil used as the substrate. The bare Ti substrate exhibits some irregularities consisting of pits and valleys, most probably formed during the nitric acid treatment.

The root mean square roughness (R_q) at the scale of 2 × 2 µm^2 was found to be 10.8 nm, and the peak-to-valley parameter (R_{pv}), which is the height between the lowest and the highest points on the scanned area, was 92.0 nm;. At the scale of 1 × 1 µm^2, they reached 5.6 nm (R_q) and 36.8 nm (R_{pv}), respectively. Figure 3c,d present 2D AFM images of the Ti foil (2 × 2 and 1 × 1 µm^2) covered by TiO$_2$. The morphology of the surface consists of quasi-spherical shaped particles ~20 nm in diameter (random agglomerated particles/clusters can also be observed). At the scale of 2 × 2 µm^2, the following corrugation parameters were observed: R_q = 21.3 nm and R_{pv} = 169.1 nm, while at the scale of 1 × 1 µm^2, they were R_q = 19.1 nm and R_{pv} = 166.4 nm. After covering the TiO$_2$ surface with lysozyme (Figure 3e,f), a high adhesion to the TiO$_2$ film was observed. The surface became completely covered with the enzymatic layer, with the TiO$_2$ particles becoming almost unrecognizable below the lysozyme layer. For the Lys/TiO$_2$/Ti sample, the following corrugation parameters were observed: R_q = 33.8 nm and R_{pv} = 272.4 nm at the scale of 2 × 2 µm^2, while at the scale of 1 × 1 µm^2, they were R_q = 27.3 nm and R_{pv} = 182.0 nm.

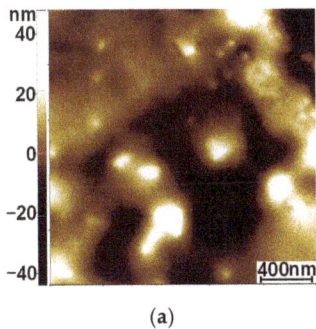
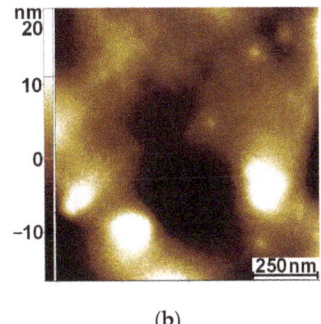

(**a**)　　　　　　　　　　　　　　(**b**)

Figure 3. *Cont.*

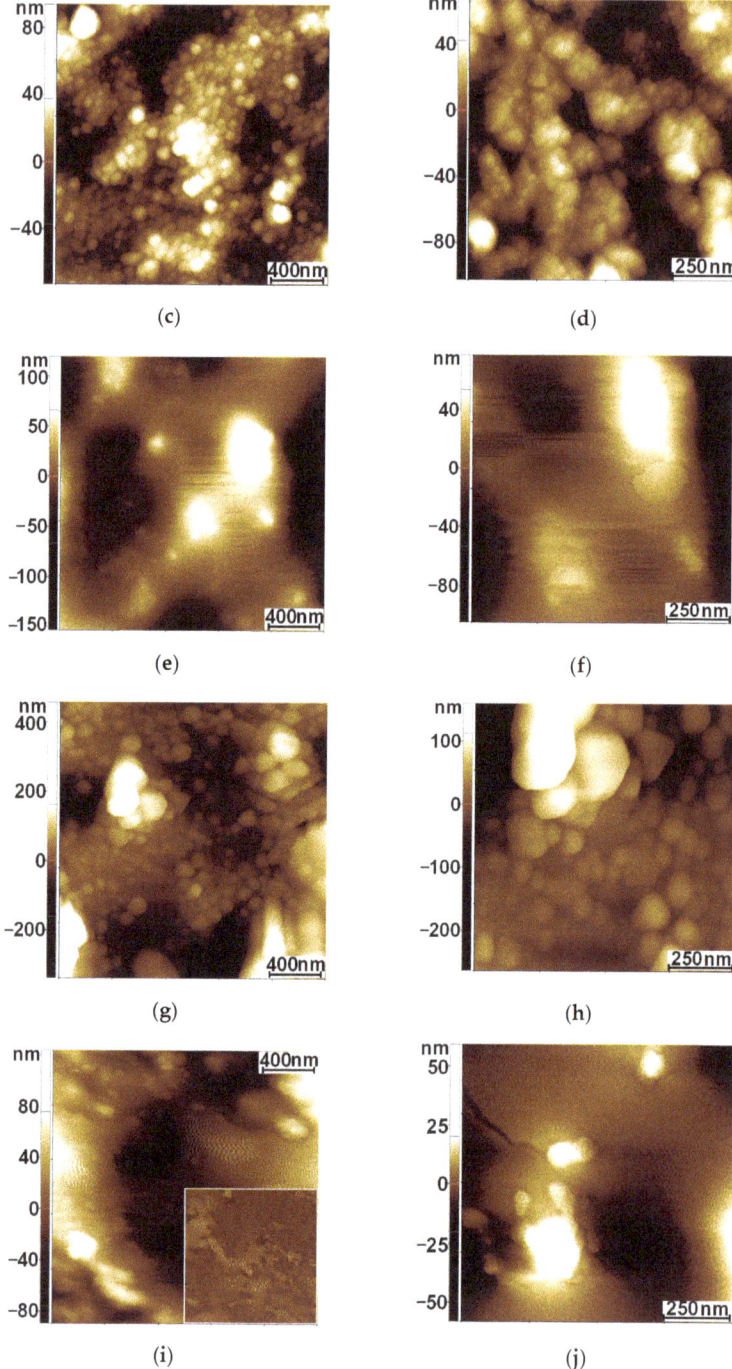

Figure 3. Cont.

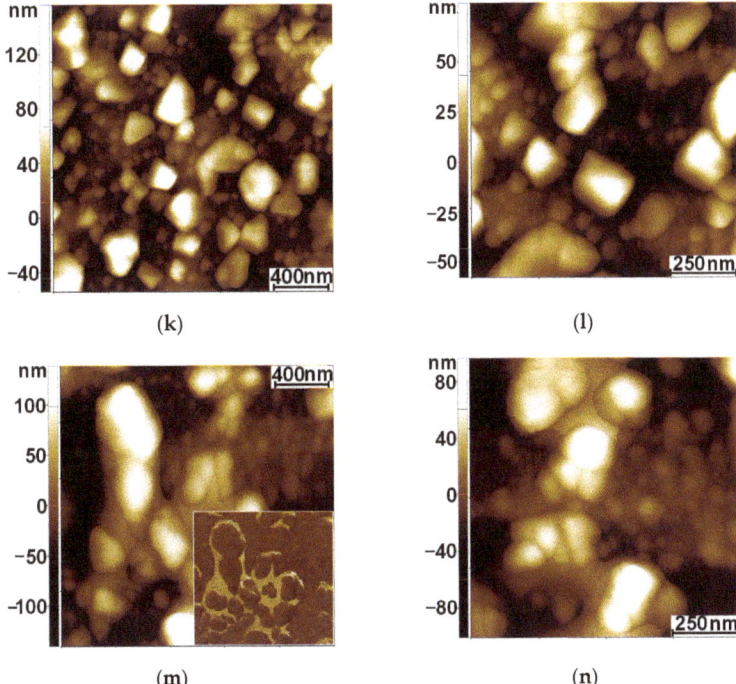

Figure 3. 2D AFM images at a scale of 2×2 µm^2 (**left** column) and 1×1 µm^2 (**right** column) for bare Ti foil (**a,b**), TiO$_2$/Ti (**c,d**), Lys/TiO$_2$ (**e,f**), Au-modified TiO$_2$ (**g,h**), Lys/Au–TiO$_2$/Ti (**i,j**), Ag-modified TiO$_2$ (**k,l**) and Lys/Ag-modified TiO$_2$ films (**m,n**). Phase contrast 2D AFM images are superimposed for Lys/Au/TiO$_2$/Ti in (**i**) and for Lys/Ag/TiO$_2$/Ti in (**m**).

Furthermore, the surface modification with noble metal NPs led to a good dispersion of Au (Figure 3g,h) and Ag NPs (Figure 3k,l) on the TiO$_2$/Ti films. The line scans collected in the fast-scan direction (not shown here) express a similar height in the vertical direction of ~250 nm (from −150 to 100 nm) for both samples. The root mean square (RMS) roughness, R_q, of the Au-modified TiO$_2$ sample, at the scale of 2×2 µm^2, was ~82.1 nm, and the peak-to-valley parameter, R_{pv} reached 793.7 nm. At the 1×1 µm^2, they were R_q = 57.5 nm and R_{pv} = 424.4 nm. Meanwhile, for the Ag-modified TiO$_2$ sample, at the scale of 2×2 µm^2, R_q equaled 28.4 nm and R_{pv} ~210.4 nm, while at the 1×1 µm^2 scale, they were R_q = 22.1 nm and R_{pv} = 137.3 nm.

The samples were further loaded with lysozyme and the resulting AFM images are presented in Figure 3i,j for the Lys/Au–TiO$_2$/Ti sample and in Figure 3m,n for the Lys/Ag–TiO$_2$/Ti sample. The images at larger scales suggested that the lysozyme loading capacity was higher for the Au-modified TiO$_2$ sample, since the corresponding image appears less clear and has "noisy" areas (see for example the phase contrast image superimposed in the bottom right corner of Figure 3i).

On the other hand, the phase contrast AFM image superimposed on the topographical one in Figure 3m suggests that lysozyme was preferentially located/agglomerated at the grain boundaries of Ag NPs, while on Au-modified TiO$_2$, it covered uniformly in larger parcels. The corrugation parameters, for lysozyme-loaded samples, at the scale of 2×2 µm^2, were found to be 43.0 nm (R_q) and 286.7 nm (R_{pv}) for Lys/Au-modified TiO$_2$, and 50.2 (R_q) and 278.0 nm (R_{pv}), respectively, for Lys/Ag-modified TiO$_2$. At the scale of 1×1 µm^2, they were R_q = 10.3 nm and R_{pv} = 117.0 nm for Lys/Au-modified TiO$_2$, while they were R_q = 50.1 nm and R_{pv} = 277.9 nm for Lys/Ag-modified TiO$_2$, demonstrating that the lysozyme exhibits a local smoothing tendency for Au as compared with Ag-modified TiO_2.

The following aspects can be summarized from the AFM analysis:
(i) The Ti foil has a defective or irregular surface; (ii) TiO_2 nanoparticles are gathered forming dense layers; (iii) metal NPs are well dispersed and faceted; (iv) lysozyme covers the upper surface in the following order: TiO_2/Ti > Au–TiO_2/Ti > Ag–TiO_2/Ti; (v) the relatively large roughness parameter values are favorable towards the adhesion of biological compounds.

3.3. XRD

The crystallinity of the films was studied using the X-Ray diffraction (XRD) method. The mean size of the ordered (crystalline) domains, L (commonly known as crystallite sizes), of the phases was calculated using the Scherrer equation: $L = K\lambda/\beta\cos\theta$, where K is a dimensionless shape factor, usually taken as 0.89; λ (nm) is the wavelength of XRD radiation (Cu); β (in radians) is the line broadening at half the maximum intensity (FWHM); and θ is the Bragg angle.

Figure 4 shows the X-Ray diffraction patterns of the TiO_2/Ti, Ag–TiO_2/Ti, and Au–TiO_2/Ti samples. The TiO_2/Ti sample contained anatase, TiO_2, and metal titanium, Ti. The reflection of metal titanium phase, Ti, identified according to ICDD file no. 44–1294, belonged to the substrate. A preferred orientation along the (002) crystal plane was observed. The anatase phase, identified according to ICDD file no. 21–1272, only presented a few broad reflections, at 2θ values of 25.26°, 48.08°, 55.04°, and 62.66°, respectively, for the (101), (200), (105) + (211), and (204) crystal planes. The crystallite size of anatase, which was only calculated for the (101) crystal plane, was around 8 nm. Both the Ag–TiO_2/Ti and Au–TiO_2/Ti samples contained anatase, TiO_2, and titanium, Ti, from the metal substrate. In the Ag–TiO_2/Ti sample, well crystalized silver, Ag, with narrow diffraction lines, was identified according to ICDD file no. 4–0783. The crystallite size for silver was around 23 nm. In the Au–TiO_2/Ti sample, gold, Au, was detected according to ICDD file no. 4–0784, respectively. The gold phase was very well crystallized, with narrow diffraction lines. The crystallite size for the gold was around 25 nm.

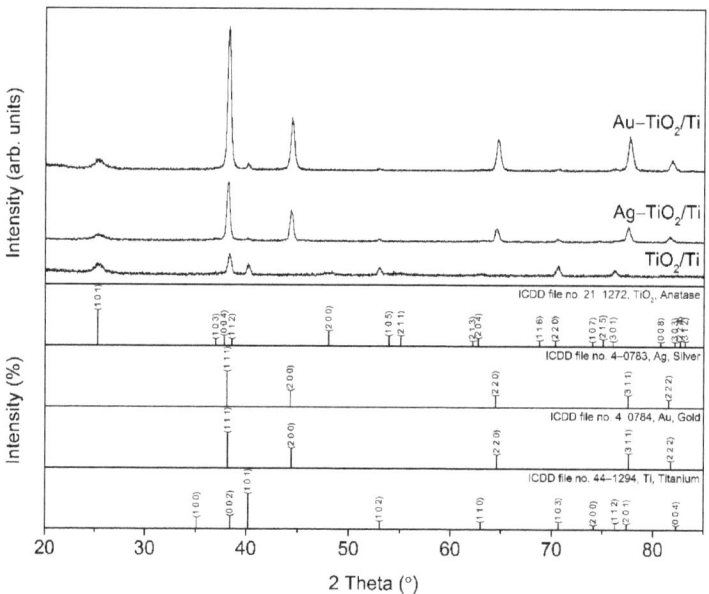

Figure 4. XRD diffractograms of the investigated samples: Au–TiO_2/Ti, Ag–TiO_2/Ti, TiO_2/Ti.

3.4. UV–Vis Spectroscopy

In order to explore the light sensitivity of the investigated samples, UV–Vis spectra were recorded. For the Au–TiO_2/Ti and TiO_2/Ti samples, Figure 5 reveals the UV light absorption around 370 nm due to the TiO_2 contribution. In the case of Ag–TiO_2/Ti, this appears to be scarcely defined and red shifted, merging with a large absorption band spanning from 390 to 800 nm.

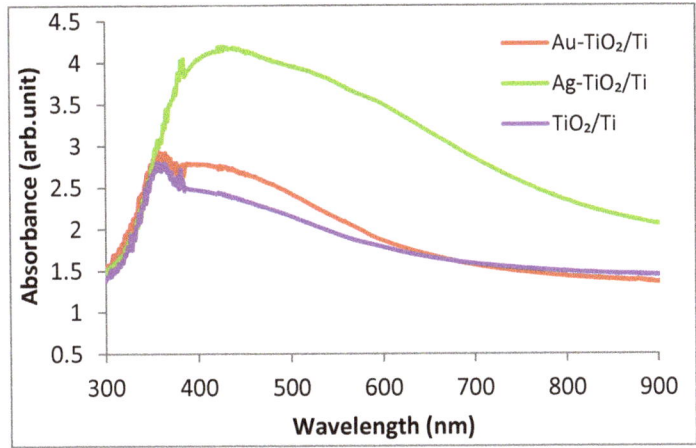

Figure 5. UV–Vis spectra of the Au–TiO_2/Ti, Ag–TiO_2/Ti, TiO_2/Ti samples ranging in the 300–900 nm domain.

Au–TiO_2/Ti also displayed an absorption band between 390 and 600 nm, which was less intense than that of the silver-containing sample, indicating the presence of Au nanoparticles but also defects in the TiO_2 nanostructured layer. These led to a long tail in the visible range for the TiO_2/Ti sample.

The literature concerning Ag-modified TiO_2 correlates the broad plasmonic peak with the presence of Ag nanoparticles with a large size distribution [30].

3.5. Photoluminescence Measurements

Generally, the photoluminescent signal of a semiconductor nanomaterial is associated with recombination of the photogenerated electron–hole pairs [31], with significant variations being induced by the modifiers. As TiO_2 is a well-known nontoxic photocatalyst, many pathways have been explored to improve its light sensitivity, especially in the visible range, including the deposition of noble metal nanoparticles. Usually, for optimal-ratio NPs/TiO_2, the PL emission decreases relative to the bare TiO_2 [32], with the recombination of photoinduced charge carriers being hindered by the electron transfer from TiO_2 to metal. From Figure 6, a broad band ranging from 400 to 470 nm with maximum located at 426 nm can be observed. For TiO_2 and Au, Ag NPs modified-TiO_2 samples, Chen et al. [33] assigned a PL signal peak at 417 nm to free excitons, with photoluminescence quenching also being observed for metal-modified materials. In Figure 6, a small PL decrease for Au–TiO_2/Ti and Ag–TiO_2/Ti can be perceived, the existence of an active interface between the metal and semiconductor being expected.

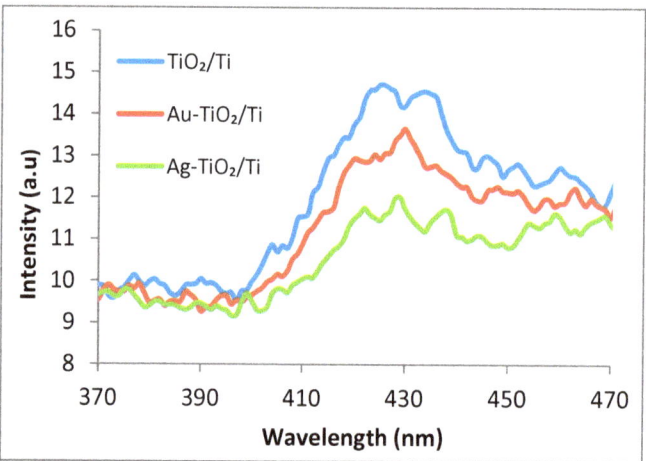

Figure 6. Photoluminescence spectra of TiO_2/Ti, Au–TiO_2/Ti and Ag–TiO_2/Ti samples registered for $\lambda_{exc} = 270$ nm.

3.6. ROS (Singlet Oxygen $^1O_2$) Identification

Nowadays, the use of singlet oxygen in photodynamic therapy is a focus of investigation [34,35], as it is a promising tool against antibiotic-resistant bacteria. There are various photosensitizers used for singlet oxygen generation, such as porphyrins and noble metal nanoparticles.

According to our previous work and the literature [36], singlet oxygen ($^1O_2$) is photogenerated by the target samples exposed to light irradiation and further triggers endoperoxide formation from the anthracene component of SOSG green reagent [37], which generates a PL emission peak centered at 535 nm for $\lambda_{exc} = 488$ nm. Figure 7 shows the intensity of the PL signals increasing with light exposure, peaking at 536 nm for both blank SOSG (a) and the samples of interest (b), (c), (d). However, higher peaks can clearly be observed for the bare TiO_2/Ti sample (b), with the metal-modified examples (c,d) generating similar signals as the blank SOSG (a small self-decomposition rate and singlet oxygen production from SOSG itself was previously reported [36,38]). Accordingly, the ability of the TiO_2/Ti sample to produce singlet oxygen under visible light exposure and its self-decontaminating behavior were verified.

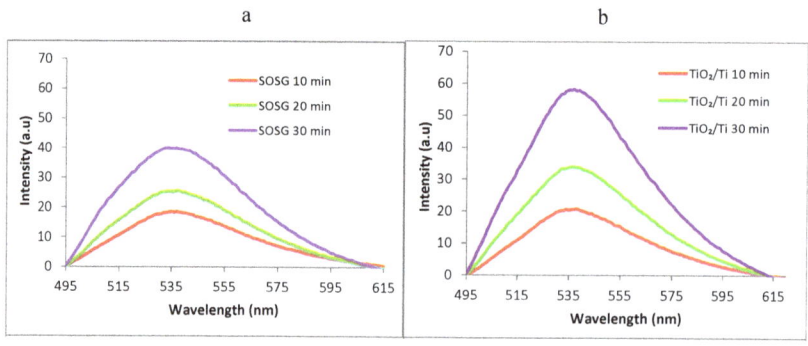

Figure 7. *Cont.*

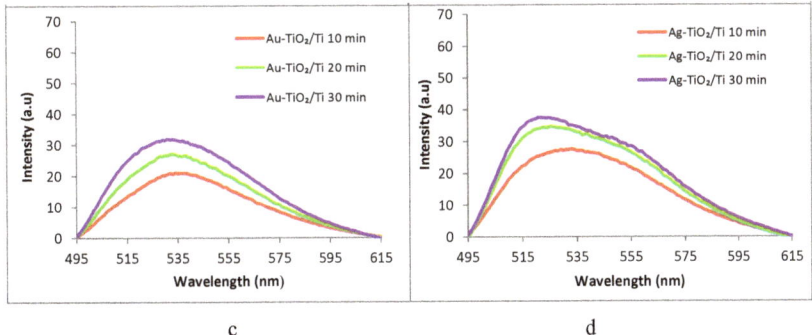

c   d

Figure 7. Time course of singlet oxygen formation after exposure to visible light (λ > 420 nm) in the absence (**a**)/presence (**b–d**) of the investigated samples, registered with SOSG singlet oxygen sensor for λ_{exc} = 488 nm.

3.7. Antibacterial Activity Assays of Inorganic Coatings TiO$_2$/Ti, Ag–TiO$_2$/Ti, and Au–TiO$_2$/Ti against M. lysodeicticus

Figure 8 clearly demonstrates the metal-modified coatings', i.e., Ag–TiO$_2$/Ti and Au–TiO$_2$/Ti, significant antimicrobial activity against *M. lysodeicticus*. The cellular viability registered for the metal-modified samples was 24% for Ag and 31% for Au modifiers. By comparison with the control sample (the C microbial cells alone), the bare TiO$_2$/Ti exhibited no antimicrobial activity. According to these results, the noble metal nanoparticles added to the TiO$_2$ layer hindered microbial growth.

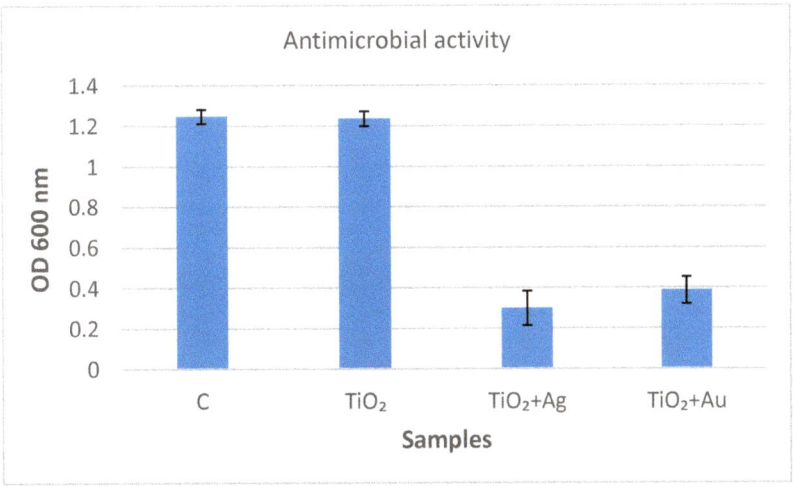

Figure 8. Microbial growth over the TiO$_2$-based coatings of titanium.

3.8. Lysozyme (Lys/TiO$_2$/Ti, Lys/Ag–TiO$_2$/Ti, Lys/Au–TiO$_2$/Ti) Activity Assays on Microbial Substrate (Micrococcus lysodeicticus)

M. lysodeicticus cell lysis in the presence of the newly developed hybrid systems was performed.

The decrease in absorbance at 450 nm measured for the *M. lysodeicticus* suspension in contact with the investigated samples allowed us to evaluate their bioactivity. Accordingly, Figure 9 shows major lysis of *M. lysodeikticus* cells in the first 10 min for the control sample (the free lysozyme in the *M. lysodeicticus* suspension) and a slight advancement during the next 24 h. A similar result was obtained for the Lys/TiO$_2$/Ti sample but for a longer incubation time (24 h).

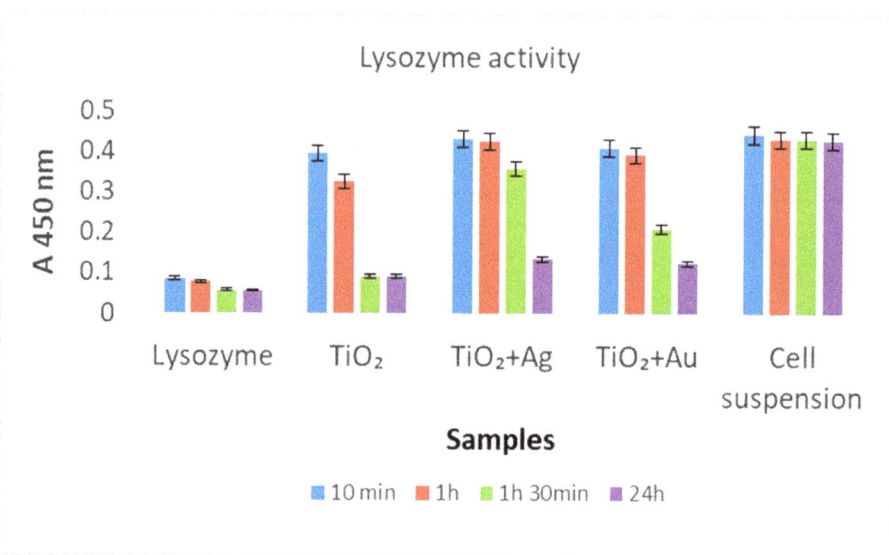

Figure 9. Activity of lysozyme (free and loaded on inorganic coatings) against *M. lysodeicticus*.

The loaded lysozyme appears to have a slower reactivity than the free enzyme, but this is significant and clearly evidenced after 24 h (as compared with the blank test carried out with the *M. lysodeikticus* cell suspension alone). Based on Figure 9, the following overall activity sequence for the target samples can be proposed: Lys/TiO$_2$/Ti > Lys/Au–TiO$_2$/Ti > Lys/Ag–TiO$_2$/Ti.

3.9. Lysozyme (Lys/TiO$_2$/Ti, Lys/Ag–TiO$_2$/Ti, Lys/Au–TiO$_2$/Ti) Activity Assays on Synthetic Substrate [4–MU–β– (GlcNAc)$_3$]

The biocatalytic assays described herein were performed in the presence of hybrid systems (Lys/TiO$_2$/Ti, Lys/Ag–TiO$_2$/Ti, Lys/Au–TiO$_2$/Ti) using a synthetic substrate, namely, 4-Methylumbelliferyl β-D-N,N′,N″-triacetylchitotrioside [4-MU-β-(GlcNAc)$_3$]. In order to evaluate the hydrolytic capacity of the loaded lysozyme, the formation of a fluorescent reaction product, namely, 7-hydroxy-4-metylcoumarin (4-methylumbelliferone), was monitored by fluorescence spectroscopy, according to previously reported data [21,22,39].

Figure 10 displays emission peaks with maxima at 450 nm assigned to the fluorescent compound 4-methylumbelliferone. This results from the hydrolysis reaction of the buffered organic substrate [4-MU-β-(GlcNAc)$_3$] subjected to incubation at 37 °C for 3 h in the presence of the previously prepared hybrid systems. Only Lys/TiO$_2$/Ti and Lys/Au–TiO$_2$/Ti exhibited well-defined peaks. Accordingly, for these inorganic coatings, a significant lysozyme loading capacity, the preservation of enzymatic activity after immobilization, and the release of fluorescent product into the buffer solution can be assumed. Unlike the above-mentioned hybrid systems, the Ag-containing sample produced insignificant amounts of 4-methylumbelliferone. Based on Figure 10, the resulting activity sequence for the samples of interest is as follows: Lys/TiO$_2$/Ti > Lys/Au–TiO$_2$/Ti > Lys/Ag–TiO$_2$/Ti.

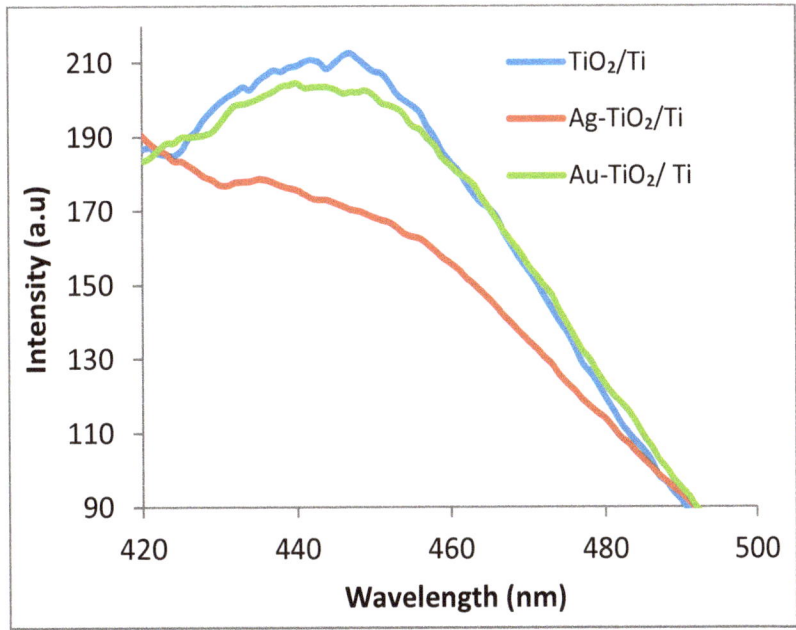

Figure 10. Fluorescence spectra of the released 7-hydroxy-4-metylcoumarin in buffer solution (λ_{exc} = 355 nm, λ_{em} = 450 nm) after 3 h of reaction.

4. Discussion

The development and structural characterization of bare and metal-modified TiO_2 coatings, appropriate for titanium dental implants, is an important area of study for engineered nanomaterials for biomedical applications. The present approach intended to simultaneously identify and test the key parameters required to produce safe and functional coatings for titanium implants, keeping in mind the natural processes triggered by the implant's presence, e.g., the formation of a thin TiO_2 layer on the titanium surface, bacterial colonization, and the activity of the lysozyme present in saliva. Therefore, it is important to understand the reactivity of the TiO_2 thin film on titanium relative to the contact environment, and to optimize the structural and functional parameters to a trigger self-disinfecting ability, intrinsic antibacterial properties, and the properties induced by the lysozyme loading capacity. In order to enhance the photosensitivity and antimicrobial properties of TiO_2, the deposition of noble metal nanoparticles was successfully performed.

SEM micrographs (Figure 1) show a sol–gel TiO_2 layer covering the titanium foil, decorated with Au/Ag nanoparticles (Figure 2, EDS spectra). The XRD analysis confirmed its anatase structure and the crystalline state of the modifiers used. Spectroscopic measurements revealed the improvement of TiO_2 light absorption after metal modification (UV–Vis). Generally, for gold nanoparticles, the data from the literature show strong plasmon resonance absorption [40]. This appears to be dependent on several light absorbing material features, such as shape, the size distribution of metallic nanoparticles, interaction between particles, and the dielectric environment [41]. In the present study, a broad absorption band was present for the Ag-TiO_2/Ti sample, which strongly decreased for the Au–TiO_2/Ti sample. This may be related to the large particle size distribution of the metallic nanoparticles. The gold and silver crystallite size identified by XRD was around 25 and 23 nm, respectively. However, SEM and AFM measurements revealed bigger, faceted surface particles. For these crystallite aggregates, a preferential growth in the (111) direction was noticed according to the main diffraction peak.

The AFM investigation also revealed the lysozyme interaction with the surface of the investigated samples. The recorded images indicated a different lysozyme coverage. This is in line with the enzymatic activity sequence of the newly developed hybrid systems.

Radical trapping measurements demonstrated singlet oxygen generation under visible light irradiation for the TiO_2/Ti sample.

By comparing the Figures 8–10, a complementary antimicrobial mechanism can be observed for the investigated materials: (a) Au and Ag nanoparticles deposited on TiO_2 trigger the antimicrobial effect of the inorganic coatings; (b) the adsorbed lysozyme, especially on bare TiO_2, preserves its enzymatic activity and could provide antibacterial protection for dental implants.

These experimental results are important since many studies are devoted to developing and improving nonaggressive antimicrobial tools, including the activity of enzymes [42]. In this sense, the key role of lysozyme in human immune defense is well recognized and has been studied; however, its mechanism of action is not fully understood. Ibrahim et al. [43] distinguish between lysozyme's bactericidal activity and its catalytic function. Therefore, further investigations on its bioactivity are needed together with the development of lysozyme-based hybrid materials.

Author Contributions: E.C., I.B. (Ion Bordeianu), M.Z., I.B. (Ioan Balint): Conceptualization; V.B., D.-I.E., J.M.C.-M., G.P., R.I., D.P. methodology, formal analysis, investigation; M.Z., I.B. (Ion Bordeianu), I.S., validation; C.A., M.A., S.P. writing—original draft preparation, E.C., C.A., M.A., S.P., M.Z., I.B. (Ioan Balint): writing—review and editing. All authors have read and agreed to the published version of the manuscript.

Funding: This research received no external funding.

Data Availability Statement: All data were reported in the paper.

Acknowledgments: The support of the Romanian Government that allowed for the acquisition of the research infrastructure under POSCCE O 2.2.1 project INFRANANOCHEM—No. 19/01.03.2009 9 is gratefully acknowledged.

Conflicts of Interest: The authors declare no conflict of interest.

References

1. Pałka, K.; Pokrowiecki, R. Porous Titanium Implants: A Review. *Adv. Eng. Mater.* **2018**, *20*, 1700648. [CrossRef]
2. Babíka, O.; Czána, A.; Holubjaka, J.; Kameníka, R.; Pilca, J. Identification of surface characteristics created by miniature machining of dental implants made of titanium based materials. *Procedia Eng.* **2017**, *192*, 1016–1021. [CrossRef]
3. Banerjee, D.; Shivapriya, P.M.; Kumar, P.; Gautam, P.K.; Misra, K.; Sahoo, A.K.; Samanta, S.K. A review on basic biology of bacterial biofilm infections and their treatments by nanotechnology-based approaches. *Proc. Natl. Acad. Sci. India Sect. B Biol. Sci.* **2019**, *90*, 243–259. [CrossRef]
4. Yang, W.E.; Hsu, M.L.; Lin, M.C.; Chen, Z.H.; Chen, L.K.; Huang, H.H. Nano/submicron-scale TiO_2 network on titanium surface for dental implant application. *J. Alloys Compd.* **2009**, *479*, 642–647. [CrossRef]
5. Cho, S.-A.; Park, K.-T. The removal torque of titanium screw inserted in rabbit tibia treated by dual acid etching. *Biomaterials* **2003**, *24*, 3611–3617. [CrossRef]
6. Iwaya, Y.; Machigashira, M.; Kanbara, K.; Miyamoto, M.; Nouguchi, K.; Izumi, Y.; Ban, S. Surface properties and biocompatibility of acid-etched titanium. *Dent. Mater. J.* **2008**, *27*, 415–421. [CrossRef] [PubMed]
7. Shemtov-Yona, K.; Rittel, D.; Dorogoy, A. Mechanical assessment of grit blasting surface treatments of dental implants. *J. Mech. Behav. Biomed. Mater.* **2014**, *39*, 375–390. [CrossRef] [PubMed]
8. Gehrke, S.A.; Ramírez-Fernandez, M.P.; Granero Marín, J.M.; Barbosa Salles, M.; Del Fabbro, M.; Calvo Guirado, J.L. A comparative evaluation between aluminium and titanium dioxide microparticles for blasting the surface titanium dental implants: An experimental study in rabbits. *Clin. Oral Implants Res.* **2018**, *29*, 802–807. [CrossRef]
9. Jemat, A.; Ghazali, M.J.; Razali, M.; Otsuka, Y. Surface modifications and their effects on titanium dental implants. *BioMed Res. Int.* **2015**, *2015*, 791725. [CrossRef]
10. Visentin, F.; El Habra, N.; Fabrizio, M.; Brianese, N.; Gerbasi, R.; Nodari, L.; Zin, V. TiO_2-HA bi-layer coatings for improving the bioactivity and service-life of Ti dental implants. *Surf. Coat. Technol.* **2019**, *378*, 125049. [CrossRef]
11. Zhao, L.; Wang, H.; Huo, K.; Cui, L.; Zhang, W.; Ni, H.; Zhang, Y.; Wu, Z.; Chu, P.K. Antibacterial nano-structured titania coating incorporated with silver nanoparticles. *Biomaterials* **2011**, *32*, 5706–5716. [CrossRef]

12. Romero-Gavilan, F.; Araújo-Gomes, N.; Sánchez-Pérez, A.M.; García-Arnáez, I.; Elortza, F.; Azkargorta, M.; Martín de Llano, J.J.; Carda, C.; Gurruchaga, M.; Suay, J.; et al. Bioactive potential of silica coatings and its effect on the adhesion of proteins to titanium implants. *Colloids Surf. B Biointerfaces* **2011**, *162*, 316–325. [CrossRef]
13. Zhang, J.; Chen, H.; Lin, T.; Yang, F.; Zhang, J.; Cai, X.; Yang, Y.; Zhang, P.; Tan, S. Fabrication of a TiO_2@Cu Core–Shell Nanorod Array as Coating for Titanium Substrate with Mechanical and Chemical Dual Antibacterial Property. *ACS Appl. Bio Mater.* **2022**, *5*, 3349–3359. [CrossRef]
14. Huang, H.; Chang, Y.Y.; Weng, J.C.; Chen, Y.C.; Lai, C.H.; Shieh, T.M. Anti-bacterial performance of Zirconia coatings on Titanium implants. *Thin Solid Films* **2013**, *528*, 151–156. [CrossRef]
15. Roknian, M.; Fattah-Alhosseini, A.; Gasht, S.O.; Keshavarz, M.K. Study of the effect of ZnO nanoparticles addition to PEO coatings on pure titanium substrate: Microstructural analysis, antibacterial effect and corrosion behavior of coatings in Ringer's physiological solution. *J. Alloys Compd.* **2018**, *740*, 330–345. [CrossRef]
16. Shibli, S.M.A.; Mathai, S. Development and bio-electrochemical characterization of a novel TiO_2–SiO_2 mixed oxide coating for titanium implants. *J. Mater. Sci. Mater. Med.* **2018**, *19*, 2971–2981. [CrossRef]
17. Kumaravel, V.; Nair, K.M.; Mathew, S.; Bartlett, J.; Kennedy, J.E.; Manning, H.G.; Whelan, B.J.; Leyland, N.S.; Pillai, S.C. Antimicrobial TiO_2 nanocomposite coatings for surfaces, dental and orthopaedic implants. *Chem. Eng. J.* **2021**, *416*, 129071. [CrossRef]
18. Negrete, O.; Bradfute, S.; Larson, S.R.; Sinha, A.; Coombes, K.R.; Goeke, R.S.; Keenan, L.A.; Duay, J.; Van Heukelom, M.; Meserole, S. *Photocatalytic Material Surfaces for SARS-CoV-2 Virus Inactivation*; No. SAND-2020-9861; Sandia National Laboratories: Livermore, CA, USA; Sandia National Laboratories: Albuquerque, NM, USA, 2020.
19. Fürst, M.M.; Salvi, G.E.; Lang, N.P.; Persson, G.R. Bacterial colonization immediately after installation on oral titanium implants. *Clin. Oral Implants Res.* **2017**, *18*, 501–508. [CrossRef]
20. Vu, T.V.; Nguyen, V.T.; Nguyen-Tri, P.; Nguyen, T.H.; Nguyen, T.V.; Nguyen, T.A. Chapter 17—Antibacterial nanocomposite coatings. In *Nanotoxicity: Prevention and Antibacterial Applications of Nanomaterials (Micro and Nano Technologies)*, 1st ed.; Rajendran, S., Mukherjee, A., Nguyen, T., Godugu, C., Shukla, R., Eds.; Elsevier: Amsterdam, The Netherlands, 2020; pp. 355–364. ISBN 978-0-12-819943-5.
21. Anastasescu, C.; Gifu, I.C.; Negrila, C.; Socoteanu, R.; Atkinson, I.; Calderon-Moreno, J.M.; Munteanu, C.; Plavan, G.; Strungaru, S.A.; Cheatham, B.; et al. Morpho-structural properties of ZnSe, TiO_2-ZnSe materials and enzymatic activity of their bioinorganic hybrids with lysozyme. *Mater. Sci. Eng. B* **2021**, *272*, 115350. [CrossRef]
22. Kao, X.K.-C.; Lin, T.-S.; Mou, C.-Y. Enhanced activity and stability of Lysozyme by immobilization in the matching nanochannels of mesoporous silica nanoparticles. *J. Phys. Chem. C* **2014**, *118*, 6734–6743. [CrossRef]
23. Soares, R.V.; Lin, T.; Siqueira, C.C.; Bruno, L.S.; Li, X.; Oppenheim, F.G.; Offner, G.; Troxler, R.F. Salivary micelles: Identification of complexes containing MG2, sIgA, lactoferrin, amylase, glycosylated proline-rich protein and lysozyme. *Arch. Oral Biol.* **2004**, *49*, 337–343. [CrossRef]
24. Silletti, E.; Vingerhoeds, M.H.; Norde, W.; van Aken, G.A. Complex formation in mixtures of lysozyme-stabilized emulsions and human saliva. *J. Colloid Interface Sci.* **2007**, *313*, 485–493. [CrossRef]
25. Franken, C.; Meijer, C.J.; Dijkman, J.H. Tissue distribution of antileukoprotease and lysozyme in humans. *J. Histochem. Cytochem.* **1989**, *37*, 493–498. [CrossRef]
26. Larsericsdotter, H.; Oscarsson, S.; Buijs, J. Thermodynamic analysis of proteins adsorbed on silica particles: Electrostatic effects. *J. Colloid Interface Sci.* **2001**, *237*, 98–103. [CrossRef]
27. Perevedentseva, E.; Cai, P.J.; Chiu, Y.C.; Cheng, C.L. Characterizing protein activities on the Lysozyme and nanodiamond complex prepared for bio applications. *Langmuir* **2011**, *27*, 1085–1091. [CrossRef]
28. Anastasescu, C.; Spataru, N.; Culita, D.; Atkinson, I.; Spataru, T.; Bratan, V.; Munteanu, C.; Anastasescu, M.; Negrila, C.; Balint, I. Chemically assembled light harvesting CuO_x-TiO_2 p–n heterostructures. *Chem. Eng. J.* **2015**, *281*, 303–311. [CrossRef]
29. Takeuchi, M.; Abe, Y.; Yoshida, Y.; Nakayama, Y.; Okazaki, M.; Akagawa, Y. Acid pretreatment of titanium implants. *Biomaterials* **2003**, *24*, 1821–1827. [CrossRef]
30. Xiong, Z.; Ma, J.; Ng, W.J.; Waite, T.D.; Zao, X.S. Silver-modified mesoporous TiO_2 photocatalyst for water purification. *Water Res.* **2011**, *45*, 2095–2103. [CrossRef] [PubMed]
31. Liqiang, J.; Yichun, Q.; Baiqi, W.; Shudan, L.; Baojiang, J.; Libin, Y.; Wei, F.; Honggang, F.; Jiazhong, S. Review of photoluminescence performance of nano-sized semiconductor materials and its relationships with photocatalytic activity. *Sol. Energ. Mat. Sol. Cells* **2006**, *90*, 1773–1787. [CrossRef]
32. Bumajdad, A.; Madkou, M. Understanding the superior photocatalytic activity of noble metals modified titania under UV and visible light irradiation. *Phys. Chem. Chem. Phys.* **2014**, *16*, 7146–7158. [CrossRef]
33. Chen, S.; Li, J.; Qian, K.; Xu, W.; Lu, Y.; Huang, W.; Yu, S. Large scale photochemical synthesis of M@TiO_2 nanocomposites (M = Ag, Pd, Au, Pt) and their optical properties, CO oxidation performance, and antibacterial effect. *Nano Res.* **2010**, *3*, 244–255. [CrossRef]
34. Zampini, G.; Planas, O.; Marmottini, F.; Gulias, O.; Agut, M.; Nonell, S.; Latterin, L. Morphology effects on singlet oxygen production and bacterial photoinactivation efficiency by different silica-protoporphyrin IX nanocomposites. *RSC Adv.* **2017**, *7*, 14422–14429. [CrossRef]

35. Maisch, T.; Baier, J.; Franz, B.; Maier, M.; Landthaler, M.; Szeimies, R.-M.; Baümler, W. The role of singlet oxygen and oxygen concentration in photodynamic inactivation of bacteria. *Proc. Natl. Acad. Sci. USA* **2007**, *104*, 7223–7228. [CrossRef]
36. Anastasescu, C.; Negrila, C.; Angelescu, D.G.; Atkinson, I.; Anastasescu, M.; Spataru, N.; Zaharescu, M.; Balint, I. Particularities of photocatalysis and formation of reactive oxygen species on insulators and semiconductors: Cases of SiO_2, TiO_2 and their composite. SiO_2-TiO_2. *Catal. Sci. Technol.* **2018**, *8*, 5657–5668. [CrossRef]
37. Ràgas, X.; Jiménez Garcia, A.; Batlori, X.; Nonell, S. Singlet oxygen photosensitisation by the fluorescent probe Singlet Oxygen Sensor Green. *Chem. Commun.* **2009**, *20*, 2920–2922. [CrossRef]
38. Kim, S.; Fujitsuka, M.; Majima, T. Photochemistry of Singlet Oxygen Sensor Green. *J. Phys. Chem. B.* **2013**, *117*, 13985–13992. [CrossRef]
39. Zeng, Y.; Wan, Y.; Zhang, D. Lysozyme as sensitive reporter for fluorometric and PCR based detection of *E. coli* and *S. aureus* using magnetic microbeads. *Microchim. Acta* **2016**, *183*, 741–748. [CrossRef]
40. Kwon, K.; Lee, K.Y.; Lee, Y.W.; Kim, M.; Heo, J.; Ahn, S.J.; Han, S.W. Controlled Synthesis of Icosahedral Gold Nanoparticles and Their Surface-Enhanced Raman Scattering Property. *J. Phys. Chem. C* **2007**, *111*, 1161–1165. [CrossRef]
41. Sancho-Parramon, J. Surface plasmon resonance broadening of metallic particles in the quasi-static approximation: A numerical study of size confinement and interparticle interaction effects. *Nanotechnology* **2009**, *20*, 235706. [CrossRef]
42. Sarbu, I.; Vassu, T.; Chifiriuc, M.C.; Bucur, M.; Stoica, I.; Petrut, S.; Rusu, E.; Moldovan, H.; Pelinescu, D. Assessment the Activity of Some Enzymes and Antibiotic Substances Sensitivity on Pathogenic Bacteria Species. *Rev. De Chim.* **2017**, *68*, 3015–3021. [CrossRef]
43. Ibrahim, H.R.; Matsuzaki, T.; Aoki, T. Genetic evidence that antibacterial activity of lysozyme is independent of its catalytic function. *FEBS Lett.* **2001**, *506*, 27–32. [CrossRef]

 *nanomaterials*

Article

Synthesis of Uniform Size Rutile TiO$_2$ Microrods by Simple Molten-Salt Method and Its Photoluminescence Activity

Hieu Minh Ngo [1,†], Amol Uttam Pawar [2,†], Jun Tang [3], Zhongbiao Zhuo [3], Don Keun Lee [2], Kang Min Ok [1] and Young Soo Kang [2,*]

[1] Department of Chemistry, Sogang University, Seoul 04107, Korea; hieungo@sogang.ac.kr (H.M.N.); kmok@sogang.ac.kr (K.M.O.)
[2] Environmental and Climate Technology, Korea Institute of Energy Technology, Naju-si 58219, Korea; amolphysics@kentech.ac.kr (A.U.P.); leedk3@kentech.ac.kr (D.K.L.)
[3] Zhejiang Coloray Technology Development Co., Ltd., No. 151, Huishan Road, Deqing County, Huzhou 313400, China; tangjun@coloray.com.cn (J.T.); zhongbiao@coloray.com.cn (Z.Z.)
* Correspondence: yskang@kentech.ac.kr
† These authors contributed equally to this work.

Abstract: Uniform-size rutile TiO$_2$ microrods were synthesized by simple molten-salt method with sodium chloride as reacting medium and different kinds of sodium phosphate salts as growth control additives to control the one-dimensional (1-D) crystal growth of particles. The effect of rutile and anatase ratios as a precursor was monitored for rod growth formation. Apart from uniform rod growth study, optical properties of rutile microrods were observed by UV−visible and photoluminescence (PL) spectroscopy. TiO$_2$ materials with anatase and rutile phase show PL emission due to self-trapped exciton. It has been observed that synthesized rutile TiO$_2$ rods show various PL emission peaks in the range of 400 to 900 nm for 355 nm excitation wavelengths. All PL emission appeared due to the oxygen vacancy present inside rutile TiO$_2$ rods. The observed PL near the IR range (785 and 825 nm) was due to the formation of a self-trapped hole near to the surface of (110) which is the preferred orientation plane of synthesized rutile TiO$_2$ microrods.

Keywords: molten-salt method; TiO$_2$ microrod; anatase; rutile; photoluminescence

1. Introduction

The uniform shape and size of nano/microparticles with selective crystal facets are always challenging for synthesis, and those uniform particles are very useful for many applications. In chromatography applications, uniform particle size shows good resolution, better packing properties, improved kinetics, and it maintains consistent performance. It also can allow even and homogeneous coating on the solid surfaces [1]. In the field of electrochemistry, Fuller et al. suggested the improved stability of Pt in the membrane due to uniform particle size distribution [2]. Similarly, in paint industries many properties also depend upon the particle size distribution, such as transparency, film color appearance, paint viscosity, color stability, and weather resistance. In short, it is understood that more uniform particle size shows better performance in many fields of applications in modern life.

TiO$_2$ is one of the versatile materials that can be used in various fields due to its unique optical and electrical properties. Hence, synthesis of uniform size and shape with selective crystal facets of TiO$_2$ particles is key in the current research field. There are different chemical methods used for the synthesis of uniform micro- and nanoparticles, for example, the micro-emulsion method [3], co-precipitation method [4], sol-gel method [5], and solvothermal/hydrothermal method with different surfactant and capping agents [6]. Kang et al. reported a truncated and rice-like one-axis {001}-oriented crystalline anatase TiO$_2$ by simple hydrothermal synthesis. It was presumed that {001} facets were highly reactive and showed better catalytic performance compared to other crystal facets [7]. Hence, shape variation can vary the active surface area, which further causes change in catalytic performance. However, not only the shape but size of nano-microparticles also shows variation in its catalytic performance. Hao et al. carried out a comparative photodegradation study of rhodamine B with TiO_2 nanoparticles for different particle sizes of 8, 16, and 150 nm. It was observed that a smaller particle size gave better performance in a photodegradation study [8]. Chen et al. reported uniform coating of TiO_2 nanoparticles with sizes of 5 to 10 nm on natural cellulose by the solvothermal method, which showed an enhanced photocatalytic performance in the field of dye degradation study [9]. However, it is not only smaller size particles that always produce better performance; many times it is completely dependent on the application. For example, in the paint industry, pigment properties mostly depend upon the light scattering which further depends upon wavelength and particle size. For efficient light scattering, the diameter of the particles should be slightly smaller than the half wavelength of incident light. For example, for visible light range from 380 to 750 nm, the particle size should be slightly less than or half of that range, which is about 200 to 300 nm and much higher than earlier examples [10]. Moreover, TiO_2 is an outstanding material with a high refractive index, which is the main reason for light scattering and one of the important parameters for white pigments. Among different phases of TiO_2, rutile TiO_2 shows a 10% higher refractive index than anatase TiO_2 [11]. Furthermore, by considering various morphologies, one-dimensional (1-D) structures such as rods, belts, wires, and fibers show excellent mechanical and electrical properties with high surface to volume ratio [12]. Finally, it was decided to synthesize rutile TiO_2 rods with a diameter approximately 300 nm with longer length (in μm) and uniform size, which will suit applications in the paint industry.

There are several reports published on the synthesis of 1-D TiO_2 with different synthesis techniques and for various applications [13–15]. The molten-salt synthesis technique is one of the promising techniques and highly accepted for rutile TiO_2 synthesis. In this technique, generally low-melting-point salt has been used as reacting media with other high-melting-point precursors. In a few studies, μm-sized rutile TiO_2 with a high aspect ratio has been synthesized by the molten-salt method [16–19]. In those studies, NaCl salt was usually used as a reacting media either alone or together with another salts such as dibasic sodium phosphate (Na_2HPO_4). This type of salt combination, known as eutectic composition, is helpful in increasing the reactivity and ion mobility in reacting media at the minimum required temperature [18]. Kim et al. reported eutectic composition of NaCl salt and sodium hexametaphosphate ($(NaPO_3)_6$) for the synthesis of 1-D TiO_2 [20]. It has been explained that the oxide material's solubility increases by Lux−Flood acid-base interaction; $(NaPO_3)_6$ produces PO_3^- ions which are responsible for a strong Lux−Flood acid that reduces the O^{2-} activity in the system. It creates the reducing atmosphere in the system, which further helps to take O_2 out of the TiO_2 crystal. Therefore, anatase TiO_2 becomes unstable and dissolves in this molten-salt method. Finally, with bond breaking and rearrangements of atoms, it was converted into rutile TiO_2 rods [20].

However, obtaining uniform size of 1D nano- or micro-particles at a large scale is still highly challenging. Kim et al. in [20] introduced extra additives (i.e., $Na_3P_4O_7$) with various combinations along with TiO_2 precursors to obtain a uniform size of rutile TiO_2 rods.

In this report, the molten-salt synthesis technique was chosen to synthesize rutile-type TiO_2 microrods with TiO_2-NaCl media to control size and shape with 1D growth in the air atmosphere. This technique shows many advantages, such as a homogeneous mixture of TiO_2-NaCl at low temperature, and an easy post-process to remove impurities. Mixtures of two different types of sodium phosphate were used as additives for the crystal growth control. Finally, synthesized rutile-type TiO_2 microrods were used for detailed study related to size-controlled morphology, crystallinity, and optical properties.

2. Experimental Procedures

2.1. Materials

Titanium (IV) oxide (rutile seed) was provided by Jiangsu Hehai Nanometer Science & Technology Co., Ltd. (Taixing, Jiangsu province China). Titanium (IV) oxide (anatase powder) (98+%) was purchased from Sigma-Aldrich, Korea. Sodium pyrophosphate decahydrate (97%) and sodium hexametaphosphate (extra pure) were purchased from Alfa-Aesar (Seoul, Korea). Sodium chloride (99.5%) was purchased from Duksan (Ansan-si, Korea). Deionized (DI) water was taken from a Milli-Q IQ 7000 system ($\rho > 18.2$ MΩ.cm). Ethanol (99.5%), acetone (95%), and hexane (95%) were purchased from Samchun (Pyeongtaek, Korea).

2.2. Preparation of TiO_2 Microrods

TiO_2 microrods were prepared by the molten-salt method [20]. Generally, TiO_2 rutile seeds and TiO_2 anatase (ratios vary from 1:1, to 1:2., 1:3, and 1:4) were mixed and ground together with a constant ratio of NaCl and sodium phosphate additives (NaCl: $(NaPO_3)_6$: $Na_3P_4O_7$ = 4:1:1). To choose the proper ratio of (NaCl: $(NaPO_3)_6$: $Na_3P_4O_7$), we performed a few experiments with different ratios and checked SEM data. The obtained SEM data are presented in Supporting Information, Figure S1. The actual weight of the precursors in each ratio is presented in the Supporting Information (Table S1). The ratio 4:1:1 showed uniform TiO_2 particles compared to other ratios. The mixture powder was pelletized using a 13 mm pelletizer (PIKE technologies) at 250 kg.cm^{-2}, for 30 s. The pelletized sample was then annealed at 900 °C for 6 h (5 °C/min heating rate) in a box furnace. After that, the pellets were ground to powder and washed with water 3 times using centrifugation at 8000 rpm, 30 min each time, then washed with ethanol and acetone to remove the water. The powder sample was collected, dried at 80 °C for 1 day, and stored at room temperature for further characterization.

2.3. Characterization

The crystal structure of the formed products was characterized by powder X-ray diffraction (Rigaku, MiniFlex 600) using Cu Kα radiation with a wavelength (λ) of 1.5406 Å. The prepared TiO_2 powder was placed on an XRD sample holder and measured spectra were in the range of 10 to 90 degree of 2 theta with scan rate 2 degree/min. The obtained spectra were plotted in the range of 20 to 80 degrees of 2 theta and are presented in Figure 1. The morphology of the nanostructures was observed by scanning electron microscopy (FE-SEM, JSM-7100F JEOL) equipped with energy dispersive X-ray spectroscopy (EDS). For the SEM measurement, TiO_2 powder was attached on the SEM holder and to obtain a high-quality picture a thin layer (~10 nm) of Pt was coated via sputter coater. Optical properties were studied using a UV–visible NIR spectrophotometer by Varian Cary with diffuse reflectance mode. In the diffuse reflectance spectroscopy (DRS), a sample holder was filled with prepared TiO_2 powder and before starting the measurement, baseline correction was done using barium sulfate. After that, each sample was measured in reflectance mode in a wavelength range 250 to 800 nm. An Hitachi F-7000 fluorescence spectrophotometer was utilized to monitor room temperature PL emission. A 355 nm excitation wavelength was used to check the emission spectra in the 400 to 900 nm wavelength range. For the measurement of excitation spectra, 825 nm emission wavelength was used, and excitation was observed in the range of 300 to 380 nm. IR spectroscopy was also observed using the Nicolate Avatar 330 FTIR instrument. For the sample measurements, attenuated total reflectance (ATR) mode was used, and background correction was done without any sample on a diamond crystal with pressure anvil. The sample powder was simply placed on the diamond crystal and pressure was applied through an anvil to start the measurements in the range of 4000 to 600 cm^{-1}.

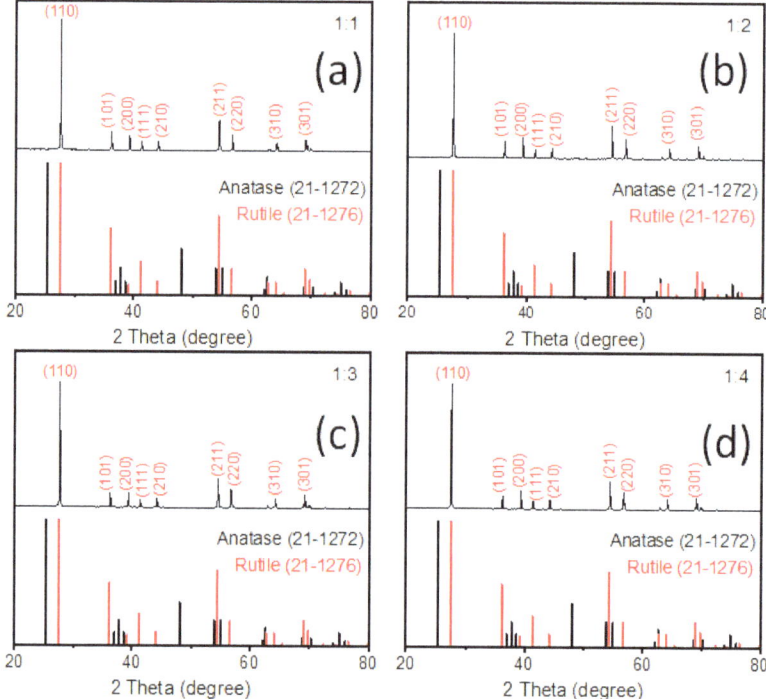

Figure 1. Typical XRD patterns of rutile TiO$_2$ microrod samples prepared by different precursor ratios of rutile and anatase TiO$_2$ (rutile, seed): TiO$_2$ (anatase) (**a**) 1:1, (**b**) 1:2, (**c**) 1:3, and (**d**) 1:4.

3. Result and Discussion

The solution of the TiO$_2$ growth process consisted of four chemicals, rutile TiO$_2$ seeds, anatase TiO$_2$, sodium chloride, and phosphate salts as an additive. The rutile seeds and anatase TiO$_2$ precursor particle size were observed as $\leq$ 50 nm and $\leq$ 150 nm, respectively, as presented in Figure S2. The rutile seeds were used as nucleation for rod growth because they are stable at high temperature (>800 °C). Sodium chloride acted as the media environment for the TiO$_2$ growth process at high temperature (>800 °C). Sodium phosphate ((NaPO$_3$)$_6$) acted as a Lux−Flood acid, which created a further reducing environment in the system that caused removal of O$_2$ from anatase TiO$_2$. Hence, anatase TiO$_2$ became unstable and dissolved, acting as a titania source for further growth of rutile TiO$_2$. The mixture of two sodium phosphate additives ((NaPO$_3$)$_6$ and Na$_3$P$_4$O$_7$) was useful to create a Lux−Flood acid/base with different Ka (acid dissociation constant) values, helping to increase the solubility and dissolution of anatase TiO$_2$ and control the aspect ratio of the TiO$_2$ rods [20]. It was expected to control the thickness of the rutile TiO$_2$ in the range of micrometers with the combination of these two phosphate salts.

To maintain a homogeneous diffusion, a large quantity of NaCl was used, four times higher than the total TiO$_2$ precursor (TiO$_2$ (rutile + anatase): NaCl = 1:4). When the ratio of TiO$_2$ (rutile) vs TiO$_2$(anatase) was kept at 1:1 and the relative ratio of additives varied (ratio of (NaPO$_3$)$_6$: Na$_3$P$_4$O$_7$ varied from 1:2 to 4:2), it was observed that the size of particles was irregular in all cases. This indicates that the ratio of anatase and rutile was still large, so that the seeds could not grow to consume all the anatase. Instead, the remaining anatase dissolved at high temperature and formed irregular rutile particles. Hence, we further varied the rutile anatase ratio from 1:1, to 1:2, 1:3, and 1:4 by keeping the NaCl: (NaPO$_3$)$_6$: Na$_3$P$_4$O$_7$ ratio constant as 8:2:2.

XRD patterns of synthesized rutile TiO$_2$ microrods with different combinations of rutile and anatase precursors are presented in Figure 1. The results were compared with standard JCPDS cards of rutile TiO$_2$ (21-1276) and anatase TiO$_2$ (21-1272). All the samples showed diffraction peaks at 2 theta values 27.5, 36.1, 39.2, 41.2, 44.05, 54.3, 56.6, 64, and 69° corresponding to crystal planes (110), (101), (200), (111), (210), (211), (220), (310), and (301) of rutile TiO$_2$ (21-1276), respectively. There was no single peak observed regarding anatase TiO$_2$ (21-1272), so it seems that all the samples were purely crystalline and had a single phase of rutile TiO$_2$ with tetragonal crystal structure. In all samples, it was observed that (110) peak intensity was comparatively much higher than the second high intense peak (211). The standard card shows first and second high intense peak ratios, i.e., (110)/(211) is equal to 1.25, and the same peaks area ratio was found in prepared TiO$_2$ microrods around 3.58, 3.64, 4.05, and 4.35 for sample 1, 2, 3, and 4, respectively, which were significantly higher than the rutile JCPDS card data. This indicates that TiO$_2$ particles grow in one axis direction with high surface area along [110], which is also known as the preferred axis (or crystal facet) orientation growth. These results are quite consistent with SEM data (Figure 2), which shows one-axis-oriented TiO$_2$ microrods in all four prepared samples. However, Figure 2a and Figure S3a in Supporting Information show samples prepared at (1:1) ratio of rutile and anatase TiO$_2$ precursors appeared with nanorods having different sizes of ~550 to 700 nm in diameter and from ~30 to 50 μm in length. All particles appeared with a smooth surface without further growth or deposition. When the amount of anatase increased and the ratio was (1:2), microrods appeared with smooth and clean surfaces (Figures 2b and S3b). The shape and size of these rods was not much different than that of the samples with a (1:1) ratio. The TiO$_2$ prepared with further increasing anatase precursors at a (1:3) ratio showed relatively uniform microrods (Figures 2c and S3c). The size of rods increased to nearly 550–700 nm in diameter with a length from 30 to 50 μm. When rutile and anatase ratios increased to 1:4, non-uniform microrods were shown with random sizes and lengths; approximately 200 nm–2 μm diameter range and 10–50 μm length range (Figures 2d and S3d). The rutile TiO$_2$ prepared with a ratio of 1:3 showed microrods comparatively uniform in shape and size. To identify other impurities, SEM-EDS measurements were carried out for all synthesized rutile TiO$_2$ samples, and are presented in Supporting Information Figure S4. It was observed that no other impurities were present in prepared samples.

Figure 3 shows the TEM, HRTEM, and SAED patterns of synthesized rutile TiO$_2$ microrods prepared with rutile and an anatase ratio 1:3. The TEM image shows a clear picture of synthesized microrods with diameter of more than 0.2 μm and length more than 2 μm. Figure 3b shows a selective area electron diffraction (SAED) pattern, and it shows clear lattice points, which are signs of single crystallinity with growth direction along [110] and [111]. In further observation, a high-resolution TEM (HRTEM) image presented in Figure 3c shows clear lattice fringes with an observed interplanar distance of 3.24 Å along the (110) plane. These data are additional confirmation of preferred orientation and growth along the [110] direction, the same as XRD.

Figure 2. SEM images of rutile TiO_2 microrod samples prepared by different ratios of TiO_2 (rutile, seed): TiO_2 (anatase) as a precursor (**a**) 1:1, (**b**) 1:2, (**c**) 1:3, and (**d**) 1:4.

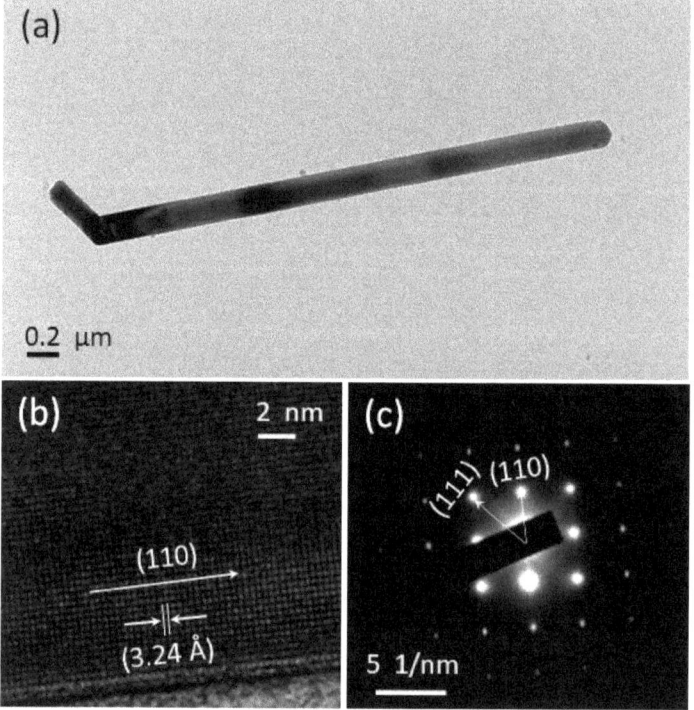

Figure 3. (**a**) TEM, (**b**) HRTEM and (**c**) SAED images of rutile TiO_2 microrod samples prepared with a 1:3 ratio of TiO_2 (rutile, seed): TiO_2 (anatase) as a precursor.

Figure 4 shows UV–visible absorption spectra of samples prepared by different ratios of rutile and anatase TiO$_2$ as precursors. Typical TiO$_2$ materials show an absorbance range until 400 nm wavelength; the same phenomenon was observed in all four synthesized samples and is presented in Figure 4. In general, nanoparticle sizes less than 100 nm show a slight variation in absorption range due to changes in band gap energy [21]. However, in this case particle sizes are quite high in the μm range, hence we did not observe any changes in light absorption. Typically, the value of band gap energy was calculated by using Tauc and Davis–Mott relation, with the equation given below (Equation (1)), [22,23].

$$h = K(h - Eg)^n \tag{1}$$

where α = absorption coefficient, h = Plank's constant, ν = frequency, K = energy independent constant, Eg = band gap energy, "n" is a nature of transmission and depends upon the materials selection rules regarding electron transition; e.g., 1/2 for allowed direct transition, 3/2 for forbidden direct transition, 2 for allowed indirect transition, and 3 for forbidden indirect transition. Rutile TiO$_2$ shows direct electronic transition, hence $n = \frac{1}{2}$ was considered in this case. However, with unknown thickness due to TiO$_2$ as powder form, finding "α" value was difficult with UV absorption spectra. To avoid this problem, diffuse reflectance spectroscopy (DRS) was used for further study. The theory of DRS is based on the Kubelka–Munk equation (Equation (2)): [24–27].

$$\frac{\alpha}{s} = \frac{(1-R)^2}{2R} = F(R) \tag{2}$$

where "α" stands for absorption coefficient, "s" for scattering coefficient and these two values changes with shape, size, and packing of materials. The reflectance of the materials is denoted as R. In practice, the measured diffuse reflectance spectrum is the ratio of the analyzed sample's reflection intensity to the standard sample's reflection intensity. In the Kubelka–Munk function, $F(R)$ is the conversion of reflectance data, which equals to the absorption coefficient (α) per unit scattering (s). Because scattering is assumed to be relatively constant for all the wavelengths, the absorption coefficient (α) is directly proportional to the $F(R)$ value. Finally, to identify the band gap energy value of synthesized rutile TiO$_2$ microrods, diffuse reflectance spectra were collected for all the samples and plotted in a graph [F(R)hν]^2 vs. hν, presented in Figure 5. Tauc plot extrapolation determines the value of band gap energy as around 3.06 eV (±0.01 eV) for all synthesized rutile TiO$_2$ microrods. The separate graphs of Tauc plot extrapolation are presented in Supporting Information as Figure S5.

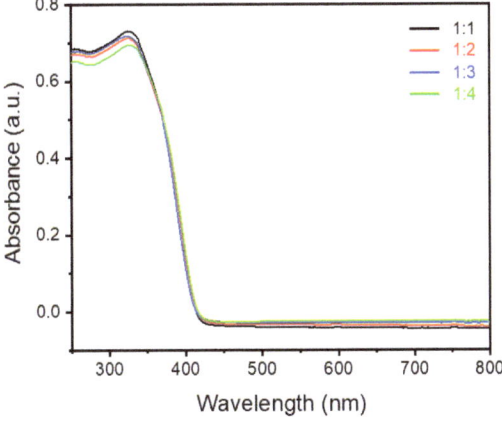

Figure 4. UV–visible absorbance spectra of rutile TiO$_2$ microrod samples prepared by different ratios of TiO$_2$ (rutile, seed): TiO$_2$ (anatase) as a precursor (black) 1:1, (red) 1:2, (blue) 1:3, and (green) 1:4.

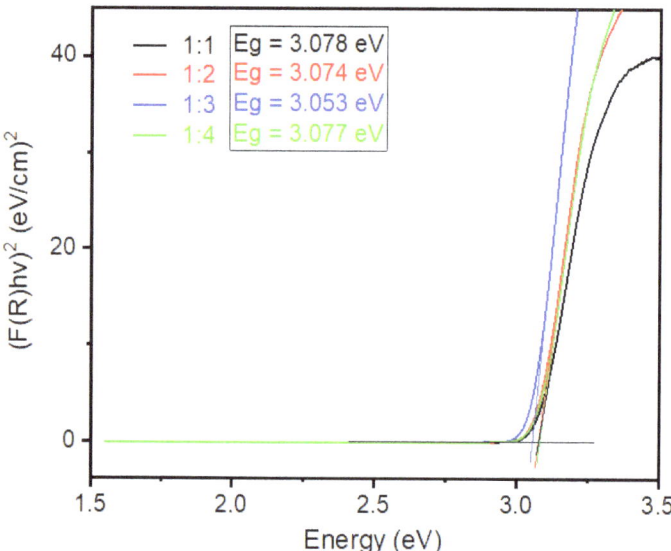

Figure 5. Band gap energy calculation using Tauc plot method for rutile TiO_2 microrod samples prepared by different ratios of TiO_2 (rutile, seed): TiO_2 (anatase) as a precursor (black) 1:1, (red) 1:2, (blue) 1:3, and (green) 1:4.

Room temperature photoluminescence (PL) measurements were done with an excitation wavelength 355 nm of a Xenon lamp. The emission spectra of all synthesized rutile TiO_2 nanorods were observed in the range between 400 to 900 nm. The excitation spectra are presented in Supporting Information as Figure S6. In the complete spectra (Figure S7), a secondary harmonic high intense peak is presented at 710 nm, which is exactly twice the excitation wavelength (355 nm). For clearer visualization, these spectra are presented in two different regions: a range from 400 nm to 675 nm (Figure 6a) and a range from 750 nm to 900 nm (Figure 6b). The peak at 430 nm is observed due to recombination of free electrons and holes near the band edge of rutile TiO_2 [28]. In the middle of visible range, two emission peaks at 470 nm and 575 nm were observed due to radiative transition of the self-trap electron-hole recombination. The self-trap state is due to the oxygen vacancy present in TiO_2. When a sample anneals at high temperature, oxygen vacancies are created inside the materials [29]. As describing in the experimental section, the sample was pelletized and annealed at 900 °C; therefore, there is a high chance of creating oxygen vacancy inside the rutile TiO_2 particles. The PL emission spectra in Figure 6a are presented in the visible range and near the UV region, which is quite similar to the anatase phase [30], but there is a possibility that rutile TiO_2 can show emission spectra similar to anatase due to oxygen vacancy [31]. Figure 6b shows typical emission spectra of rutile TiO_2, which shows PL emission at 825 nm and a small hump at 785 nm. The rutile TiO_2 shows an emission range near the IR (NIR) region, due to 1) radiative recombination of trapped hole with free electrons or 2) radiatively recombination of trapped electrons with a free hole [30]. In the rutile TiO_2 crystal structure, along the (110) and (100) plane are shown threefold coordinated oxygen atoms. When light is incident on rutile TiO_2, electrons are excited to conduction band level and holes are created in the valence band, those generated holes transferred towards the surface, but due to threefold coordinated oxygen atoms few of those holes are trapped near to the (110) or (100) plane/surface [32–34]. That trapped energy level, known as a self-trapped hole (STH), is quite low in energy state and matched with the near-IR range [30]. As we have observed in XRD, SEM, and TEM samples, synthesized rutile type TiO_2 microrod samples in this report show preferred orientation along the [110] direction and rod-type morphology. Hence, they have a larger site of self-trapped holes, which are responsible for PL emission at 825 nm and 785 nm wavelengths.

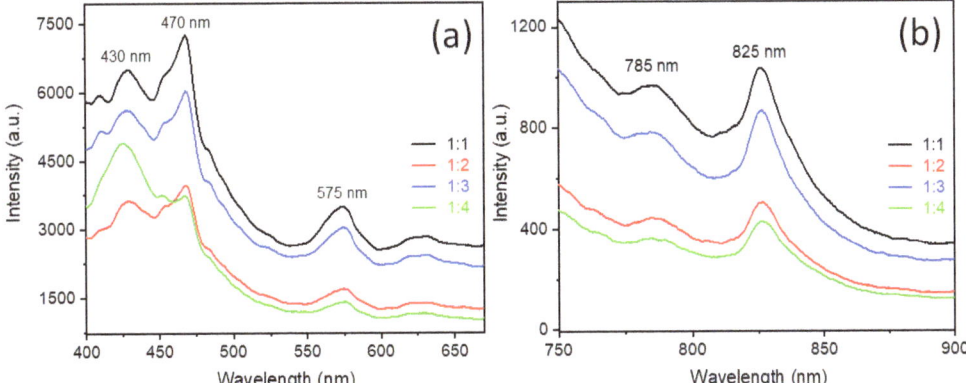

Figure 6. Photoluminescence emission spectra of synthesized rutile TiO_2 prepared with different precursors ratio of rutile: anatase TiO_2 : (black) 1:1, (red) 1:2, (blue) 1:3, and (green) 1:4. (**a**) PL emission range from 400 to 675 nm and (**b**) PL emission range from 750 nm to 900 nm at excitation wavelength 355 nm.

In the synthesis experiments, sodium phosphate was used as an additive, therefore it could be present as impurities in the final product. Hence, FTIR spectra were measured to identify any type of impurities of synthesized rutile TiO_2 microrods with different ratios of titania precursors. The observed data (Figure 7) show typical spectrum of rutile TiO_2 microrods with the hydroxyl group (OH) on the surface. It is well known that air moisture (hydroxyl group) can be easily attached on the surface of rutile TiO_2 microrods, and it is called a surface hydroxyl group [35]. Hence, the broad peaks at around 3420 cm^{-1} show stretching vibration of the OH group. A peak position at around 1629 cm^{-1} represents OH bending vibration on the rutile TiO_2 microrod surface [36]. The peak appearing at 1012 cm^{-1} is a characteristic peak of O-O stretching vibration [36]. However, no impurity peaks regarding phosphate were observed in all synthesized TiO_2 microrods.

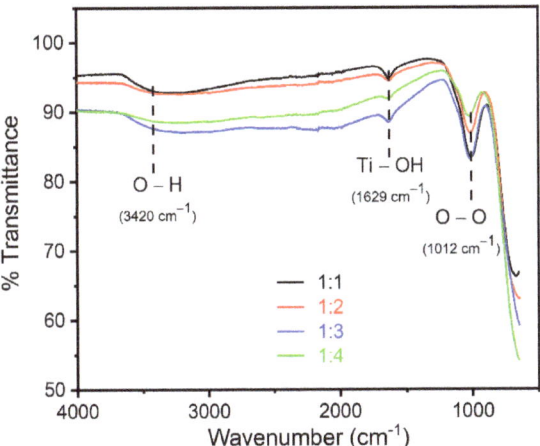

Figure 7. FTIR spectra for rutile TiO_2 microrod samples prepared by different ratios of TiO_2 (rutile, seed): TiO_2 (anatase) as a precursor (black) 1:1, (red) 1:2, (blue) 1:3, and (green) 1:4.

4. Conclusions

- It is concluded that micrometer-length TiO_2 rods were successfully synthesized by using the molten-salt method using TiO_2(rutile: anatase)/NaCl/Na$(PO_3)_6$/Na$_3$P$_4$O$_7$ as precursors.
- In the molten-salt precursors, rutile TiO_2 acted as nuclei for rod formations and anatase TiO_2 acted as a source of titanium (Ti) for rutile rod growth in the presence of NaCl as reacting media with eutectic composition of sodium phosphates (Na$(PO_3)_6$/Na$_3$P$_4$O$_7$).
- For proper eutectic composition, five different ratios of NaCl/Na$(PO_3)_6$/Na$_3$P$_4$O$_7$ were used and among them only 4:1:1 showed TiO_2 rods significantly controlled in size and length.
- By keeping the constant ratio of sodium chloride and sodium phosphate (NaCl: Na$(PO_3)_6$:Na$_3$P$_4$O$_7$) as 4:1:1, variation of TiO_2 precursors (rutile:anatase) was studied and it was found that a (1:3) ratio produced comparatively uniform size and length of TiO_2 rods.
- The synthesized rutile TiO_2 showed various emission wavelengths, such as 430, 470, 575, 785, and 825 nm at 355 nm excitation wavelength. Photoluminescence emission was observed due to oxygen vacancy generated at high temperature annealing (900 °C).

Supplementary Materials: The following supporting information can be downloaded at: https://www.mdpi.com/article/10.3390/nano12152626/s1. Figure S1. Synthesized rutile TiO_2 micro-rods by changing additives ratio ((NaPO$_3$)$_6$:Na$_3$P$_4$O$_7$) to (a) 4:0, (b) 3:1, (c) 2:2, (d) 1:3 and (e) 0:4 and keeping other precursors ratio constant (rutile : anatase : NaCl = 0.5:1.5:8); Figure S2. Low and high magnification SEM images of TiO_2 precursors (a,a') anatase TiO_2 and (b,b') rutile TiO_2; Figure S3. SEM images with lower magnification of rutile TiO_2 micro-rods samples prepared by different ratios of TiO_2 (rutile, seed):TiO_2 (anatase) as a precursor (a) 1:1, (b) 1:2, (c) 1:3 and (d) 1:4; Figure S4. SEM and EDS data of synthesized rutile TiO_2 rods with different ratios of TiO_2 (rutile, seed):TiO_2 (anatase) as a precursor (a,a') 1:1, (b,b') 1:2, (c,c') 1:3 and (d,d') 1:4; Figure S5. Band gap energy calculation by using Tauc plot method for TiO_2 sample prepared by different ratios of TiO_2 (rutile, seed):TiO_2 (anatase) as a precursor (a) 1:1, (b) 1:2, (c) 1:3 and (d) 1:4; Figure S6. Photoluminescence excitation spectra of synthesized rutile TiO_2 micro-rods with different combination of rutile:anatase precursors such as (black) 1:1, (red) 1:2, (blue) 1:3 and (green) 1:4. for the emission wavelength 825 nm. Figure S7. Complete photoluminescence spectra of synthesized rutile TiO_2 micro-rods with different combination of rutile : anatase precursors such as (black) 1:1, (red) 1:2, (blue) 1:3 and (green) 1:4. Secondary harmonic generation peak observed at 710 nm which started from 675 nm and end up at 750 nm. This secondary harmonic peak is exactly twice of excitation wavelength 355 nm. Table S1. Actual weight and related ratios of precursors used for rutile TiO2 synthesis by molten salt method.

Author Contributions: Conceptualization, Y.S.K. and H.M.N.; methodology, H.M.N.; validation, Y.S.K. and H.M.N.; formal analysis, H.M.N. and A.U.P.; investigation, H.M.N. and A.U.P.; resources, J.T., Z.Z., Y.S.K.; data curation, A.U.P., D.K.L.; writing—original draft preparation, H.M.N. and A.U.P.; writing—review and editing, A.U.P., D.K.L. and Y.S.K.; supervision, J.T., Z.Z., K.M.O., Y.S.K.; project administration, J.T., Z.Z., Y.S.K.; funding acquisition, J.T., Z.Z., Y.S.K. All authors have read and agreed to the published version of the manuscript.

Funding: This research was funded by Zhejiang Coloray Technology Development Co., Ltd. (Huzhou, Zhejiang, China).

Institutional Review Board Statement: Not applicable.

Informed Consent Statement: Not applicable.

Data Availability Statement: Not applicable.

Acknowledgments: This project has been supported by Zhejiang Coloray Technology Development Co., LTD (Huzhou, Zhejiang, China). This work was supported by the KENTECH Research Grant (KRG2022-01-005).

Conflicts of Interest: The authors declare that they have no known competing financial interests or personal relationships that could have appeared to influence the work reported in this paper.

References

1. Wedd, M.; Ward-Smith, S.; Rawle, A. Particle Size Analysis. In *Encyclopedia of Analytical Science*; Elsevier: Amsterdam, The Netherlands, 2019; pp. 144–157.
2. Trogadas, P.; Fuller, T.F. The Effect of Uniform Particle Size Distribution on Pt Stability. *ECS Trans.* **2011**, *41*, 761–773. [CrossRef]
3. Nan, J.; Huang, C.; Tian, L.; Shen, C. Effects of micro-emulsion method on microwave dielectric properties of 0.9Al$_2$O$_3$-0.1TiO$_2$ ceramics. *Mater. Lett.* **2019**, *249*, 132–135. [CrossRef]
4. Shivaraj, B.; Prabhakara, M.C.; Naik, H.S.B.; Naik, E.I.; Viswanath, R.; Shashank, M.; Swamy, B.E.K. Optical, bio-sensing, and antibacterial studies on Ni-doped ZnO nanorods, fabricated by chemical co-precipitation method. *Inorg. Chem. Commun.* **2021**, *134*, 109049. [CrossRef]
5. Lee, B.T.; Han, J.K.; Gain, A.K.; Lee, K.H.; Saito, F. TEM microstructure characterization of nano TiO$_2$ coated on nano ZrO$_2$ powders and their photocatalytic activity. *Mater. Lett.* **2006**, *60*, 2101–2104. [CrossRef]
6. Jiang, H.; Liu, Y.; Li, J.; Wang, H. Synergetic effects of lanthanum, nitrogen and phosphorus tri-doping on visible-light photoactivity of TiO$_2$ fabricated by microwave-hydrothermal process. *J. Rare Earths* **2016**, *34*, 604–613. [CrossRef]
7. Nguyen, C.K.; Cha, H.G.; Kang, Y.S. Axis-Oriented, Anatase TiO$_2$ Single Crystals with Dominant {001} and {100} Facets. *Cryst. Growth Des.* **2011**, *11*, 3947–3953. [CrossRef]
8. Hao, W.C. Comparison of the photocatalytic activity of TiO$_2$ powder with different particle size. *J. Mater. Sci. Lett.* **2002**, *21*, 1807. [CrossRef]
9. Chu, S.; Miao, Y.; Qian, Y.; Ke, F.; Chen, P.; Jiang, C.; Chen, X. Synthesis of uniform layer of TiO$_2$ nanoparticles coated on natural cellulose micrometer-sized fibers through a facile one-step solvothermal method. *Cellulose* **2019**, *26*, 4757–4765. [CrossRef]
10. Veronovski, N. TiO$_2$ Applications as a Function of Controlled Surface Treatment. In *Titanium Dioxide*; Chapter 21; IntechOpen: London, UK, 2018.
11. Wijnhoven, J.E.G.J.; Bechger, L.; Vos, W.L. Fabrication and Characterization of Large Macroporous Photonic Crystals in Titania. *Chem. Mater.* **2001**, *13*, 4486–4499. [CrossRef]
12. Liu, W.; Wang, Y.; Ge, M.; Gao, Q. One-dimensional light-colored conductive antimony-doped tin oxide@TiO$_2$ whiskers: Synthesis and applications. *J. Mater. Sci. Mater. Electron.* **2018**, *29*, 619–627. [CrossRef]
13. Wang, X.; Li, Z.; Shi, J.; Yu, Y. One-Dimensional Titanium Dioxide Nanomaterials: Nanowires, Nanorods, and Nanobelts. *Chem. Rev.* **2014**, *114*, 9346–9384. [CrossRef]
14. Lee, K.; Mazare, A.; Schmuki, P. One-Dimensional Titanium Dioxide Nanomaterials: Nanotubes. *Chem. Rev.* **2014**, *114*, 9385–9454. [CrossRef]
15. Xia, Y.; Yang, P.; Sun, Y.; Wu, Y.; Mayers, B.; Gates, B.; Yin, Y.; Kim, F.; Yan, H. One-Dimensional Nanostructures: Synthesis, Characterization, and Applications. *Adv. Mater.* **2003**, *15*, 353–389. [CrossRef]
16. Roy, B.; Fuierer, P.A. Influence of Sodium Chloride and Dibasic Sodium Phosphate SaltMatrices on the Anatase–Rutile Phase Transformation and Particle Sizeof Titanium Dioxide Powder. *J. Am. Ceram. Soc.* **2010**, *93*, 436–444. [CrossRef]
17. Tseng, L.; Luo, X.; Tan, T.T.; Li, S.; Yi, J. Doping concentration dependence of microstructure and magnetic behaviours in Co-doped TiO$_2$ nanorods. *Nanoscale Res. Lett.* **2014**, *9*, 673. [CrossRef]
18. Liu, B.; Chen, H.M.; Liu, C.; Andrews, S.C.; Hahn, C.; Yang, P. Large-Scale Synthesis of Transition-Metal-Doped TiO$_2$ Nanowires with Controllable Overpotential. *J. Am. Chem. Soc.* **2013**, *135*, 9995–9998. [CrossRef]
19. Roy, B.; Ahrenkiel, S.P.; Fuierer, P.A. Controlling the Size and Morphology of TiO$_2$ Powder by Molten and Solid Salt Synthesis. *J. Am. Ceram. Soc.* **2008**, *91*, 2455–2463. [CrossRef]
20. Beyene, A.M.; Baek, C.; Jung, W.K.; Ragupathy, P.; Kim, D.K. Understanding the role of oxygen ion (O^{2-}) activity in 1-D crystal growth of rutile TiO$_2$ in molten salts. *CrystEngComm* **2018**, *20*, 487–495. [CrossRef]
21. Egerton, T.A. UV-Absorption—The Primary Process in Photocatalysis and Some Practical Consequences. *Molecules* **2014**, *19*, 18192–18214. [CrossRef]
22. Tauc, J. *Amorphous and Liquid Semiconductors*; Plenum Press: New York, NY, USA, 1974; Volume 159.
23. Sánchez-Vergara, M.E.; Álvarez-Bada, J.R.; Perez-Baeza, C.O.; Loza-Neri, E.A.; Torres-García, R.A.; Rodríguez-Gómez, A.; Alonso-Huitron, J.C. Morphological and Optical Properties of Dimetallo-Phthalocyanine-Complex Thin Films. *Adv. Mater. Phys. Chem.* **2014**, *4*, 20–28. [CrossRef]
24. Kubelka, P. New Contributions to the Optics of Intensely Light-Scattering Materials. Part I. *J. Oct. Soc. Am.* **1948**, *38*, 448. [CrossRef] [PubMed]
25. Barcelo, D. *Modern Fourier Transform Infrared Spectroscopy*; Wilson & Wilson's: New York, NY, USA, 2001.
26. Murphy, A.B. Band-gap determination from diffuse reflectance measurements of semiconductor films, and application to photoelectrochemical water-splitting. *Sol. Energy Mater. Sol. Cell* **2007**, *91*, 1326. [CrossRef]
27. Fochs, P.D. The Measurement of the Energy Gap of Semiconductors from their Diffuse Reflection Spectra. *Proc. Phys. Soc.* **1956**, *B69*, 70. [CrossRef]
28. Selman, A.M.; Hassan, Z. Structural and Photoluminescence Studies of Rutile TiO$_2$ Nanorods Prepared by CBD Method on Si Substrates. *Am. J. Mater. Sci.* **2015**, *5*, 16–20.

29. Nasralla, N.H.S.; Yeganeh, M.; Astuti, Y.; Piticharoenphun, S.; Šiller, L. Systematic study of electronic properties of Fe-doped TiO_2 nanoparticles by X-ray photoemission spectroscopy. *J. Mater. Sci. Mater. Electron.* **2018**, *29*, 17956. [CrossRef]
30. Pallotti, D.K.; Passoni, L.; Maddalena, P.; Fonzo, F.D.; Lettieri, S. Photoluminescence Mechanisms in Anatase and Rutile TiO_2. *J. Phys. Chem. C* **2017**, *121*, 9011–9021. [CrossRef]
31. Nasralla, N.H.S.; Yeganeh, M.; Šiller, L. Photoluminescence study of anatase and rutile structures of Fe-doped TiO_2 nanoparticles at different dopant concentrations. *Appl. Phys. A Mater. Sci. Process.* **2020**, *126*, 192. [CrossRef]
32. Nakamura, R.; Nakato, Y. Primary Intermediates of Oxygen Photoevolution Reaction on TiO_2 (Rutile) Particles, Revealed by in Situ FTIR Absorption and Photoluminescence Measurements. *J. Am. Chem. Soc.* **2004**, *126*, 1290–1298. [CrossRef]
33. Nakamura, R.; Okamura, T.; Ohashi, N.; Imanishi, A.; Nakato, Y. Molecular Mechanisms of Photoinduced Oxygen Evolution, PL Emission, and Surface Roughening at Atomically Smooth (110) and (100) N-TiO_2 (Rutile) Surfaces in Aqueous Acidic Solutions. *J. Am. Chem. Soc.* **2005**, *127*, 12975–12983. [CrossRef]
34. Imanishi, A.; Okamura, T.; Ohashi, N.; Nakamura, R.; Nakato, Y. Mechanism of Water Photooxidation Reaction at Atomically Flat TiO_2 (Rutile) (110) and (100) Surfaces: Dependence on Solution pH. *J. Am. Chem. Soc.* **2007**, *129*, 11569–11578. [CrossRef]
35. Lee, J.S.; Kim, J.H.; Lee, Y.J.; Jeong, N.C.; Yoon, K.B. Manual Assembly of Microcrystal Monolayers on Substrates. *Angew. Chem. Int. Ed.* **2007**, *46*, 3087–3090. [CrossRef]
36. Foratirad, H.; Baharvandi, H.R.; Maragheh, M.G. Chemo-Rheological Behavior of Aqueous Titanium Carbide Suspension and Evaluation of the Gelcasted Green Body Properties. *Mater. Res.* **2017**, *20*, 175–182. [CrossRef]

Article

Construction of Spindle-Shaped Ti^{3+} Self-Doped TiO$_2$ Photocatalysts Using Triethanolamine-Aqueous as the Medium and Its Photoelectrochemical Properties

Zunfu Hu [1], Qi Gong [1], Jiajia Wang [1], Xiuwen Zheng [2], Aihua Wang [3] and Shanmin Gao [2,3,*]

1. School of Materials Science and Engineering, Linyi University, Linyi 276005, China; huzunfu@lyu.edu.cn (Z.H.); gongqi72909@163.com (Q.G.); wangjiajia@lyu.edu.cn (J.W.)
2. School of Chemistry & Chemical Engineering, Linyi University, Linyi 276005, China; zhengxiuwen@lyu.edu.cn
3. Yantai North China Microwave Technology Co., Ltd., Yantai 264025, China; 13953598088@163.com
* Correspondence: gaoshanmin@lyu.edu.cn

Abstract: To enhance the utilization efficiency of visible light and reduce the recombination of photo-generated electrons and holes, spindle-shaped TiO$_2$ photocatalysts with different Ti^{3+} concentrations were fabricated by a simple solvothermal strategy using low-cost, environmentally friendly TiH$_2$ and H$_2$O$_2$ as raw materials and triethanolamine-aqueous as the medium. The photocatalytic activities of the obtained photocatalysts were investigated in the presence of visible light. X-ray diffraction (XRD), Raman spectra, transmission electron microscope (TEM), X-ray photoelectron spectroscopy (XPS), and Fourier transform infrared (FT-IR) spectra were applied to characterize the structure, morphologies, and chemical compositions of as-fabricated Ti^{3+} self-doped TiO$_2$. The concentration of triethanolamine in the mixed solvent plays a significant role on the crystallinity, morphologies, and photocatalytic activities. The electron–hole separation efficiency was found to increase with the increase in the aspect ratio of as-fabricated Ti^{3+} self-doped TiO$_2$, which was proved by transient photocurrent response and electrochemical impedance spectroscopy.

Keywords: titanium dioxide; Ti^{3+} self-doped; spindle-shaped; triethanolamine; photocatalytic

1. Introduction

As one of the most promising strategies to deal with the worldwide environmental and energy crises, photocatalytic (PC) technology takes particular attention of being a green technology that can be applied under solar conditions. PC activity is largely determined by the separation efficiency of photogenerated electrons and holes with high chemical energy [1,2]. Due to its earth abundance, nontoxic high-photoelectrical conversion performance [3], and strong chemical stability against photo and chemical corrosion, TiO$_2$ has been extensively applied as photocatalyst for solar hydrogen production [4], photovoltaic power generation [5,6], sewage purification [7–9], air purification, and other fields [10,11]. Hence, TiO$_2$-based photocatalysts with high PC activity have received continuous attention. However, two defects, wide band-gap (3.2 eV for pure TiO$_2$) and rapid charge recombination rate, limit the large-scale application of TiO$_2$-based photocatalysts [12,13]. Hence, PC activity may be enhanced by comprehensive studies on band engineering and charge carrier dynamics to solve the above problems of pure TiO$_2$ photocatalyst [14,15].

Many strategies have been developed during the past few decades to enhance the absorption of visible light [1,16]. Heteroatom doping is the most basic way of changing the electronic structure of functional materials; for instance C [17], S [18,19], and N doping [20,21], acting as electronic donation, and receipt have long been actively used to modify wide band-gap TiO$_2$ photocatalysts for visible light absorption. In addition, surface modification strategies also were applied to improve the PC activity of TiO$_2$, such as transition metal doping, dye sensitization, inorganic combination, and nanostructure-tuning [22]. These efforts usually suffer from the leakage of harmful metal ions, thermal instability, and the increase in deleterious recombination centers for photogenerated charges [23–26]. These vital drawbacks have increased the limited use of TiO_2-based materials in developing photocatalysts. Therefore, reasonable modification strategies play a major role in improving PC properties of TiO_2-based photocatalysts.

As a typical n-type semiconductor, TiO_2 possesses intrinsic oxygen deficiency. The introduction of Ti^{3+} and oxygen vacancies (OVs) was employed to broaden the visible light absorption range and promote the charge separation [13,27,28]. Hence, Ti^{3+} self-doped TiO_2 nanomaterials have been widely investigated for their extended visible light absorption and high conductivity [29,30]. Inducting Ti^{3+} and OVs would create new defect states below the conduction band of TiO_2, thus increasing its PC activity [31–34]. Thus far, various strategies for introducing Ti^{3+} are often used; examples are thermal annealing at a high temperature in various reducing atmospheres, such as H_2, CO, H_2/Ar, or vacuum [35–37]. Nevertheless, the above strategies face the problems of high equipment requirements and high manufacturing costs. Hence, developing a facile strategy to construct Ti^{3+} self-doped TiO_2 is still a potential research topic.

Bulk and surface studies cannot elucidate the shape dependence effect, which is regarded as the origin of the higher performance of smaller particles. Serial organic surfactants, such as PVP [38], PEG [39], CTAB [40], thiophenol [41], and TEA [42] have been employed to tune the size and morphology and to prevent agglomeration. The introduction of surfactant could also stabilize the active adsorption sites that are conducive to the formation of smaller TiO_2-based photocatalysts [43,44]. In our previous work, reduced TiO_2 photocatalysts with different OVs contents were prepared by hydrothermal method, and the effects of hydrothermal treatment temperature and time consumed on the structure, morphology, and properties of the obtained samples were studied [45,46]. However, the solvent effect has an important impact on the morphology of the product, thus affecting the PC performance of the products. As a soluble, inexpensive, and readily available organic surfactant, triethanolamine ($N(CH_2CH_2OH)_3$, TEA) plays a vital role in the regulation of crystallinity, morphology, and electronic structure of the TiO_2-based photocatalysts [47–53]. Herein, to optimize the PC performance of TiO_2, a facile mixed solvothermal method was developed to fabricate spindle-shaped Ti^{3+} self-doped TiO_2 with TEA as structure-directing and capping agent.

2. Construction of Spindle-Shaped TiO_2 Photocatalysts with Ti^{3+} and OVs

Ti^{3+} self-doped TiO_2 photocatalysts were prepared with TiH_2 as Ti source, H_2O_2 as oxidant, and TEA as structure-directing agent. The entire synthesis process is illustrated in Scheme 1. The sol obtained by oxidizing TiH_2 with H_2O_2 was evenly divided into five parts, and then different amounts of TEA and H_2O were added, respectively. The solvothermal reaction subsequently was conducted at 180 °C for 20 h. The products were cleaned with deionized water and absolute ethanol three times, and centrifugate was collected. After drying at 60 °C, light blue photocatalysts were obtained.

The chemicals, detailed preparation process, characterization, PC, and photoelectrochemical measurement experiment of the spindle-shaped TiO_2 with Ti^{3+} self-doped are detailed in the Supplementary Materials.

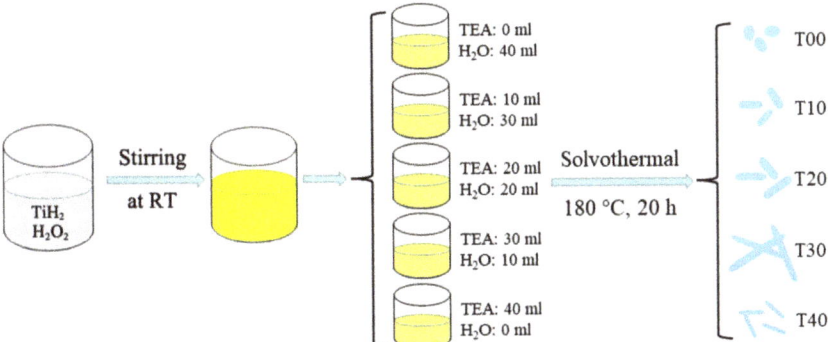

Scheme 1. Schematic diagram of the synthesis route of Ti^{3+} self-doped TiO_2 photocatalysts.

3. Results and Discussion

Ti^{3+} self-doped TiO_2 nanoparticles were prepared with different concentrations of TEA (0, 10, 20, 30, and 40 mL) via the solvothermal method at 180 °C for 20 h. The corresponding TiO_2 nanoparticles were denoted as T00, T10, T20, T30, and T40, respectively. The morphology, structure, and composition of nanomaterials are known to have a direct impact on their PC performance. X-ray diffraction (XRD) was applied to study the structure and crystallization of obtained photocatalysts. As depicted in Figure 1a, those peaks located at 25.3°, 37.9°, 48.0°, 54.1°, 54.9°, 62.7°, and 68.9° represented the (101), (004), (200), (105), (211), (204), and (216) crystal planes of pure TiO_2, maintaining the standard card of anatase TiO_2 (JCPDS No. 21–1272) [16,54]. The peaks of TiH_2 were not found in the XRD results, which proved that TiH_2 was converted into anatase TiO_2. Furthermore, with the increase in TEA content from 10 to 40 mL, the intensities of diffraction peaks decreased gradually, suggesting a decreased crystallinity for TiO_2. It might be attributed to the formation of defects with high content of TEA (Figure 1b). With the increase in TEA amount, the half-width was gradually increasing, indicating that the crystallinity of the sample was deteriorated. The above XRD results show that the content of TEA had a direct effect on the crystallinity of TiO_2 photocatalysts.

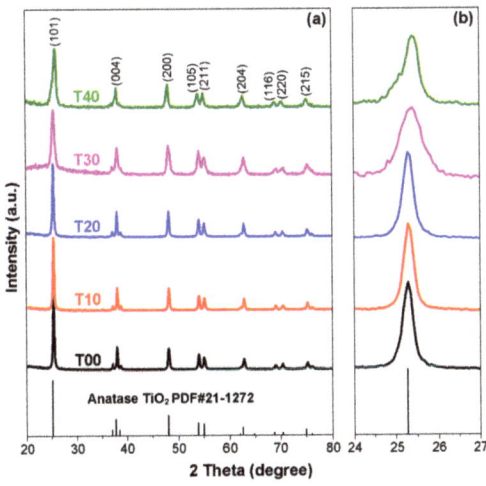

Figure 1. (**a**) XRD spectra of TiO_2 photocatalysts obtained by solvothermal treatment of the sol at different triethanolamine concentrations and (**b**) enlarged XRD results.

As a simple, effective, and high-sensitivity detection technology, Raman spectra, originating from the vibration of molecular bonds, were employed to explore the significant structural changes in TiO_2. As shown in Figure 2a, these peaks located at 145, 398, 514, and 636 cm^{-1} could index, respectively, to Eg, B1g, A1g, and Eg lattice vibration modes, indicating that all these nanomaterials were mainly anatase TiO_2 [55]. As illustrated in Figure 2b, the strongest Raman bands around 145 cm^{-1} shifted to a lower wavenumber along with the peak broadening as the content of TEA increased from 10 to 40 mL, which might be attributed to the increase in the aspect ratio of spindle-shaped TiO_2 nanoparticles. At the same time, the shift and broadening content reached the highest with 40 mL TEA, indicating that there existed a large number of defects in T40.

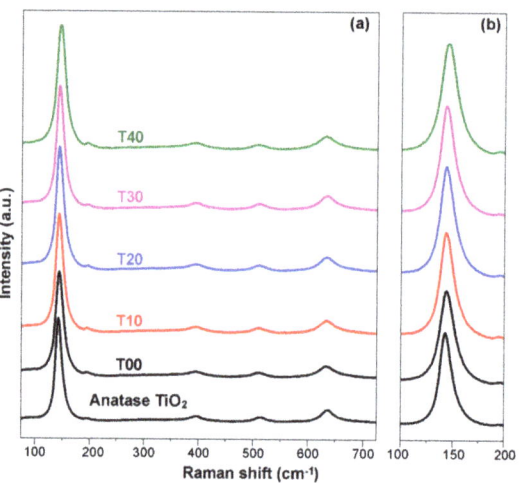

Figure 2. (a) Raman spectra of TiO_2 photocatalysts obtained by hydrothermal treatment the sol at different triethanolamine concentrations and (b) enlarged Raman results.

Transmission electron microscopy (TEM) and HRTEM were employed to investigate the representative microstructures and crystalline properties of as-prepared nanomaterials. As shown in Figure 3, as-obtained samples were spindle-shaped particles. It was easy to find that the TEA concentration in the medium significantly affected the morphology of TiO_2 nanomaterials. In absence of TEA, the obtained nanomaterials were a spherical-like particle (Figure S1a,b). The morphologies of as-obtained TiO_2 nanomaterials become spindle-shaped in the presence of TEA. With the increase in TEA amount, the spindle-shaped nanorods became narrower and longer (Figure 3a–c). However, when the amount of TEA reached to 40 mL, the obtained sample become fragmented (Figure 3e), which further indicated that the amount of TEA had an important effect on the morphology of the TiO_2 nanomaterials. The HRTEM results demonstrated that the crystallinity of the TiO_2 photocatalysts decreased gradually with the increase in TEA (Figures 3d,f and S1c,d). The crystal lattices of T00 were measured at 0.351 nm), which matched the (101) crystal plane of anatase TiO_2. While with increase in TEA, the crystal lattices of T10, T20, T30, and T40 decreased from 0.351 nm to 0.346 nm (Figures S1c,d and 3d,f), indicating that the modification of TEA would affect the crystallinity of TiO_2 nanomaterials.

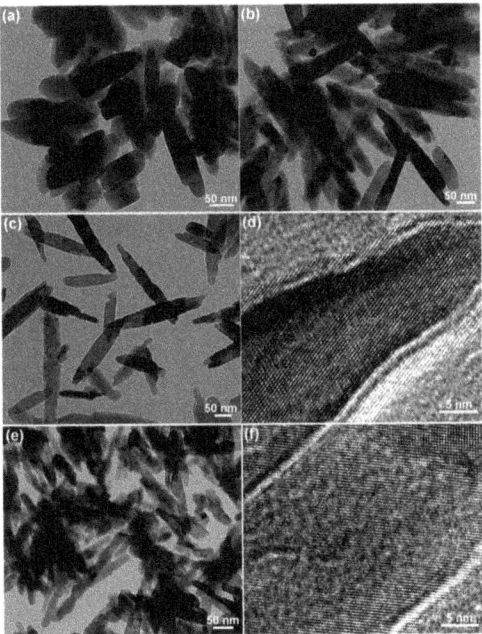

Figure 3. TEM and HRTEM pictures of TiO$_2$ photocatalysts obtained with different content of TEA: (**a**) T10, (**b**) T20, (**c**,**d**) T30, and (**e**,**f**) T40.

The XPS analysis is applied to further study the surface chemical environment of as-obtained TiO$_2$ photocatalysts. The survey XPS patterns showed that as-fabricated TiO$_2$ photocatalysts all contain O, C, and Ti (Figure S2). With the increase in TEA amounts in the medium, the N element binding energy peak appeared, indicating that the product contains N element on its surface. As illustrated in Figure 4a, the high-resolution XPS spectrum of Ti 2p for T00, T40 showed two peaks, which can be fitted into four peaks at 457.9, 458.6, 463.4, and 464.3 eV, assigning to Ti^{3+} 2p$_{3/2}$, Ti^{4+} 2p$_{3/2}$, Ti^{3+} 2p$_{1/2}$, and Ti^{4+} 2p$_{1/2}$, respectively [56,57]. Similarly, the XPS spectrum of O 1s for T00 and T40 also showed two peaks, which can be fitted into three peaks at 529.6, 530.5, and 532.2 eV, assigning to Ti–O bond, O–H, and OVs [58,59]. Compared with T00, the content of Ti^{3+} and OVs in the sample with TEA was greatly increased, indicating that the addition of TEA could not only form a one-dimensional spindle product but also regulate the content of Ti^{3+} and OVs in the product. Through comparative analysis of the binding energy peaks of Ti^{3+} 2p$_{3/2}$ in all obtained samples, the content of Ti^{3+} was found to increase with the increase in TEA amount in the mixed solvent, which further demonstrated the influence of solvent effect on the product composition (Supporting Information, Figure S3). On one hand, a one-dimensional structure was conducive to the separation of photogenerated electrons and holes [55]. On the other hand, the introduction of Ti^{3+}/OVs could lower the bandgap of TiO$_2$, which facilitated the application of visible light and the improvement of PC performance. The C 1s spectra of T00, shown in Figure 4c, was deconvolved into one peak located at 284.7 eV, assigning to the C–C bond with sp^2 orbital [60]. As comparison, C 1s spectra of T20 were fitted into two peaks at 284.7 and 288.5 eV, the latter of which was assigned as C–N bond. Similarly, the C 1s spectra of T40 were deconvolved into three peaks at 284.7, 286.1, and 288.5 eV, among which the peak at 286.1 eV was assigned to C–H with sp^3 hybridization, indicating the introduction of TEA. Moreover, high-resolution N 1s spectra of T20 and T40 showed one more peak at 399.8 eV compared to the N 1s spectra of T00, which was assigned to the C–N–C bond (Figure 4d), further confirming the introduction of TEA on the surface of TiO_2. A binding energy peak of Ti–C or Ti–N bond was not in all samples, which indicated that the C or N doping is not formed during the solvothermal treatment, and TEA only attached to the surface of TiO_2 by physical adsorption. In addition, the FT-IR analysis results also proved that TEA anchored onto TiO_2 photocatalysts (Supporting Information, Figure S4).

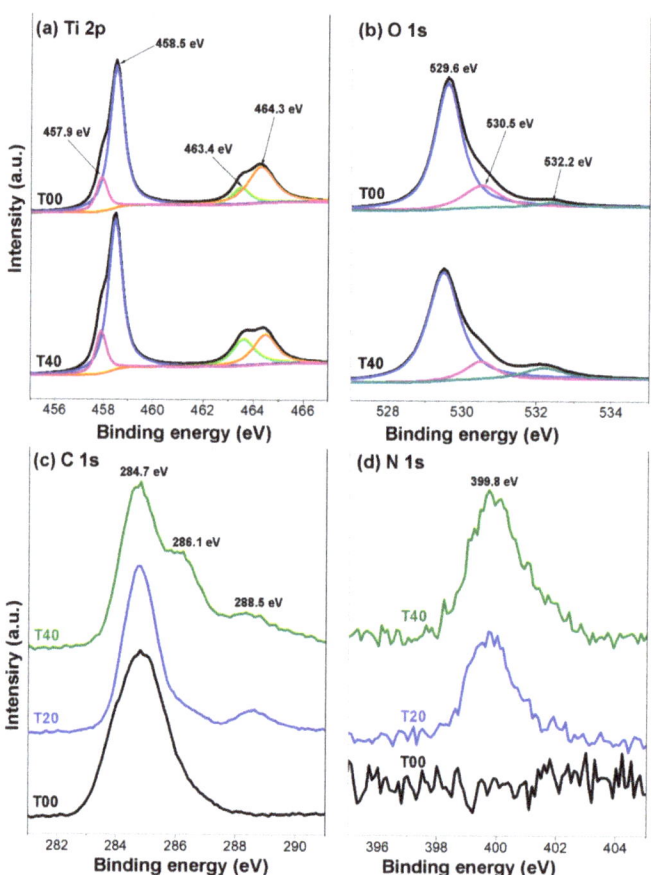

Figure 4. High-resolution XPS spectrum of different samples.

The optical performance of as-fabricated TiO_2 photocatalysts was investigated by UV-vis diffuse reflectance spectra (DRS). As illustrated in Figure 5a, with TEA as a structure-directing and capping agent, the photo-response performance of TiO_2 photocatalysts was greatly improved, extending to visible and infrared light regions. Furthermore, T40 exhibited the highest UV-vis absorption property than the other samples, indicating that the introduction Ti^{3+}/OVs could produce a novel vacancy band located below the conduction band edge of anatase TiO_2. As illustrated in Figure 5b, the band-gap energy of T40 was calculated to be 2.93 eV via plotting Kubelka–Munk formula against the photon energy. The band energy of T30, T20, T10, T00, and anatase TiO_2 were calculated to be 3.00, 3.03, 3.08, 3.05, and 3.12 eV, which were larger than that of T40.

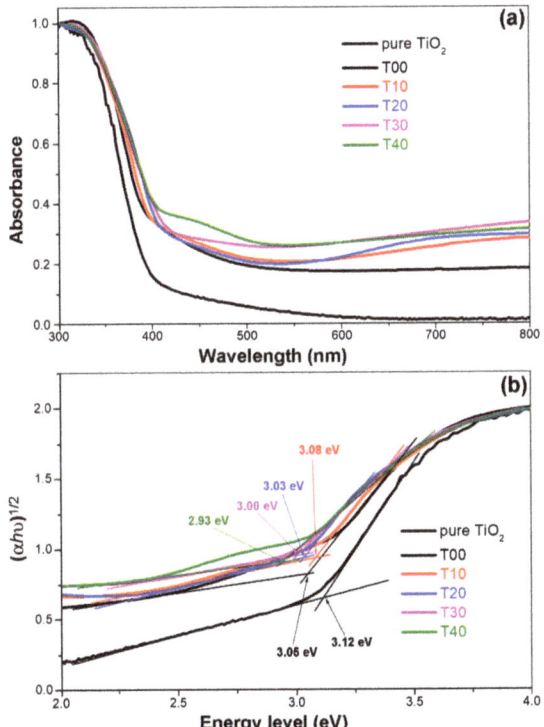

Figure 5. (a) UV-vis diffuse reflectance spectrum of TiO_2 photocatalysts; (b) Plotting curves of Kubelka–Munk against photon energy over pure TiO_2, T00, T10, T20, T30 and T40, respectively.

Photoelectrochemical transient photocurrent response under visible light can feasibly investigate the photogenerated electrons transfer property [61]. The instantaneous photocurrent data of those photocatalysts were investigated under visible light irradiation at an interval of 50 s. As displayed in Figure 6a, the photocurrent density of T10 was measured as 1.3 µA cm^{-2}, while that of T20 and T30 were 2.8 and 3.6 µA cm^{-2}, respectively. The photocurrent disappeared immediately after the visible light was removed, indicating that the photocurrent was completely generated by the photoelectrodes. Charge separation and transfer of photocatalysts were obtained by electrochemical impedance spectroscopy [62]. Figure 6b shows that the radius of the arch for T30 was smallest among those prepared TiO_2 photocatalysts, indicating that T30 exhibited the lowest impedance. The above experimental results showed that Ti^{3+} self-doped and spindle structure were beneficial to the separation and transfer efficiency of photogenerated electron-holes.

The PC performance of as-prepared TiO_2 photocatalysts was investigated by the degradation of Rhodamine B (RhB) when exposed to visible light. The time-dependent curve of concentration and spectrum during RhB degradation is shown in Figure 7a. Figure 7a also shows that the photodegradation of RhB dyes was not observed without photocatalyst. This indicates that RhB cannot be degraded under visible light while T30 sample showed the strongest PC activity, which was mainly attributed to its aspect ratio, allowing the generation of photogenerated electrons and holes and electron transfer to the surface. The stability of photocatalyst had an important influence on its practical application. Recycling experiment was employed to investigate the stability of as-prepared photocatalysts. As illustrated in Figure 7b, after six cycles, the PC performance of T30 had almost no loss, indicating that as-obtained possessed high PC stability and high practical application had potential.

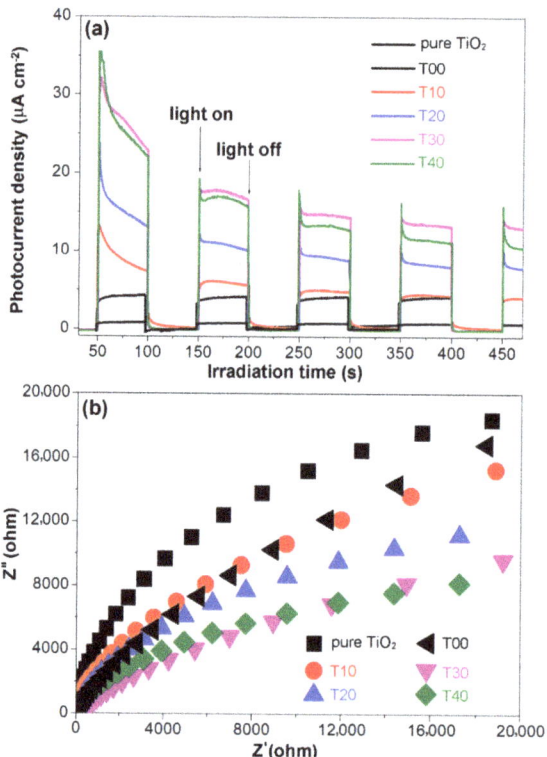

Figure 6. Transient photocurrent response (**a**) and EIS spectrum (**b**) of Nyquist plots acquired at open-circuit potential exposed to visible light illumination for different samples.

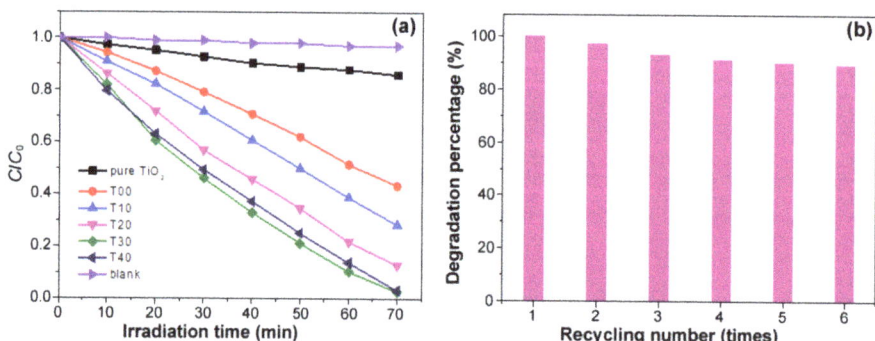

Figure 7. (**a**) Photocatalytic performance of the different samples; (**b**) recyclability of T30 in degradation of Rhodamine B.

Generally, the PC activity is determined directly by the separation efficiency of photogenerated electrons and holes. In addition, it is widely recognized that photocatalysts with a larger specific surface can supply more surface active sites for the adsorption of RhB molecules, resulting in an enhanced PC performance. The BET surface results show that T00 sample has the largest surface area, while T30 sample has the smallest surface (Supporting Information, Figure S5); however, T30 has the best PC performance, which indicates that the specific surface areas are not the main factor affecting the PC performance. In addition,

Figure 6 shows that T30 sample has the largest photocurrent and the smallest impedance, indicating that T30 has the higher separation effect of photogenerated electrons and holes. Therefore, the enhanced PC activity of Ti^{3+} self-doped TiO_2 photocatalysts is attributed to the following reasons: (i) The introduction of Ti^{3+}/OVs defects are used to reduce the band-gap of TiO_2 into the visible light range and subsequently the new defect states appear below conduction band, and (ii) the spindle-shaped nanoparticles are beneficial to the separation of photogenerated electron holes and electron transport [55]. In addition, the introduction of Ti^{3+}/OVs enhances the charge separation and diffusion while suppressing charge carrier recombination. When exposed to visible light, photogenerated electrons could change from valence band to conduction band, promoting the migration of electrons and holes to the surface TiO_2 nanoparticles. As displayed in Figure 8, photogenerated electrons react with dissolved O_2 to produce superoxide anion radicals ($\bullet O_2^-$), while holes oxidize the surface OH^- groups and H_2O molecules to generate $\bullet OH$. The highly reductive $\bullet O_2^-$ and oxidizing $\bullet OH$ will decompose RhB molecules into small micromolecules, such as H_2O and CO_2. Nevertheless, excessive Ti^{3+} would lead to the formation of the recombination center for photogenerated electron–hole pairs, which were harmful to the improvement of photocatalytic performance of TiO_2.

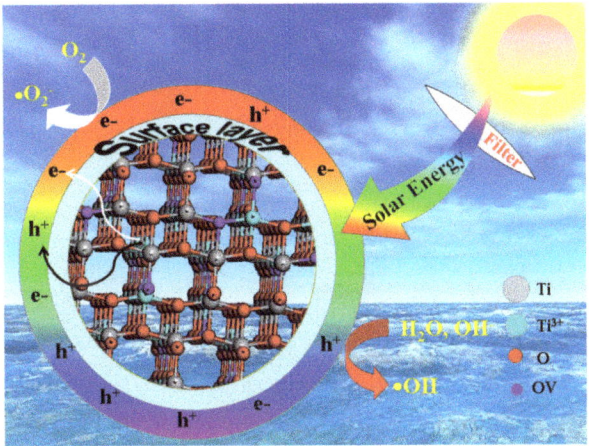

Figure 8. Schematic illustration of the visible-light photocatalytic mechanism of the Ti^{3+} self-doped TiO_2 nanoparticles.

4. Conclusions

In conclusion, spindle-shaped Ti^{3+} self-doped TiO_2 photocatalysts were fabricated by a simple solvothermal strategy with triethanolamine as a structure-directing agent. The morphology and photochemical properties of TiO_2 photocatalysts could be effectively controlled by adjusting the amount of trimethylamine. In addition, the crystallinity, photogenerated electrons, and holes of the TiO_2 catalysts could also be effectively controlled by changing the amount of trimethylamine. As-prepared TiO_2 photocatalysts were also employed for the photocatalytic degradation of rhodamine B. It was found that T30 exhibited the highest catalytic efficiency, which was mainly due to the large aspect ratio. Hence, the Ti^{3+} self-doped TiO_2 photocatalysts with large aspect ratio will be a feasible alternative strategy for water treatment and organic degradation in the future.

Supplementary Materials: The following supporting information can be downloaded at: https://www.mdpi.com/article/10.3390/nano12132298/s1, Materials and Methods. Figure S1: TEM and HRTEM images of the samples obtained at different triethanolamine concentrations; Figure S2: Survey XPS spectra of the obtained samples at different TEA amount; Figure S3: High-resolution XPS spectrum of Ti^{3+} 2p3/2 for different samples; Figure S4: FTIP spectra for pure TiO_2 and obtained samples with different TEA amount; Figure S5: N_2 adsorption-desorption isotherms and the corresponding BET surface area of the different samples.

Author Contributions: Conceptualization, Z.H., X.Z. and S.G.; methodology, Z.H. and J.W.; validation, X.Z. and A.W.; investigation, Z.H., Q.G. and J.W.; resources, X.Z. and S.G.; data curation, Z.H., Q.G., J.W. and A.W.; writing—original draft preparation, Z.H. and S.G.; writing—review and editing, Z.H. and S.G.; visualization, S.G.; supervision, X.Z. and S.G.; project administration, S.G.; funding acquisition, Z.H. and S.G. All authors have read and agreed to the published version of the manuscript.

Funding: This work was financially supported by the Natural Science Foundation of Shandong Province (Grant Number: ZR2019MB019, ZR2020QB170).

Institutional Review Board Statement: Not applicable.

Informed Consent Statement: Not applicable.

Data Availability Statement: Data presented in this article are available at request from the corresponding author.

Conflicts of Interest: The authors declare no competing interests.

References

1. Kang, X.; Song, X.Z.; Han, Y.; Cao, J.; Tan, Z. Defect-engineered TiO_2 Hollow Spiny Nanocubes for Phenol Degradation under Visible Light Irradiation. *Sci. Rep.* **2018**, *8*, 5904. [CrossRef] [PubMed]
2. Fang, L.; Chen, J.; Zhang, M.; Jiang, X.; Sun, Z. Introduction of Ti^{3+} ions into heterostructured TiO_2 nanotree arrays for enhanced photoelectrochemical performance. *Appl. Surf. Sci.* **2019**, *490*, 1–6. [CrossRef]
3. Ma, X.; Wang, C.; Wu, F.; Guan, Y.; Xu, G. TiO_2 nanomaterials in photoelectrochemical and electrochemiluminescent biosensing. *Top. Curr. Chem.* **2020**, *378*, 28. [CrossRef] [PubMed]
4. Li, J.; Cushing, S.K.; Zheng, P.; Senty, T.; Meng, F.; Bristow, A.D.; Manivannan, A.; Wu, N. Solar hydrogen generation by a CdS-Au-TiO_2 sandwich nanorod array enhanced with Au nanoparticle as electron relay and plasmonic photosensitizer. *J. Am. Chem. Soc.* **2014**, *136*, 8438–8449. [CrossRef]
5. Pradhan, S.C.; Velore, J.; Hagfeldt, A.; Soman, S. Probing photovoltaic performance in copper electrolyte dye-sensitized solar cells of variable TiO_2 particle size using comprehensive interfacial analysis. *J. Mater. Chem. C* **2022**, *10*, 3929–3936. [CrossRef]
6. Xiu, Z.; Guo, M.; Zhao, T.; Pan, K.; Xing, Z.; Li, Z.; Zhou, W. Recent advances in Ti^{3+} self-doped nanostructured TiO_2 visible light photocatalysts for environmental and energy applications. *Chem. Eng. J.* **2020**, *382*, 123011. [CrossRef]
7. Deng, X.; Zhang, H.; Guo, R.; Ma, Q.; Cui, Y.; Cheng, X.; Xie, M.; Cheng, Q. Effect of Ti^{3+} on enhancing photocatalytic and photoelectrochemical properties of TiO_2 nanorods/nanosheets photoelectrode. *Sep. Purif. Technol.* **2018**, *192*, 329–339. [CrossRef]
8. Cheng, D.; Li, Y.; Yang, L.; Luo, S.; Yang, L.; Luo, X.; Luo, Y.; Li, T.; Gao, J.; Dionysiou, D.D. One-step reductive synthesis of Ti^{3+} self-doped elongated anatase TiO_2 nanowires combined with reduced graphene oxide for adsorbing and degrading waste engine oil. *J. Hazard. Mater.* **2019**, *378*, 120752. [CrossRef]
9. Hao, W.; Li, X.; Qin, L.; Han, S.; Kang, S.-Z. Facile preparation of Ti^{3+} self-doped TiO_2 nanoparticles and their dramatic visible photocatalytic activity for the fast treatment of highly concentrated Cr(vi) effluent. *Catal. Sci. Technol.* **2019**, *9*, 2523–2531. [CrossRef]
10. Wajid Shah, M.; Zhu, Y.; Fan, X.; Zhao, J.; Li, Y.; Asim, S.; Wang, C. Facile Synthesis of Defective TiO_{2-x} Nanocrystals with High Surface Area and Tailoring Bandgap for Visible-light Photocatalysis. *Sci. Rep.* **2015**, *5*, 15804. [CrossRef]
11. Liu, G.; Han, K.; Ye, H.; Zhu, C.; Gao, Y.; Liu, Y.; Zhou, Y. Graphene oxide/triethanolamine modified titanate nanowires as photocatalytic membrane for water treatment. *Chem. Eng. J.* **2017**, *320*, 74–80. [CrossRef]
12. Cai, J.; Huang, Z.a.; Lv, K.; Sun, J.; Deng, K. Ti powder-assisted synthesis of Ti^{3+} self-doped TiO_2 nanosheets with enhanced visible-light photoactivity. *RSC Adv.* **2014**, *4*, 19588–19593. [CrossRef]
13. Xing, H.; Wen, W.; Wu, J.-M. Enhanced UV photoactivity of Ti^{3+} self-doped anatase TiO_2 single crystals hydrothermally synthesized using Ti-H_2O_2-HF reactants. *J. Photochem. Photobiol. A Chem.* **2019**, *382*, 111958. [CrossRef]
14. Naldoni, A.; Altomare, M.; Zoppellaro, G.; Liu, N.; Kment, S.; Zboril, R.; Schmuki, P. Photocatalysis with Reduced TiO_2: From Black TiO_2 to Cocatalyst-Free Hydrogen Production. *ACS Catal.* **2019**, *9*, 345–364. [CrossRef] [PubMed]
15. Yin, G.; Huang, X.; Chen, T.; Zhao, W.; Bi, Q.; Xu, J.; Han, Y.; Huang, F. Hydrogenated Blue Titania for Efficient Solar to Chemical Conversions: Preparation, Characterization, and Reaction Mechanism of CO_2 Reduction. *ACS Catal.* **2018**, *8*, 1009–1017. [CrossRef]
16. Hamisu, A.; Gaya, U.I.; Abdullah, A.H. Bi-template assisted sol-gel synthesis of photocatalytically-active mesoporous anatase TiO_2 nanoparticles. *Appl. Sci. Eng. Prog.* **2021**, *14*, 313–327. [CrossRef]
17. Wang, Y.; Sun, L.; Chudal, L.; Pandey, N.K.; Zhang, M.; Chen, W. Fabrication of Ti^{3+} Self-doped TiO_2 via a Facile Carbothermal Reduction with Enhanced Photodegradation Activities. *ChemistrySelect* **2019**, *4*, 14103–14110. [CrossRef]

18. Li, M.; Xing, Z.; Jiang, J.; Li, Z.; Kuang, J.; Yin, J.; Wan, N.; Zhu, Q.; Zhou, W. In-situ Ti^{3+}/S doped high thermostable anatase TiO_2 nanorods as efficient visible-light-driven photocatalysts. *Mater. Chem. Phys.* **2018**, *219*, 303–310. [CrossRef]
19. Huang, Z.; Gao, Z.; Gao, S.; Wang, Q.; Wang, Z.; Huang, B.; Dai, Y. Facile synthesis of S-doped reduced TiO_{2-x} with enhanced visible-light photocatalytic performance. *Chin. J. Catal.* **2017**, *38*, 821–830. [CrossRef]
20. Sun, M.; Yao, Y.; Ding, W.; Anandan, S. N/Ti^{3+} co-doping biphasic TiO_2/Bi_2WO_6 heterojunctions: Hydrothermal fabrication and sonophotocatalytic degradation of organic pollutants. *J. Alloy. Compd.* **2020**, *820*, 153172. [CrossRef]
21. Cao, Y.; Xing, Z.; Shen, Y.; Li, Z.; Wu, X.; Yan, X.; Zou, J.; Yang, S.; Zhou, W. Mesoporous black Ti^{3+}/N-TiO_2 spheres for efficient visible-light-driven photocatalytic performance. *Chem. Eng. J.* **2017**, *325*, 199–207. [CrossRef]
22. Qiao, M.; Wu, S.; Chen, Q.; Shen, J. Novel triethanolamine assisted sol–gel synthesis of N-doped TiO_2 hollow spheres. *Mater. Lett.* **2010**, *64*, 1398–1400. [CrossRef]
23. Si, L.; Huang, Z.; Lv, K.; Tang, D.; Yang, C. Facile preparation of Ti^{3+} self-doped TiO_2 nanosheets with dominant {001} facets using zinc powder as reductant. *J. Alloy. Compd.* **2014**, *601*, 88–93. [CrossRef]
24. Pan, Y.; Wen, M. Noble metals enhanced catalytic activity of anatase TiO_2 for hydrogen evolution reaction. *Int. J. Hydrog. Energy* **2018**, *43*, 22055–22063. [CrossRef]
25. Wang, L.; Luo, H.; Zhou, X.; Wei, A.; Zhou, K.; Chen, Z.; Zhang, D. Enhanced permittivity and energy density of P (VDF-HFP)-based capacitor using core-shell structured $BaTiO_3$@TiO_2 fillers. *Ionics* **2018**, *24*, 3975–3982. [CrossRef]
26. Wang, J.; Wang, X.; Yan, J.; Tan, Q.; Liang, G.; Qu, S.; Zhong, Z. Enhanced Photoelectrochemical Properties of Ti^{3+} Self-Doped Branched TiO_2 Nanorod Arrays with Visible Light Absorption. *Materials* **2018**, *11*, 1971. [CrossRef]
27. Li, Y.; Ye, X.; Cao, S.; Yang, C.; Wang, Y.; Ye, J. Oxygen-Deficient Dumbbell-Shaped Anatase TiO_{2-x} Mesocrystals with Nearly 100% Exposed {101} Facets: Synthesis, Growth Mechanism, and Photocatalytic Performance. *Chem. A Eur. J.* **2019**, *25*, 3032–3041. [CrossRef]
28. Kuang, J.; Xing, Z.; Yin, J.; Li, Z.; Tan, S.; Li, M.; Jiang, J.; Zhu, Q.; Zhou, W. Ti^{3+} self-doped rutile/anatase/TiO_2(B) mixed-crystal tri-phase heterojunctions as effective visible-light-driven photocatalysts. *Arab. J. Chem.* **2020**, *13*, 25682578. [CrossRef]
29. Zhang, X.; Hu, W.; Zhang, K.; Wang, J.; Sun, B.; Li, H.; Qiao, P.; Wang, L.; Zhou, W. Ti^{3+} Self-Doped Black TiO_2 Nanotubes with Mesoporous Nanosheet Architecture as Efficient Solar-Driven Hydrogen Evolution Photocatalysts. *ACS Sustain. Chem. Eng.* **2017**, *5*, 6894–6901. [CrossRef]
30. Ren, R.; Wen, Z.; Cui, S.; Hou, Y.; Guo, X.; Chen, J. Controllable Synthesis and Tunable Photocatalytic Properties of Ti^{3+}-doped TiO_2. *Sci. Rep.* **2015**, *5*, 10714. [CrossRef]
31. Kang, X.; Liu, S.; Dai, Z.; He, Y.; Song, X.; Tan, Z. Titanium dioxide: From engineering to applications. *Catalysts* **2019**, *9*, 191. [CrossRef]
32. Nguyen, T.D.; Li, J.; Lizundia, E.; Niederberger, M.; Hamad, W.Y.; MacLachlan, M.J. Black titania with nanoscale helicity. *Adv. Funct. Mater.* **2019**, *29*, 1904639. [CrossRef]
33. Zhen, D.; Gao, C.; Yang, D.; Zhu, X.; Grimes, C.A.; Liu, Y.; Cai, Q. Blue Ti^{3+} self-doped TiO_2 nanosheets with rich {001} facets for photocatalytic performance. *New J. Chem.* **2019**, *43*, 5759–5765. [CrossRef]
34. Motola, M.; Čaplovičová, M.; Krbal, M.; Sopha, H.; Thirunavukkarasu, G.K.; Gregor, M.; Plesch, G.; Macak, J.M. Ti^{3+} doped anodic single-wall TiO_2 nanotubes as highly efficient photocatalyst. *Electrochim. Acta* **2020**, *331*, 135374. [CrossRef]
35. Yu, X.; Fan, X.; An, L.; Li, Z.; Liu, J. Facile synthesis of Ti^{3+}-TiO_2 mesocrystals for efficient visible-light photocatalysis. *J. Phys. Chem. Solids* **2018**, *119*, 94–99. [CrossRef]
36. Pellegrino, F.; Morra, E.; Mino, L.; Martra, G.; Chiesa, M.; Maurino, V. Surface and Bulk Distribution of Fluorides and Ti^{3+} Species in TiO_2 Nanosheets: Implications on Charge Carrier Dynamics and Photocatalysis. *J. Phys. Chem. C* **2020**, *124*, 3141–3149. [CrossRef]
37. Wang, P.; Jia, C.; Li, J.; Yang, P. Ti^{3+}-doped TiO_2(B)/anatase spheres prepared using thioglycolic acid towards super photocatalysis performance. *J. Alloy. Compd.* **2019**, *780*, 660–670. [CrossRef]
38. Zhao, X.; Guo, K.; Zhang, K.; Duan, S.; Chen, M.; Zhao, N.; Xu, F.J. Orchestrated Yolk-Shell Nanohybrids Regulate Macrophage Polarization and Dendritic Cell Maturation for Oncotherapy with Augmented Antitumor Immunity. *Adv. Mater.* **2022**, *34*, 2108263. [CrossRef]
39. Lei, J.; Zhang, M.; Wang, S.; Deng, L.; Li, D.; Mu, C. One-Pot Approach for the Synthesis of Water-Soluble Anatase TiO_2 Nanoparticle Cluster with Efficient Visible Light Photocatalytic Activity. *J. Phys. Chem. C* **2018**, *122*, 26447–26453. [CrossRef]
40. Mousavi, S.M.; Zarei, M.; Hashemi, S.A.; Ramakrishna, S.; Chiang, W.H.; Lai, C.W.; Gholami, A. Gold nanostars-diagnosis, bioimaging and biomedical applications. *Drug Metab. Rev.* **2020**, *52*, 1–20. [CrossRef]
41. Tian, L.; Feng, H.; Dai, Z.; Zhang, R. Resorufin-based responsive probes for fluorescence and colorimetric analysis. *J. Mater. Chem. B* **2021**, *9*, 53–79. [CrossRef] [PubMed]
42. Dlamini, N.N.; Rajasekhar Pullabhotla, V.S.R.; Revaprasadu, N. Synthesis of triethanolamine (TEA) capped CdSe nanoparticles. *Mater. Lett.* **2011**, *65*, 1283–1286. [CrossRef]
43. Kang, K.-H.; Lee, D.-K. Synthesis of magnesium oxysulfate whiskers using triethanolamine as a morphology control agent. *J. Ind. Eng. Chem.* **2014**, *20*, 2580–2583. [CrossRef]
44. Prakash, T.; Navaneethan, M.; Archana, J.; Ponnusamy, S.; Muthamizhchelvan, C.; Hayakawa, Y. Chemical synthesis of highly size-confined triethylamine-capped TiO_2 nanoparticles and its dye-sensitized solar cell performance. *Bull. Mater. Sci.* **2018**, *41*, 40. [CrossRef]

45. Liu, X.; Gao, S.M.; Xu, H.; Lou, Z.Z.; Wang, W.J.; Huang, B.B.; Dai, Y. Green synthetic approach for Ti^{3+} self-doped TiO_{2-x} nanoparticles with efficient visible light photocatalytic activity. *Nanoscale* **2013**, *5*, 1870–1875. [CrossRef]
46. Wang, X.T.; Li, Y.M.; Liu, X.; Gao, S.M.; Huang, B.B.; Dai, Y. Preparation of Ti^{3+} self-doped TiO_2 nanoparticles and their visible light photocatalytic activity. *Chin. J. Catal.* **2015**, *36*, 389–399. [CrossRef]
47. Haque, M.A.; Mahalakshmi, S. Triethanolamine-assisted synthesis of cadmium sulfide nanoclusters. *Res. Chem. Intermed.* **2014**, *41*, 5205–5215. [CrossRef]
48. Du, M.; Yin, X.; Gong, H. Effects of triethanolamine on the morphology and phase of chemically deposited tin sulfide. *Mater. Lett.* **2015**, *152*, 40–44. [CrossRef]
49. Poortavasoly, H.; Montazer, M.; Harifi, T. Aminolysis of polyethylene terephthalate surface along with in situ synthesis and stabilizing ZnO nanoparticles using triethanolamine optimized with response surface methodology. *Mater. Sci. Eng. C Mater. Biol. Appl.* **2016**, *58*, 495–503. [CrossRef]
50. Bai, Q.; Lavenas, M.; Vauriot, L.; Le Trequesser, Q.; Hao, J.; Weill, F.; Delville, J.P.; Delville, M.H. Hydrothermal Transformation of Titanate Scrolled Nanosheets to Anatase over a Wide pH Range and Contribution of Triethanolamine and Oleic Acid to Control the Morphology. *Inorg. Chem.* **2019**, *58*, 2588–2598. [CrossRef]
51. Rezaie, A.B.; Montazer, M. Polyester modification through synthesis of copper nanoparticles in presence of triethanolamine optimized with response surface methodology. *Fibers Polym.* **2017**, *18*, 434–444. [CrossRef]
52. Liang, J.; Li, L.; Luo, M.; Wang, Y. Fabrication of Fe_3O_4 octahedra by a triethanolamine-assisted hydrothermal process. *Cryst. Res. Technol.* **2011**, *46*, 95–98. [CrossRef]
53. Peng, S.; Dan, M.; Guo, F.; Wang, H.; Li, Y. Template synthesis of $ZnIn_2S_4$ for enhanced photocatalytic H_2 evolution using triethanolamine as electron donor. *Colloids Surf. A Physicochem. Eng. Asp.* **2016**, *504*, 18–25. [CrossRef]
54. Ramanathan, S.; Moorthy, S.; Ramasundaram, S.; Rajan, H.K.; Vishwanath, S.; Selvinsimpson, S.; Durairaj, A.; Kim, B.; Vasanthkumar, S. Grape seed extract assisted synthesis of dual-functional anatase TiO_2 decorated reduced graphene oxide composite for supercapacitor electrode material and visible light photocatalytic degradation of bromophenol blue dye. *ACS Omega* **2021**, *6*, 14734–14747. [CrossRef] [PubMed]
55. Xiong, H.; Wu, L.; Liu, Y.; Gao, T.; Li, K.; Long, Y.; Zhang, R.; Zhang, L.; Qiao, Z.A.; Huo, Q.; et al. Controllable Synthesis of Mesoporous TiO_2 Polymorphs with Tunable Crystal Structure for Enhanced Photocatalytic H_2 Production. *Adv. Energy Mater.* **2019**, *9*, 1901634. [CrossRef]
56. Yin, G.; Wang, Y.; Yuan, Q. Ti^{3+}-doped TiO_2 hollow sphere with mixed phases of anatase and rutile prepared by dual-frequency atmospheric pressure plasma jet. *J. Nanoparticle Res.* **2018**, *20*, 1–12. [CrossRef]
57. Abdullah, S.; Sahdan, M.; Nafarizal, N.; Saim, H.; Embong, Z.; Rohaida, C.C.; Adriyanto, F. Influence of substrate annealing on inducing Ti^{3+} and oxygen vacancy in TiO_2 thin films deposited via RF magnetron sputtering. *Appl. Surf. Sci.* **2018**, *462*, 575–582. [CrossRef]
58. Duan, Y.; Grah, P.A.; Cai, F.; Yuan, Z. Enhanced photoelectrochemical performance for hydrogen generation via introducing Ti^{3+} and oxygen vacancies into TiO_2 nanorod arrays. *J. Mater. Sci. Mater. Electron.* **2018**, *29*, 20236–20246.
59. Pan, J.; Dong, Z.; Wang, B.; Jiang, Z.; Zhao, C.; Wang, J.; Song, C.; Zheng, Y.; Li, C. The enhancement of photocatalytic hydrogen production via Ti^{3+} self-doping black TiO_2/g-C_3N_4 hollow core-shell nano-heterojunction. *Appl. Catal. B Environ.* **2019**, *242*, 92–99. [CrossRef]
60. Ming, H.; Zhang, H.; Ma, Z.; Huang, H.; Lian, S.; Wei, Y.; Liu, Y.; Kang, Z. Scanning transmission X-ray microscopy, X-ray photoelectron spectroscopy, and cyclic voltammetry study on the enhanced visible photocatalytic mechanism of carbon–TiO_2 nanohybrids. *Appl. Surf. Sci.* **2012**, *258*, 3846–3853. [CrossRef]
61. Yu, J.; Wang, B. Effect of calcination temperature on morphology and photoelectrochemical properties of anodized titanium dioxide nanotube arrays. *Appl. Catal. B Environ.* **2010**, *94*, 295–302. [CrossRef]
62. Khan, M.; Ansari, S.; Pradhan, D.; Ansari, M.; Han, D.; Lee, J.; Cho, M. Band gap engineered TiO_2 nanoparticles for visible light induced photoelectrochemical and photocatalytic studies. *J. Mater. Chem. A* **2014**, *2*, 637–644. [CrossRef]

Article

Efficient Charge Transfer Channels in Reduced Graphene Oxide/Mesoporous TiO₂ Nanotube Heterojunction Assemblies toward Optimized Photocatalytic Hydrogen Evolution

Zhenzi Li [†], Decai Yang [†], Hongqi Chu, Liping Guo, Tao Chen, Yifan Mu, Xiangyi He, Xueyan Zhong, Baoxia Huang, Shiyu Zhang, Yue Gao, Yuxiu Wei, Shijie Wang * and Wei Zhou *

Shandong Provincial Key Laboratory of Molecular Engineering, School of Chemistry and Chemical Engineering, Qilu University of Technology (Shandong Academy of Sciences), Jinan 250353, China; zzli@qlu.edu.cn (Z.L.); decaiyang@qlu.edu.cn (D.Y.); hongqichu@qlu.edu.cn (H.C.); lipingguo@qlu.edu.cn (L.G.); taochen@qlu.edu.cn (T.C.); yifanmou@qlu.edu.cn (Y.M.); xiangyihe@qlu.edu.cn (X.H.); xueyanzhong@qlu.edu.cn (X.Z.); baoxiahuang@qlu.edu.cn (B.H.); shiyuzhang@qlu.edu.cn (S.Z.); yuegao123@qlu.edu.cn (Y.G.); yuxiuwei@qlu.edu.cn (Y.W.)
* Correspondence: wsj0924@qlu.edu.cn (S.W.); wzhou@qlu.edu.cn (W.Z.)
† These authors contributed equally to this work.

Abstract: Interface engineering is usually considered to be an efficient strategy to promote the separation and migration of photoexcited electron-hole pairs and improve photocatalytic performance. Herein, reduced graphene oxide/mesoporous titanium dioxide nanotube heterojunction assemblies (rGO/TiO₂) are fabricated via a facile hydrothermal method. The rGO is anchored on the surface of TiO₂ nanosheet assembled nanotubes in a tightly manner due to the laminated effect, in which the formed heterojunction interface becomes efficient charge transfer channels to boost the photocatalytic performance. The resultant rGO/TiO₂ heterojunction assemblies extend the photoresponse to the visible light region and exhibit an excellent photocatalytic hydrogen production rate of 932.9 $\mu mol^{-1}\ g^{-1}$ under simulated sunlight (AM 1.5G), which is much higher than that of pristine TiO₂ nanotubes (768.4 $\mu mol\ h^{-1}\ g^{-1}$). The enhancement can be ascribed to the formation of a heterojunction assembly, establishing effective charge transfer channels and favoring spatial charge separation, the introduced rGO acting as an electron acceptor and the two-dimensional mesoporous nanosheets structure supplying a large surface area and adequate surface active sites. This heterojunction assembly will have potential applications in energy fields.

Keywords: TiO₂; photocatalysis; mesoporous structure; assembly; heterojunction

1. Introduction

In recent years, the energy crisis and environmental pollution have become two major topics in the field of science and technology research [1,2]. The development and utilization of new technology has become the primary task of current research. Among various options, photocatalysis is favored for its low carbon footprint and use of renewable resources [3–5]. The most important thing is that it can directly use solar energy to generate new energy, such as hydrogen energy, and plays an important role in the degradation of pollutants and nitrogen fixation [6]. Therefore, it is considered to be an efficient strategy to solve the current increasingly serious environmental problems, and it has good application prospects.

In 1972, Honda and Fujishima [7] discovered that titanium dioxide (TiO₂) could cause photocatalytic water-splitting. Since then, many scholars have begun to conduct a large number of experimental studies on TiO₂ [8]. So far, TiO₂ has obviously become the most widely studied photocatalyst due to its advantages of high stability and activity, super hydrophilicity, low cost and environmental friendliness in the field of photocatalysis [9,10]. However, from the perspective of the mechanism, it can be found that TiO₂ has three fatal defects, which seriously limit its development. Firstly, TiO_2 has a large band gap energy (3.2 eV), which means that the photogenerated electron-hole can only be activated by ultraviolet light with an energy higher than 3.2 eV (wavelength less than 387 nm), which only accounts for ~5% of sunlight. It was clear that visible light with a relatively long wavelength (almost 43% of sunlight) could not stimulate the photocatalytic activity of TiO_2 [11,12]. Secondly, the excited photogenerated electrons and holes in TiO_2 have a strong oxidability and reducibility, which leads to a rapid recombination and inefficient quantum efficiency. Studies have shown that most (about 90%) of the excited photogenerated electrons and holes of TiO_2 have recombined before photocatalysis [13]. Thirdly, traditional TiO_2 materials have a low specific surface area, which restricts the photocatalytic activity of TiO_2 obviously [14,15].

In recent years, some methods have been adopted to improve the photocatalytic performance of TiO_2. Firstly, the surface area of TiO_2 could be improved by adjusting the morphology [16]. In addition, an excellent morphology can also improve the probability of charge transport by shortening the transmission path [17]. Secondly, TiO_2 is modified by doping, metal deposition, and recombination to improve its activity. Doping can be used for improving the charge transfer efficiency and can be used for dragging down the bandgap [18]. Metal deposition can either form a Schottky barrier or an Ohmic contact, improving the efficiency of charge separation [19], and the introduction of metal can also broaden the absorption of light into the visible light region [20]. Recombination can improve charge separation efficiency through electron transfer [21]. Among various methods, heterojunction construction is a simple and practical method to effectively promote charge separation [22–24]. Graphene, as a representative of two-dimensional carbon materials, is composed of a single layer of carbon atoms with an sp^2 hybrid orbital in the shape of a hexagonal honeycomb [25]. Graphene oxide (GO) is prepared by embedding O atoms into the c-scaffold of graphene to make sp^2 and sp^3 domains exist simultaneously in the structure, which promotes the expansion of multiple interactions. On this basis, reduced graphene oxide (rGO) can provide more interfacial polarization sites and absorption sites [26]. Thus, rGO has obvious application prospects in many fields including energy and environment, along with thermal and electrical properties [27–29]. In recent years, some reports had been reported on the application of rGO/TiO_2 composite materials in the field of photocatalysis. Balsamo et al. successfully prepared a TiO_2-rGO composite structure using the one-pot method by solar irradiation, which is simple and green. The degradation rate of 2,4-dichlorophenoxyacetic acid after irradiation for 3 h reached 97%. The enhanced absorption of the composite is due to the interaction between the TiO_2 and rGO [30]. Zouzelka et al. optimized rGO/TiO_2 composites by electrophoretic deposition. The presence of rGO can, to some extent, promote the photocatalytic degradation of 4-chlorophenol. Compared with TiO_2, the photocatalytic degradation rate of the composite material is several times higher, and this enhanced performance is due to the charge transfer promoting the formation of hydroxyl radical [31]. Firstly, rGO could change the band gap width of TiO_2 to expand its light response range. Secondly, it can be used as an electron transporter, which can effectively prevent photogenerated electron hole recombination and significantly prolong the lifetime of the electron-hole pair. In addition, a large specific surface area can provide more active sites for photocatalytic reactions [32]. Therefore, rGO/TiO_2 composite materials maybe have good application prospects in the field of photocatalysis due to their high charge separation efficiency. Although these rGO/TiO_2 composites indeed improved the photocatalytic performance obviously, they are still far from being practical applications. The question of how to further promote the charge separation of rGO/TiO_2-based materials via interface engineering is still a great challenge.

Here, novel reduced graphene oxide/mesoporous titanium dioxide nanotube heterojunction assemblies (rGO/TiO_2) are fabricated via a facile hydrothermal method by utilizing mesoporous TiO_2 nanosheet-assembled nanotubes as the host. For one thing, the morphology of TiO_2 was regulated into the nanotube structure assembled by the mesoporous nanosheet, which increased the specific surface area of the material and exposed more active sites. Besides, a tubular structure increased the multiple reflection of sunlight inside the tube and improved the utilization rate of light. For another thing, rGO, as an electron acceptor, was introduced to form rGO/TiO_2 heterojunction assemblies, which contributed to improve the spatial charge separation efficiency and extend the photoresponse to the visible light region. The rGO/TiO_2 heterojunction assemblies significantly improved the photocatalytic activity, and the photocatalytic hydrogen production rate was up to 932.9 umol h^{-1} g^{-1}, 1.2 times higher than that of pristine TiO_2. The novel rGO/TiO_2 heterojunction assemblies may provide new insights for the fabrication of TiO_2-based photocatalysts with a high performance.

2. Materials and Methods

2.1. Chemicals

Titanium oxysulfate ($TiOSO_4$), potassium permanganate ($KMnO_4$), sodium nitrate ($NaNO_3$) and sulfuric acid (H_2SO_4) were purchased from Aladdin Chemical (Shanghai, China). Ethanol (EtOH), ether ($C_2H_5OC_2H_5$) and glycerol ($C_3H_5(OH)_3$) were purchased from Tianjin Kermio Chemical (Tianjin, China). None of the chemicals were further purified.

2.2. Preparation of TiO_2 Nanotube

The typical method for the synthesis of a TiO_2 nanotube is based on the literature, and the specific steps are as follows [33]. 1 g $TiOSO_4$ was added to 18 mL EtOH with intense stirring for 30 min, after which 9 mL $C_3H_5(OH)_3$ and 9 mL $C_2H_5OC_2H_5$ were added drop by drop, stirred overnight, and then transferred to the Teflon reactor at 170 °C for 10 h. After the reaction, the obtained precursor was collected by repetitive centrifugation at 8000 rpm for 10 min and washed several times by deionized water and ethanol and dried at 60 °C for 12 h. After drying, the samples were transferred to a Muffle furnace and kept at 600 °C for 4 h. Finally, the TiO_2 nanotube sample was obtained.

2.3. Preparation of rGO/TiO_2

The formation process of rGO/TiO_2 is shown in Figure 1a. rGO/TiO_2 samples are synthesized on the basis of TiO_2, and rGO is prepared by the hummer method using $KMnO_4$, $NaNO_3$ and H_2SO_4 as raw materials [34,35]. The prepared 5 mg rGO was added to 18 mL EtOH, stirred and ultrasonicated for 20 min, after which 1 g $TiOSO_4$ was added. Then, 9 mL $C_3H_5(OH)_3$ and 9 mL $C_2H_5OC_2H_5$ were added drop by drop, stirred overnight, and transferred to the Teflon-lined autoclave at 170 °C for 10 h. After the reaction, the resulting product was centrifuged and washed with deionized water and anhydrous ethanol 3 times and dried for later use. After drying, the samples were transferred to a Muffle furnace for calcination at 600 °C for 4 h. Finally, rGO/TiO_2 samples were obtained. According to the feeding ratio, the content of rGO on TiO_2 is 1 wt%.

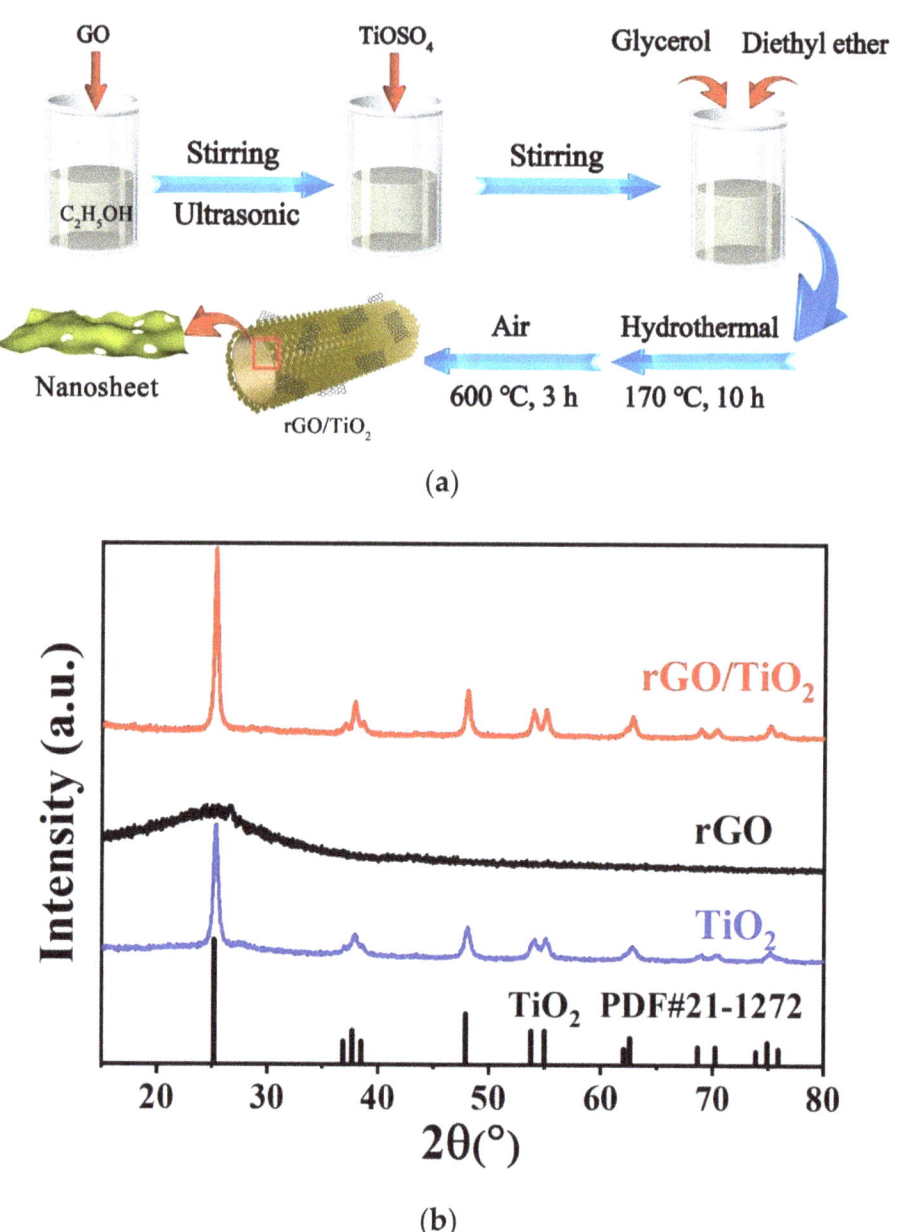

Figure 1. *Cont.*

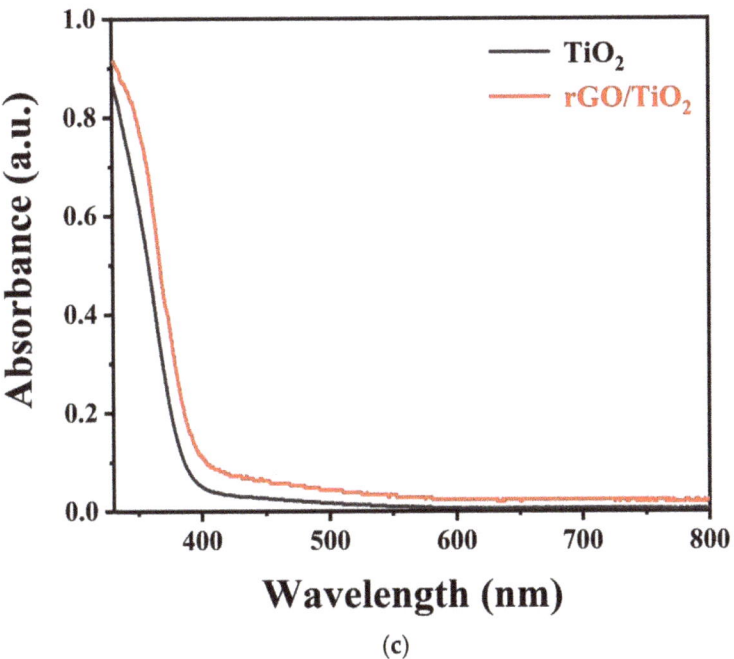

(c)

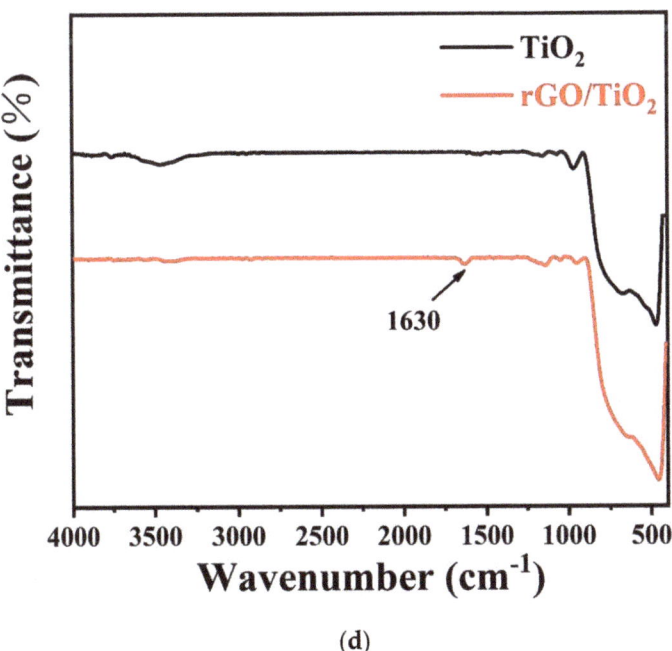

(d)

Figure 1. Schematic representation of the formation of (**a**) rGO/TiO2 heterojunction assembly photocatalysts; (**b**) X-ray diffraction (XRD) patterns, (**c**) UV–visible diffuse reflectance spectra, and (**d**) FTIR of TiO_2 and rGO/TiO_2, respectively.

3. Results and Discussion

3.1. Crystal Structure and Optical Absorption of rGO/TiO$_2$

Figure 1b shows the X-ray diffraction (XRD) patterns of GO, TiO$_2$ and GO/TiO$_2$. The multiple diffraction peaks located at 25.3, 37.8, 48.0, 55.1, 62.8, 70.2 and 75.0° can be assigned to the (101), (004), (200), (211), (204), (116) and (220) planes of anatase TiO$_2$ (JCPDS 21-1272), respectively [36]. This indicates the successful preparation of anatase TiO$_2$. The characteristic peak positions of rGO (Figure 1b and Figure S1) are about 26° and 42.5°, which is consistent with the literature [37]. Figure 1b shows the XRD pattern of the rGO/TiO$_2$ composite material, which is basically consistent with that of TiO$_2$, indicating that the introduction of rGO does not destroy the crystal structure of TiO$_2$. This also implies the stability of the material. However, the increased strength of the (101) crystal plane preliminarily indicates the successful introduction of rGO. The reason for this stems from the coincidence of the position of the TiO$_2$ (101) crystal plane (25.3°) and rGO (26°), and no single peak of rGO can be observed. As shown in Figure 1c, UV-vis was applied to obtain the absorption property of the materials. TiO$_2$ has a strong light absorption capacity when the wavelength is less than 400 nm, which is the inherent absorption property of crystal anatase TiO$_2$. The absorption intensity of rGO/TiO$_2$ was higher than that of TiO$_2$, indicating the enhancement of the composites. The photoresponse region of the material extended the UV absorption to the visible light region, which was attributed to the modification of rGO. Figure 1d shows FTIR of TiO$_2$ and rGO/TiO$_2$, the principal vibration of TiO$_2$ corresponding to the wide band in the range of 400–800 cm^{-1}. The significant band at 670 cm^{-1} corresponds to the bending vibration mode of the Ti–O–Ti bond. For the rGO/TiO$_2$ heterostructure, the wide absorption at low frequencies is similar to the spectrum of the TiO$_2$ and can therefore be attributed to the vibration of the TiO$_2$. In addition, the peak broadening of rGO/TiO$_2$ relative to TiO$_2$ below 1000 cm^{-1} may be due to the superposition of the peak existing in the Ti–O–C vibration (798 cm^{-1}) on Ti–O–Ti. The emerging absorption band at around 1630 cm^{-1} can be associated with the bone vibration of the graphene sheet, indicating the successful loading of rGO [38]. The weak peak was due to the low amount of rGO on the TiO$_2$ surface (about 1 wt%).

3.2. Morphology of rGO/TiO$_2$ Assembly

A scanning electron microscope (SEM, Hitachi, S-4800, Tokyo, Japan) and transmission electron microscope (TEM, JEOL, JEM-2100, Tokyo, Japan) were used to observe the microstructure of TiO$_2$ and rGO/TiO$_2$. It can be clearly seen that TiO$_2$ shows a very uniform tubular structure, assembled by many sheets, with a diameter of about 700 nm (Figure 2a,b). Compared with TiO$_2$, the surface of rGO/TiO$_2$ becomes slightly rough, and the diameter of the tube increases slightly (Figure 2c,d). The sheet structure can increase the specific surface area and expose more active sites. The tubular structure formed by the self-assembly of multiple nanosheets can further increase the specific surface area of the material. Secondly, the tubular structure can reflect and utilize the sunlight efficiently, which improves the utilization rate of light and is conducive to improving the photocatalytic activity. In addition, the heterostructure formed by TiO$_2$ and rGO is beneficial to the electron transfer between materials, which reduces the electron-hole recombination rate and improves the charge separation efficiency and photocatalytic activity. TEM images of rGO/TiO$_2$, as shown in Figure 2e, illustrate a nanosheet-assembled tubular structure, and the light-colored part on the surface is a layer of rGO, which is consistent with the results obtained by SEM. The rGO could be anchored on the surface of the nanosheets due to the laminated effect, thus forming efficient heterojunctions. The HRTEM images (Figure 2f,g) of rGO/TiO$_2$ show that the edge of the sample is a layer of rGO and that the inner layer is TiO$_2$. As shown in Figure 2g, clear lattice fringes can be seen with a lattice spacing of about 0.35 nm corresponding to the (101) crystal plane of anatase TiO$_2$. Meanwhile, it can be clearly seen that the lattice spacing of rGO is 0.34 nm, which further indicates the successful preparation of the rGO/TiO$_2$ heterojunction assembly material. Elemental mapping (Figure 2h–j) shows that O, C and

Ti elements in rGO/TiO$_2$ are evenly dispersed. The above information demonstrates the successful manufacturing of the rGO/TiO$_2$ heterojunction assembly.

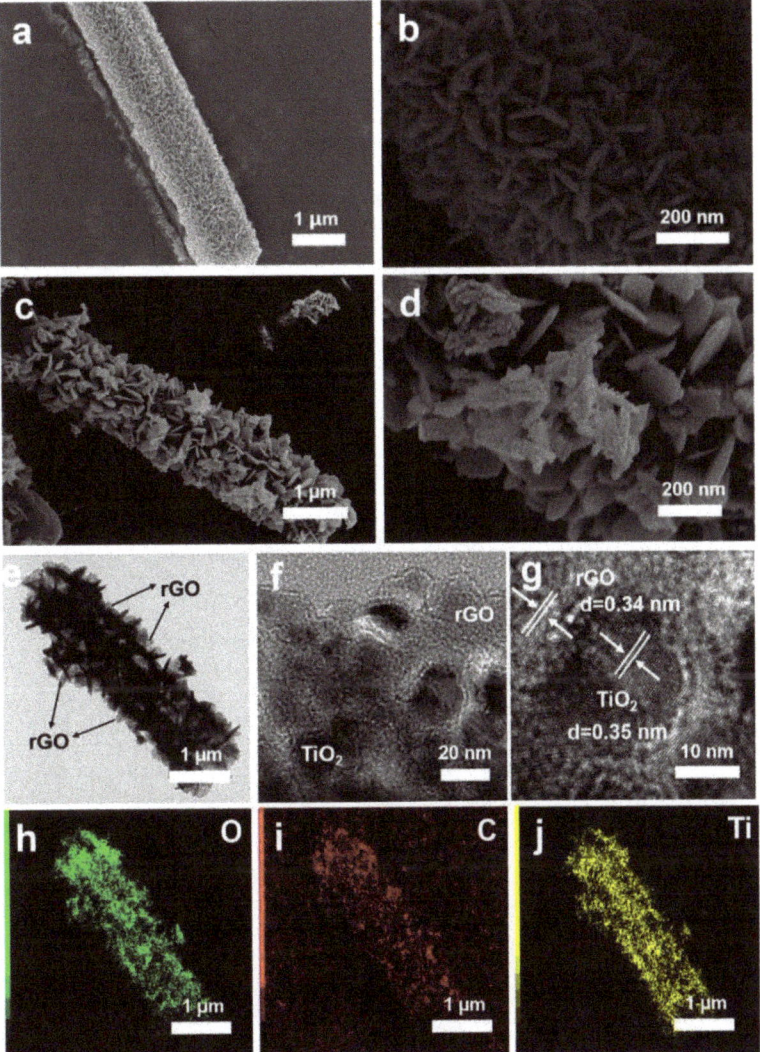

Figure 2. The SEM images of (**a,b**) TiO$_2$ and (**c,d**) rGO/TiO$_2$, and the (**e**) TEM, (**f,g**) HRTEM images and (**h–j**) elemental mappings of the rGO/TiO$_2$ heterojunction assembly.

3.3. Photocatalytic Performance of rGO/TiO$_2$

The photocatalytic performance of the catalysts for hydrogen production under favorable conditions of the catalyst Pt was conducted. As shown in Figure 3a, the hydrogen production performance of rGO is nearly 0, indicating that the presence of rGO does not contribute to hydrogen production, which is consistent with the reports in the literature. The hydrogen production quantity of rGO/TiO$_2$ is 932.9 umol h^{-1} g^{-1}, which is 1.2 times higher than that of original TiO$_2$ (768.4 umol h^{-1} g^{-1}). We can clearly see the excellent photocatalytic performance of rGO/TiO$_2$, which is significantly better than in the previous literature (Table S1). This is mainly attributed to the formation of a heterojunction between

TiO$_2$ and rGO, which promotes the spatial charge separation and inhibits the recombination of electron-hole pairs. Secondly, the design of an rGO sheet supported on a TiO$_2$ sheet is more conducive to external proton reduction. This 2D/2D design can also maximize the interface contact area, integrate the advantages of each 2D component, and facilitate the photocatalytic reaction. Finally, the nanosheet-assembled tubular structure provides a large specific surface area and sufficient surface active sites for the photocatalytic reaction. Under sunlight irradiation (AM 1.5G), the special microstructure can realize multiple reflections within the tubular structure, improving the utilization rate of sunlight. To investigate the stability of catalysts, cyclic tests were performed (Figure 3b). Five cycles of rGO/TiO$_2$ were tested, and the hydrogen production rates were almost identical for each cycle. Besides, the amount of hydrogen production was not attenuated, indicating that rGO/TiO$_2$ possessed a high stability, which was favorable for practical applications.

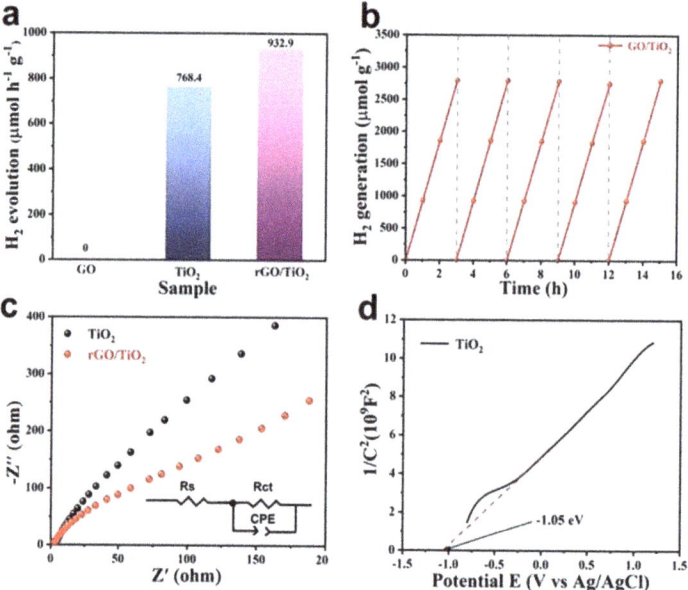

Figure 3. (a) Photocatalytic H$_2$ evolution rate of rGO, TiO$_2$ and rGO/TiO$_2$, (b) recycling tests of rGO/TiO$_2$, (c) the Nyquist plots of electrochemical impedance under AM 1.5G of TiO$_2$ and rGO/TiO$_2$, and (d) the Mott–Schottky plots of TiO$_2$.

Figure 3c shows the electrochemical impedance spectra (EIS, Shanghai Chenhua Instrument Co., Ltd., CHI760E, Shanghai, China) of TiO$_2$ and rGO/TiO$_2$ under AM 1.5G light irradiation. It can be clearly seen that the arc radius of the EIS diagram of rGO/TiO$_2$ is smaller than that of TiO$_2$, indicating that rGO/TiO$_2$ has a lower electrochemical impedance under light irradiation. This result clearly indicates that there is a positive synergistic effect between TiO$_2$ and rGO on the enhancing electron migration, which has a more efficient charge carrier separation and faster electron transfer to reduce the electron-hole recombination rate and promote the photocatalytic hydrogen generation [39]. In order to analyze the semiconductor type and band structure of TiO$_2$, the Mott–Schottky curve was tested for TiO$_2$. As shown in Figure 3d, the slope of the curve is positive, which is typical for n-type semiconductors [40]. At the same time, the flat band potential of TiO$_2$ (−1.05 eV) can be obtained from the slope of the curve, and the conduction band of TiO$_2$ is −0.81 eV [41]. Combined with UV-vis, the valence band position is determined at 2.31 eV.

3.4. Photocatalytic Mechanism of rGO/TiO$_2$

Based on the above, the mechanism diagram of photocatalytic hydrogen production of the rGO/TiO$_2$ heterojunction was made, as shown in Scheme 1. When the rGO/TiO$_2$ catalyst is irradiated by light, the electrons on TiO$_2$ are excited to the conduction band (CB), leaving holes in the valence band (VB). Meanwhile, TiO$_2$ will come into contact with rGO to form a built-in electric field. Since rGO is an electron acceptor, electrons will transfer from TiO$_2$ CB to rGO. Additionally, the reduction reaction H$^+$ converts to H$_2$ on rGO. The improved photocatalytic performance is attributed to an electron derived from the rGO/TiO$_2$ heterojunction, which reduces the recombination and improves the spatial charge separation efficiency. Such high photocatalytic properties are inseparable from the structure itself. The nanosheet-assembled hollow tubular structures provide a large specific surface area and adequate surface active sites for photocatalytic reactions. In addition, the nanotube structure can improve the utilization rate of light energy. The reason for this is that light can be reflected and used multiple times in the cavity. These factors all contribute to improving the photocatalytic hydrogen evolution of the rGO/TiO$_2$ heterojunction assembly.

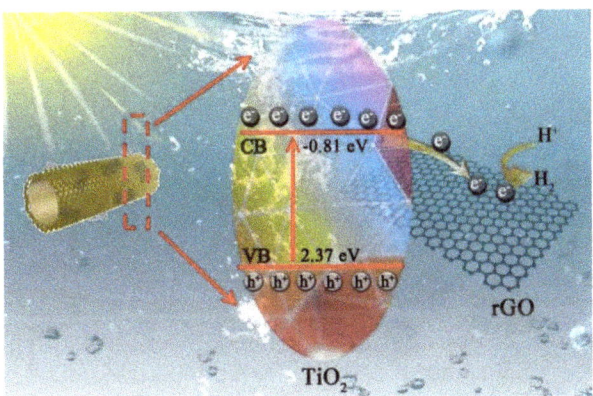

Scheme 1. The mechanism diagram of photocatalytic hydrogen production of the rGO/TiO$_2$ heterojunction assembly.

4. Conclusions

In summary, an rGO/TiO$_2$ heterojunction assembly was prepared by a step hydrothermal method, and the H$_2$ production rate was much faster than that of the original TiO$_2$ nanotubes under AM 1.5G irradiation. This was mainly decided by the following two aspects. Firstly, the rGO heterostructure with TiO$_2$ promoted the electron transfer, reduced the electron recombination rate, and improved the spatial charge efficiency. The design of rGO sheets supported by TiO$_2$ sheets is more conducive to the reduction of external protons. Secondly, the advantage of the special structure was that the nanosheet-assembled hollow tube provided a large specific surface area. The combination of the two 2D structures maximizes the interface contact area and exposes enough reaction sites on the surface, which is favorable for photocatalytic hydrogen production. This novel strategy provides new ideas for the fabrication of other heterojunction photocatalysts with an efficient solar energy conversion.

Supplementary Materials: The following supporting information can be downloaded at: https://www.mdpi.com/article/10.3390/nano12091474/s1. Characterizations, Photocatalytic activity, Photoelectrochemical measurements [42–47]. Figure S1: XRD patterns of rGO, Table S1: Comparison with other TiO$_2$ and rGO composite catalysts.

Author Contributions: Conceptualization, Z.L., S.W. and W.Z.; methodology, D.Y.; software, L.G.; validation, T.C., Y.M. and X.H.; formal analysis, X.Z.; investigation, H.C.; resources, B.H.; data curation, S.Z., Y.W. and Y.G.; writing—original draft preparation, Z.L. and D.Y.; writing—review and editing, S.W. and W.Z.; visualization, W.Z.; supervision, S.W. and W.Z.; project administration, W.Z.; funding acquisition, W.Z. All authors have read and agreed to the published version of the manuscript.

Funding: This work was financially supported by the National Natural Science Foundation of China (Grant Number: 21871078, 52172206); the Shandong Province Natural Science Foundation (Grant Number: ZR2021MB016), and the Development plan of Youth Innovation Team in Colleges and Universities of Shandong Province.

Data Availability Statement: Data presented in this article are available at request from the corresponding author.

Conflicts of Interest: The authors declare no conflict of interest.

References

1. Uddin, N.; Zhang, H.; Du, Y.; Jia, G.; Wang, S.; Yin, Z. Structural-phase catalytic redox reactions in energy and environmental applications. *Adv. Mater.* **2020**, *32*, 1905739. [CrossRef] [PubMed]
2. Jeong, G.; Sasikala, S.; Yun, T.; Lee, G.; Lee, W.; Kim, S. Nanoscale assembly of 2d materials for energy and environmental applications. *Adv. Mater.* **2020**, *32*, 1907006. [CrossRef] [PubMed]
3. Fang, B.; Xing, Z.; Sun, D.; Li, Z.; Zhou, W. Hollow semiconductor photocatalysts for solar energy conversion. *Adv. Powder Mater.* **2022**, *1*, 100021. [CrossRef]
4. Wang, W.; Deng, C.; Xie, S.; Li, Y.; Zhang, W.; Sheng, H.; Chen, C.; Zhao, J. Photocatalytic c–c coupling from carbon dioxide reduction on copper oxide with mixed-valence copper(I)/copper(II). *J. Am. Chem. Soc.* **2021**, *143*, 2984–2993. [CrossRef] [PubMed]
5. Jian, S.; Tian, Z.; Hu, J.; Zhang, K.; Zhang, L.; Duan, G.; Yang, W.; Jiang, S. Enhanced visible light photocatalytic efficiency of La-doped ZnO nanofibers via electrospinning-calcination technology. *Adv. Powder Mater.* **2022**, *1*, 100004. [CrossRef]
6. Ma, B.; Blanco, M.; Calvillo, L.; Chen, L.; Chen, G.; Lau, T.-C.; Dražić, G.; Bonin, J.; Robert, M.; Granozzi, G. Hybridization of Molecular and Graphene Materials for CO_2 Photocatalytic Reduction with Selectivity Control. *J. Am. Chem. Soc.* **2021**, *143*, 8414–8425. [CrossRef]
7. Fujishima, A.; Honda, K. Electrochemical Photolysis of Water at a Semiconductor Electrode. *Nature* **1972**, *238*, 37–38. [CrossRef]
8. Gong, H.; Wang, L.; Zhou, K.; Zhang, D.; Zhang, Y.; Adamaki, V.; Bowen, C.; Sergejevs, A. Improved photocatalytic performance of gradient reduced TiO_2 ceramics with aligned pore channels. *Adv. Powder Mater.* **2021**, *1*, 100025. [CrossRef]
9. Qiu, P.; Li, W.; Thokchom, B.; Park, B.; Cui, M.; Zhao, D.; Khim, J. Uniform core–shell structured magnetic mesoporous TiO_2 nanospheres as a highly efficient and stable sonocatalyst for the degradation of bisphenol-A. *J. Mater. Chem. A* **2015**, *3*, 6492–6500. [CrossRef]
10. Hao, Z.; Chen, Q.; Dai, W.; Ren, Y.; Zhou, Y.; Yang, J.; Xie, S.; Shen, Y.; Wu, J.; Chen, W.; et al. Oxygen-deficient blue TiO_2 for ultrastable and fast lithium storage. *Adv. Energy Mater.* **2020**, *10*, 1903107. [CrossRef]
11. Wang, H.; Hu, X.; Ma, Y.; Zhu, D.; Li, T.; Wang, J. Nitrate-Group-Grafting-Induced Assembly of Rutile TiO_2 Nanobundles for Enhanced Photocatalytic Hydrogen Evolution. *Chin. J. Catal.* **2020**, *41*, 95–102. [CrossRef]
12. Yu, L.; Zhang, G.; Liu, C.; Lan, H.; Liu, H.; Qu, J. Interface Stabilization of Undercoordinated Iron Centers on Manganese Oxides for Nature-inspired Peroxide Activation. *ACS Catal.* **2018**, *8*, 1090–1096. [CrossRef]
13. Skinner, D.; Colombo, D.; Cavaleri, J.; Bowman, R. Femtosecond Investigation of Electron Trapping in Semiconductor Nanoclusters. *J. Phys. Chem.* **1995**, *99*, 7853–7856. [CrossRef]
14. Lan, K.; Liu, Y.; Zhang, W.; Liu, Y.; Elzatahry, A.; Wang, R.; Xia, Y.; Al-Dhayan, D.; Zheng, N.; Zhao, D. Uniform Ordered Two-Dimensional Mesoporous TiO_2 Nanosheets from Hydrothermal-Induced Solvent-Confined Monomicelle Assembly. *J. Am. Chem. Soc.* **2018**, *140*, 4135–4143. [CrossRef]
15. Zhao, Y.; Zhao, Y.; Shi, R.; Wang, B.; Waterhouse, G.I.N.; Wu, L.; Tung, C.; Zhang, T. Tuning Oxygen Vacancies in Ultrathin TiO_2 Nanosheets to Boost Photocatalytic Nitrogen Fixation up to 700 nm. *Adv. Mater.* **2019**, *31*, 1806482. [CrossRef]
16. Wu, J.; Qiao, P.; Li, H.; Xu, Y.; Yang, W.; Yang, F.; Lin, K.; Pan, K.; Zhou, W. Engineering surface defects on two-dimensional ultrathin mesoporous anatase TiO_2 nanosheets for efficient charge separation and exceptional solar-driven photocatalytic hydrogen evolution. *J. Mater. Chem. C* **2020**, *8*, 3476. [CrossRef]
17. Bae, D.; Pedersen, T.; Seger, B.; Malizia, M.; Kuznetsov, A.; Hansen, O.; Chorkendorffa, I.; Vesborg, P. Back-illuminated Si photocathode: A combined experimental and theoretical study for photocatalytic hydrogen evolution. *Energy Environ. Sci.* **2015**, *8*, 650–660. [CrossRef]
18. Yi, L.; Ci, S.; Luo, S.; Shao, P.; Hou, Y.; Wen, Z. Scalable and Low-cost Synthesis of Black Amorphous Al-Ti-O Nanostructure for High-efficient Photothermal Desalination. *Nano Energy* **2017**, *41*, 600–608. [CrossRef]
19. Bae, D.; Kanellos, G.; Faasse, G.M.; Dražević, E.; Venugopal, A.; Smith, W.A. Design principles for efficient photoelectrodes in solar rechargeable redox flow cell applications. *Commun. Mater.* **2020**, *1*, 17. [CrossRef]

20. Abed, J.; Rajput, N.S.; El Moutaouakil, A.; Jouiad, M. Recent Advances in the Design of Plasmonic Au/TiO$_2$ Nanostructures for Enhanced Photocatalytic Water Splitting. *Nanomaterials* **2020**, *10*, 2260. [CrossRef]
21. Song, W.; Ji, J.; Guo, K.; Wang, X.; Wei, X.; Cai, Y.; Tan, W.; Li, L.; Sun, J.; Tang, C.; et al. Solid-phase impregnation promotes Ce doping in TiO$_2$ for boosted denitration of CeO$_2$/TiO$_2$ catalysts. *Chin. Chem. Lett.* **2022**, *33*, 935–938. [CrossRef]
22. Zhu, Y.; Shah, M.; Wang, C. Insight into the role of Ti^{3+} in photocatalytic performance of shuriken-shaped BiVO$_4$/TiO$_{2-x}$ heterojunction. *Appl. Catal. B Environ.* **2017**, *203*, 526–532. [CrossRef]
23. Wang, H.; Liu, J.; Xiao, X.; Meng, H.; Wu, J.; Guo, C.; Zheng, M.; Wang, X.; Guo, S.; Jiang, B. Engineering of SnO$_2$/TiO$_2$ heterojunction compact interface with efficient charge transfer pathway for photocatalytic hydrogen evolution. *Chin. Chem. Lett.* **2022**, *in press*. [CrossRef]
24. Jia, M.; Yang, Z.; Xiong, W.; Cao, J.; Xiang, Y.; Peng, H.; Jing, Y.; Zhang, C.; Xu, H.; Song, P. Magnetic heterojunction of oxygen-deficient Ti^{3+}-TiO$_2$ and Ar-Fe$_2$O$_3$ derived from metal-organic frameworks for efficient peroxydisulfate (PDS) photo-activation. *Appl. Catal. B Environ.* **2021**, *298*, 120513. [CrossRef]
25. Novoselov, K.S.; Geim, A.K.; Morozov, S.V.; Jiang, D.; Zhang, Y.; Dubonos, S.V.; Grigorieva, I.V.; Firsov, A.A. Electric Field Effect in Atomically Thin Carbon Films. *Science* **2004**, *306*, 666–669. [CrossRef]
26. Sharma, D.; Menon, A.; Bose, S. Graphene templated growth of copper sulphide 'flowers' can suppress electromagnetic interference. *Nanoscale Adv.* **2020**, *2*, 3292–3303. [CrossRef]
27. Wang, X.; Meng, G.; Zhu, C.; Huang, Z.; Qian, Y.; Sun, K.; Zhu, X. A generic synthetic approach to large-scale pristine-graphene/metal-nanoparticles hybrids. *Adv. Funct. Mater.* **2013**, *23*, 5685. [CrossRef]
28. Qiu, B.; Xing, M.; Zhang, J. Mesoporous TiO$_2$ nanocrystals grown in situ on graphene aerogels for high photocatalysis and lithium-ion batteries. *J. Am. Chem. Soc.* **2014**, *136*, 5852–5855. [CrossRef]
29. Xu, C.; Wang, C.; He, X.; Lyu, M.; Wang, S.; Wang, L. Processable graphene oxideembedded titanate nanofiber membranes with improved filtration performance. *J. Hazard. Mater.* **2017**, *325*, 214–222. [CrossRef] [PubMed]
30. Balsamo, S.; Fiorenza, R.; Condorelli, M.; Pecoraro, R.; Brundo, M.; Presti, F.; Sciré, S. One-Pot Synthesis of TiO$_2$-rGO Photocatalysts for the Degradation of Groundwater Pollutants. *Materials* **2021**, *14*, 5938. [CrossRef]
31. Zouzelka, R.; Remzova, M.; Plšek, J.; Brabec, L.; Rathousky, J. Immobilized rGO/TiO$_2$ Photocatalyst for Decontamination of Water. *Catalysts* **2019**, *9*, 708. [CrossRef]
32. Wang, S.; Cai, J.; Mao, J.; Li, S.; Shen, J.; Gao, S.; Huang, J.; Wang, X.; Parkin, I.P.; Lai, Y. Defective black Ti3$^+$ self-doped TiO$_2$ and reduced graphene oxide composite nanoparticles for boosting visible-light driven photocatalytic and photoelectrochemical activity. *Appl. Surf. Sci.* **2019**, *467–468*, 45–55. [CrossRef]
33. Zhang, X.; Hu, W.; Zhang, K.; Wang, J.; Sun, B.; Li, H.; Qiao, P.; Wang, L.; Zhou, W. Ti^{3+} Self-Doped Black TiO$_2$ Nanotubes with Mesoporous Nanosheet Architecture as Efficient Solar-Driven Hydrogen Evolution Photocatalysts. *ACS Sustain. Chem. Eng.* **2017**, *5*, 6894–6901. [CrossRef]
34. Marcano, D.C.; Kosynkin, D.V.; Berlin, J.M.; Sinitskii, A.; Sun, Z.; Slesarev, A.; Alemany, L.B.; Lu, W.; Tour, J.M. Improved synthesis of graphene oxide. *ACS Nano* **2010**, *4*, 4806–4814. [CrossRef]
35. Cote, L.J.; Kim, F.; Huang, J. Langmuir-blodgett assembly of graphite oxide single layers. *J. Am. Chem. Soc.* **2009**, *131*, 1043–1049. [CrossRef]
36. Yu, C.; Li, M.; Yang, D.; Pan, K.; Yang, F.; Xu, Y.; Yuan, L.; Qu, Y.; Zhou, W. NiO nanoparticles dotted TiO$_2$ nanosheets assembled nanotubes P-N heterojunctions for efficient interface charge separation and photocatalytic hydrogen evolution. *Appl. Surf. Sci.* **2021**, *568*, 150981. [CrossRef]
37. Chen, C.; Yang, S.; Ding, J.; Wang, G.; Zhong, L.; Zhao, S.; Zang, Y.; Jiang, J.; Ding, L.; Zhao, Y.; et al. Non-covalent self-assembly synthesis of AQ2S@rGO nanocomposite for the degradation of sulfadiazine under solar irradiation: The indispensable effect of chloride. *Appl. Catal. B Environ.* **2021**, *298*, 120495. [CrossRef]
38. Sang, Y.; Zhao, Z.; Tian, J.; Hao, P.; Jiang, H.; Liu, H.; Claverie, J.P. Enhanced Photocatalytic Property of Reduced Graphene Oxide/TiO$_2$ Nanobelt Surface Heterostructures Constructed by an In Situ Photochemical Reduction Method. *Small* **2014**, *18*, 3775–3782. [CrossRef]
39. Zhang, Y.; Fu, F.; Zhou, F.; Yang, X.; Zhang, D.; Chen, Y. Synergistic effect of RGO/TiO$_2$ nanosheets with exposed (001) facets for boosting visible light photocatalytic activity. *Appl. Surf. Sci.* **2020**, *510*, 145451. [CrossRef]
40. Kronik, L.; Shapira, Y. Surface photovoltage phenomena: Theory, experiment, and applications. *Surf. Sci. Rep.* **1999**, *37*, 1–206. [CrossRef]
41. Yang, F.; Li, H.; Pan, K.; Wang, S.; Sun, H.; Xie, Y.; Xu, Y.; Wu, J.; Zhou, W. Engineering Surface N-Vacancy Defects of Ultrathin Mesoporous Carbon Nitride Nanosheets as Efficient Visible-Light-Driven Photocatalysts. *Sol. RRL* **2020**, *5*, 2000610. [CrossRef]
42. Gupta, B.; Melvin, A.; Matthews, T.; Dhara, S.; Dash, S.; Tyagi, A. Facile gamma radiolytic methodology for TiO$_2$-rGO synthesis: Effect on photo-catalytic H$_2$ evolution. *Int. J. Hydrogen Energy* **2015**, *40*, 5815–5823. [CrossRef]
43. Bharad, P.; Sivaranjani, K.; Gopinath, C. A rational approach towards enhancing solar water splitting: A case study of Au-RGO/N-RGO-TiO$_2$. *Nanoscale* **2015**, *7*, 11206–11215. [CrossRef]
44. Ma, J.; Dai, J.; Duan, Y.; Zhang, J.; Qiang, L.; Xue, J. Fabrication of PANI-TiO$_2$/rGO hybrid composites for enhanced photocatalysis of pollutant removal and hydrogen production. *Renew. Energ.* **2020**, *156*, 1008–1018. [CrossRef]
45. Tudu, B.; Nalajala, N.; Reddy, K.; Saikia, P.; Gopinath, C. Electronic integration and thin film aspects of Au–Pd/rGO/TiO$_2$ for improved solar hydrogen generation. *ACS Appl. Mater. Inter.* **2019**, *11*, 32869–32878. [CrossRef]

46. Agegnehu, A.; Pan, C.; Tsai, M.; Rick, J.; Su, W.; Lee, J.; Hwang, B. Visible light responsive noble metal-free nanocomposite of V-doped TiO$_2$ nanorod with highly reduced graphene oxide for enhanced solar H$_2$ production. *Int. J. Hydrogen Energy* **2016**, *41*, 6752–6762. [CrossRef]
47. Wei, X.; Cao, J.; Fang, F. A novel multifunctional Ag and Sr^{2+} co-doped TiO$_2$@rGO ternary nanocomposite with enhanced p-nitrophenol degradation, and bactericidal and hydrogen evolution activity. *RSC Adv.* **2018**, *8*, 31822–31829. [CrossRef]

MDPI
St. Alban-Anlage 66
4052 Basel
Switzerland
www.mdpi.com

Nanomaterials Editorial Office
E-mail: nanomaterials@mdpi.com
www.mdpi.com/journal/nanomaterials

Disclaimer/Publisher's Note: The statements, opinions and data contained in all publications are solely those of the individual author(s) and contributor(s) and not of MDPI and/or the editor(s). MDPI and/or the editor(s) disclaim responsibility for any injury to people or property resulting from any ideas, methods, instructions or products referred to in the content.

www.ingramcontent.com/pod-product-compliance
Lightning Source LLC
LaVergne TN
LVHW070705100526
838202LV00013B/1035